the geography collection world wide

a core text for A-level

a collaboration between

Richard Walker	Head of Geography, Bedford School
Ian Selmes	Head of Geography, Oakham School
Ken Grocott	Head of Geography, Brighton College
Anne Fielding Smith	Head of Humanities, Brighton Hove and Sussex Sixth Form College
Roger Smith	Principal Lecturer in Geography, University of Brighton
Gary Phillips	Head of Geography, Oundle School
John Lifford	Head of Geography, North Devon College
Keith Grimwade	Head of Geography, Hinchingbrooke School
Steve Burton	Humanities Co-ordinator, Barnwell School

**series editor
Ian Selmes**

the geography collection is a series of texts designed for students of the new A-level geography syllabuses and has been written by a team of eminent geography teachers. The series provides an enlightened approach to the discipline, opening up the world of the geographer through contemporary text, clarifying illustrations, integrated questions and decision-making exercises. Each of the nine sections of the Core Text and the Option Books contain guidelines for geographical skills and project work together with a glossary of important terms. A feature of the series are the themes that run through each section and Option Book, giving extra flexibility in the way that students can use the texts.

Acknowledgements

The authors and publishers would like to thank the following for permission to reproduce copyright materials in this book. Every effort has been made to trace and acknowledge all copyright holders but if any have been overlooked the publishers will be pleased to make the necessary arrangements.

AEA Technology, Figure 4.28; Allen and Unwin, *Elements of Geographical Hydrology*, B J Knapp, Figure 3.6d; Bath Evening Chronicle, Figure 3.31; Blackwell, *Manufacturing Industry*, D Horsfall, Figure 6.2; Butterworth, *Sediments*, David Briggs; Figure 3.20; Cambridge Evening News, Figure 6.29; Cambridge University Press and B Ellis, *The West Midlands*, B Ellis, Figures 6.17, 6.18; Cambridge Weekly News, Figure 6.28; City and Industrial Development Corporation of Maharashtra Ltd, Figure 8.47; © The Economist, March 1992, Figure 9.8; Geest plc, Figure 7.14; *Geomagazine*, Peter McClure, Figures 6.12, 6.14, 6.21; The Geographical Association, *Geography*, Oakey, Figure 6.5a, Keeble Figure 6.6, 6.23, Hoggart, Figure 6.9, Mounfield, Figure 6.19, Fielding, Figure 8.16; The Guardian ©, adapted from a diagram by Paddy Allen, Figure 3.8, Figures 8.17, 8.18, 8.38; HMSO, © Crown Copyright, Figure 4.17; Institute of Geological and Nuclear Sciences Limited, Figure 1.1, inside front cover; London Weather Centre, Figure 4.22; Longman Education, *The Location of Manufacturing Industry*, J Bale, Figures 6.16, 6.20, *The United States: A Contemporary Human Geography*, Knox et al, Figures 6.31abc; Macdonald Publishers, *The Water of Leith*, Figure 3.6b; G McDonald and K Irvine for material in sections 1 and 2; Ordnance Survey, © Crown Copyright, Figures 7.30, 8.8; Oxford University Press, *Flooding and Flood Hazard in the UK*, Malcolm Newson, Figure 3.34; Penguin Books Ltd, *The Third Revolution*, Paul Harrison, Figure 9.6; Philip Allan, *Geography Review*, May 1990, Figures 4.16, 4.17, *Geography Review*, Figure 6.13; Population Reference Bureau inc, Figure 9.3; Reed Book Services and George Philip, *Philips World Handbook and Modern School Atlas*, Figures 6.1, 6.24; Resources for Learning Ltd, by kind permission of Anglian Water, Figure 3.24; Routledge, *Changing Geography of the UK*, Johnston and Gardiner, Figure 6.26; NRSC Ltd/Science Photo Library, Figure 3.35 inside back cover; Severn Trent, Figure 3.29; Stanley Thornes, *Geofile*, Figures 6.22, 6.25; The Sunday Telegraph, Figures 3.25, 6.4; Thomas Nelson, *Hydrology: Measurements and Application*, Malcolm Newson, Figure 3.26, *North and South America*, David Waugh, Figure 6.31d; © Times Newspapers Limited 1992 and the British Library, Figure 4.7; V H Winston & Sons, *Non-Metropolitan Industrialisation*, Lonsdale & Seyler, Figure 6.5b; United Nations Population Fund, Figures 9.1, 9.2, 9.24; Yorkshire Dales National Park, *Landscapes for Tomorrow*, Figure 7.27; Warren Springs Laboratories, Figure 4.28.

The author and publishers also thank the following for permission to reproduce photographs in this book.
The Evening Argus, Figures 4.19, 4.20; GSF Picture Library, Figures 3.18, 4.18, 5.19, 5.20, 5.27; Midland Examining Group, Figure 1.33; NERC Satellite Station, Dundee University, Figure 4.21; Panos Pictures, 5.22; Panos Pictures/David Reed, Figure 5.25; Robert Harding, Figures 8.1abc, 8.6, 8.36, 8.37; Still Pictures/Adrian Harris, Figure 8.32; Still Pictures/Heldur Metocmy, Figure 9.31; Still Pictures/Jorgen Schytte, Figure 9.32; Still Pictures/Mark Edwards, Figures 8.35, 9.7; Still Pictures/Mikkel Ostergaard, Figure 8.27; Still Pictures/Oliver Gillie, Figure 5.34; Sylvia Cordaiy, Figures 5.11, 8.1i, 9.25; Topham, Figures 4.25, 8.31; Adam Woolfitt, Figure 6.15.

All other photos supplied by the authors.

A catalogue record for this title is available from the British Library

ISBN 0 340 61865 5

First published 1995
Impression number 10 9 8 7 6 5 4 3 2 1
Year   1999 1998 1997 1996 1995

Copyright © 1995 Richard Walker, Ian Selmes, Ken Grocott, Anne Fielding Smith, Roger Smith, Gary Phillips, John Lifford, Keith Grimwade and Steve Burton.

All rights reserved. No part of this publication may be reproduced or transmitted in any form or by any means, electronic or mechanical, including photocopy, recording, or any information storage and retrieval system, without permission in writing from the publisher or under licence from the Copyright Licensing Agency Limited. Further details of such licences (for reprographic reproduction) may be obtained from the Copyright Licensing Agency Limited, of 90 Tottenham Court Road, London W1P 9HE.

Typeset by Wearset, Boldon, Tyne and Wear.
Printed in Malaysia for Hodder & Stoughton Educational, a division of Hodder Headline Plc, 338 Euston Road, London NW1 3BH by Colorcraft Ltd.

contents

Section 1 Broken up and downhill — page 5
Denudation and weathering;
Slope form and mass movement processes;
DME: Mam Tor and the A625.

Section 2 All iced up — page 19
Ice and its distribution; Temporal variation;
Moving ice; Landforms and erosion; Glacial
deposition; Periglacial processes; DME: a new UK ski resort.

Section 3 Water on the move — page 33
The hydrosphere; Investigating drainage basins;
Dynamic drainage channels; Creating river valleys;
Monitoring and managing rivers; A vital resource;
DME: Bath floods.

Section 4 In the air — page 63
The atmosphere – a strange brew; Furnaces and freezers;
Elevator going up; That sinking feeling; The human
radiator; Bubbles and balloons; Wind in the willows;
DME: Urban air pollution.

Section 5 The living world — page 97
Ecosystems; Soils and soil processes;
Biomes; Ecosystems in the field; DME: soil erosion in the
South Downs.

Section 6 Economic activity — page 129
Manufacturing industry; Service industries;
DME: four-wheel drive in the USA.

Section 7 Food for thought — page 161
A world view; Food issues in ELDCs; Food issues in EMDCs;
DME: rural planning in the Atlas Mountains of Morocco.

Section 8 Life in the city — page 191
Urban settlement; World pattern of urbanisation;
Managing urban areas; Alternative approaches to
urban development; DME: an out-of-town superstore for
Huntingdon.

Section 9 All the people — page 223
Global population change; The demographic wedge;
Density and distribution; People on the move;
Population problems and solutions; DME: population change
in three countries.

Index — page 256

thematic contents

Transport
landslides on Mam Tor (15, 17), trans-alaskan pipeline (23), on rivers (35, 40), Trade Winds (73), 1987 storm (77), Trans-Alaskan pipeline in the Tundra (121), transport costs and industry (131, 133), technology and industry (135, 154, 155), cars (137), transport and city growth (195, 197), counterurbanisation (197, 208), migration (246).

Trade
Gulf of Bothnia (23), Trade Winds (73, 74), ocean currents (75), hardwoods (112), softwoods (123), raw materials for industry (131), trade and industrialisation (140, 147), inward investment (142), trade barriers (147, 150, 151), managing famine (166), agribusiness (171), trade agreements (172), trade and urban growth (202, 206, 211, 214), in skills (244, 245).

Hazards
landslide (1, 5, 17), weathering on buildings (12), mass movements (15), raindrops (17), Little Ice Age (22), permafrost (29), solifluction (30), floods (34, 48, 50, 51, 57, 60, 74), drought (35), water pollution (54–60), wind (74), depressions (77–81, 83), anticyclones and drought (83), anticyclones as pollution traps (84), air pollution (86), soil leaching (101), soil erosion (113, 116, 179), deforestation (112), salinisation (119, 167), wetland drainage (123), industrial decentralisation, (154), famine (35, 164, 166, 244), agricultural impacts (179), homelessness (204, 206), urbanisation (206, 212, 214, 216), population growth (223, 227).

Green
limestone quarrying (12), climatic cycles (22), Aral Sea (54), biotic index (55), water quality (57), urban air pollution (65, 69, 85, 87), Greenhouse Effect (65, 101), ozone (68), vegetation and microclimate (91–2), nutrient recycling (101, 102), soil fertility (108, 118), sustainable development (113, 117, 118), controlling soil erosion (117, 118), conservation area (104, 113, 118, 121, 123, 181, 182), greenfield sites for industry (139, 154), sustainable farming practices (169), self help housing (209), urban renewal (212), population growth and the environment (225, 227, 228).

Political
hydropolitics (38, 39, 54), UK water supply (54), controlling urban air pollution (85), managing forests (113), CAP and soil erosion (116, 118), economic development (133), government industrial policies (134, 142, 143, 145, 147, 153, 154,), UK cars (142), industrialisation processes (147, 149), land tenure and land use (165), market manipulation (178), CAP (178), urban planning (197, 198, 208, 209, 212, 215, 216), census accuracy (229, 235), population policies (229, 235, 236, 246, 251, 252), migrations (242, 246, 247).

Skills
slope analysis (14), morphological mapping (15), transects (18, 124, 220), photo interpretation (19), sediment analysis (27), drainage basin analysis (36, 40, 43), Landsat images (42), sediment analysis (48), discharge (50), flood prediction (50, 57), water quality (55), satellite weather image and synoptic chart interpretation (80), tephigrams (90) Spearmans Rank Correlation (92, 186, 197, 243), sampling (93, 220), soil texturing (106, 107, 124), quadrats (124), transects (124), chi-squared test (125, 164, 220), soil analysis (125), spatial correlation (164), scattergraph (13, 172, 186), climograph (174), index calculation (177), bi-polar analysis (192), choropleth mapping (197, 240), urban modelling (201, 204), dependency ratio (234), population pyramids (234, 236), Lorenz Curve (240), A-level projects (18, 31, 61, 93, 124, 125, 158, 189, 220, 254).

SECTION 1

Broken up and downhill

by Richard Walker

> **KEY IDEAS**
>
> - **What is the cycle of denudation?**
> - **How does weathering operate?**
> - **What effects does weathering have on the landscape?**
> - **Which factors influence the effectiveness of weathering?**
> - **How can the different slope processes be identified?**
> - **Why are slope profiles different in different areas?**

ALL BROKEN UP

The land's surface with its variety of different landforms is the product of **denudation**. These landforms provide the characteristic landscapes of various parts of the world and also give clues as to how these landscapes were created. Being a geographer involves detective work in unmasking these features and in deducing the processes at work.

Denudation really starts with the **tectonic forces** that produce rocks and that cause them to be exposed to the atmosphere. The focus of this section is on the subsequent breakdown of rock which is known as **weathering** and on the **mass movement** on slopes. Together these processes define the shape of slopes characteristic to particular environments.

Cycle of denudation

Figure 1.1 (see inside front cover) gives evidence of the processes of denudation. It shows Mount Cook in South Island, New Zealand, which at 3764 m is the highest peak in the Southern Alps. In December 1991, a huge rock avalanche took place here taking a large block out of the eastern slopes of the mountain. The 400 m long summit remained intact but substantially undermined. What then is the background to this event?

It is important to recognise that the Southern Alps are on a plate margin (see Figure 1.2, page 6) at the meeting point of the Pacific Plate and the Indo-Australian Plate. The two plates are experiencing great compression along a major **fault** line known as the Alpine Fault. The Southern Alps are one of the results of these tectonic forces with a great section of **sedimentary** and **metamorphic** rocks thrust up to heights of 3750 m and still rising today.

This compression leads to incredible stress within the rocks which in turn fractures, twists and buckles the layers of sedimentary and metamorphic rocks. The resulting faults and **joints** divide up the rock into fragments of very different sizes. With the continuous upward movement, these faults grow and weaken the overall structure of the rocks which would otherwise show great mechanical strength.

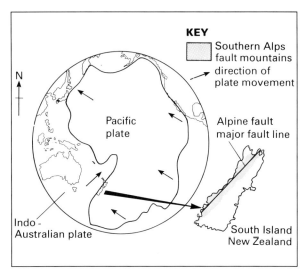

Figure 1.2 *New Zealand on the plate margins*

At this stage, the rocks are more open to a third effect, the process of weathering. The availability of a large surface area along these faults allows water, as the key component to weathering, to further weaken the structure by physical and chemical processes. The scree slopes in Figure 1.1 reflect the ability of weathering to break off fragments of rock to form the fan-shaped scree deposits shown.

The results of this faulting and the weathering processes as well as the continued upward movements of the Alps are seen in the many steep slopes created with fundamentally weakened rock. From time to time, mass movement occurs on these slopes. The weathered material moves under the influence of gravity. Sometimes there is a much greater degree of slope failure, as captured in this photograph (Figure 1.1), and whole sections of the hillside collapse. The speed of such processes can be rapid as in Figure 1.1, but often it can be slow and imperceptible.

Lastly the land is under attack from **erosion** processes. Here the agent of erosion is the Tasman Glacier which was almost entirely covered by debris even before the rock avalanche. The 17 km long glacier sculpts the land, carving deep valleys and transporting the material downslope. This erosion process can severely undermine the slopes and may in itself have contributed to the avalanche.

Much of the material will eventually end up in the sea, transported down by ice and other agents of erosion. Here in its temporary resting place, the material is deposited and over time will become the new sedimentary rock. This type of rock may then be pushed up by tectonic forces to form mountain chains such as the Himalayas or the Andes.

The photo in Figure 1.1 highlights several of the key physical processes at work in the natural world.

As new land is created or pushed up, so there are forces operating to wear it down again. The next section focuses on the process of weathering in more detail.

> **Q1.** *Use the glossary to find out the differences between the main types of rock. Draw Figure 1.3 in your notes and complete the boxes with each rock type and label the arrows with the processes.*
>
> **Q2.** *If an agent of erosion is involved in breaking down, transporting away and depositing material, what are the other agents of erosion besides ice?*
>
> **Q3.** *The processes are said to be part of the cycle of denudation. In what sense is it a cycle? Draw this cycle as a diagram incorporating the main processes mentioned in the text.*

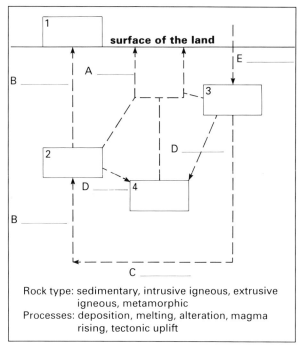

Figure 1.3 *Rock types and interactions*

Types of weathering

One of the signs of a good geographer is to know the difference between weathering and erosion. At its simplest, weathering involves the breaking down of rock without movement or in situ, whereas erosion implies the action and movement of an agent of erosion such as ice. Agents of erosion bring about the destruction and removal of the rocks, leading to the deposition of sediment and thus formation of new sedimentary rocks.

There are normally three main categories of weathering which can be identified.

1. Physical — The mechanical break up of rock into small pieces.
2. Chemical — The chemical alteration or decomposition of minerals within the rock.
3. Biological — The effects of both plants and animals on rocks and their minerals.

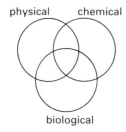

A Freeze-Thaw: expansion of water by freezing process within the cracks in the rocks. The crack widens and thawing allows the water to enter further into the rock. Eventually fragments of rock break off.

B Plant Roots: root growth along lines of weakness, especially joints, widens these cracks to break off fragments of rock.

C Carbonation: water absorbs CO_2 from the air and soil to produce a weak carbonic acid which attacks various minerals, especially calcium.

D Oxidation: oxygen combines with minerals, especially iron, to form soluble minerals. This process can be seen in reverse when oxygen is released (known as reduction).

E Salt crystallisation: expansion of crystals, especially salts, during evaporation exerts stresses within rock sometimes leading to **granular disintegration**.

F Chelation: decaying vegetation gives off organic acids which affect certain minerals making them soluble.

G Insolation: expansion and contraction of minerals, or, outside layers of rock, due to heating during the day and cooling at night.

H Hydration: absorption of water by various minerals leading to swelling, physical stress and eventually the break up of rock.

I Animal acids: animals such as sea urchins on shorelines release acid which attack minerals within the rock.

J Hydrolysis: ions within water react with ions in the mineral to produce decomposition.

This list is not exhaustive as the later case studies will show.

Figure 1.4 *Classifying weathering processes*

Q4. On a piece of A4 paper draw three overlapping circles as shown in Figure 1.4 and label the circles, physical, chemical and biological. Now read the list of processes in Figure 1.4 and fit them into the appropriate part of the diagram. Notice there are key overlap areas.

Q5. Compare your classification with those of other students. Discuss similarities and differences. If parts of the diagram have no processes, what does that suggest about the classification scheme?

A weathering landscape

Having identified the key weathering processes, a helpful way to understand them is through their effects on landforms and scenery. To do this, a case study of the Matopos District of southern Zimbabwe has been chosen (see Figure 1.5). This region shows many of the processes operating which can then be applied to other landscapes in order to understand the factors that affect them.

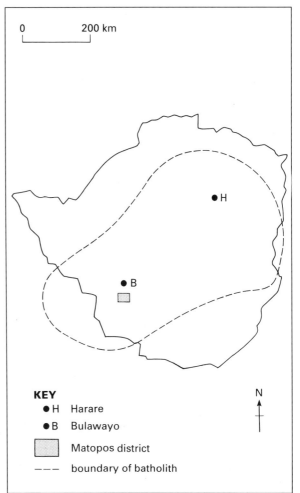

Figure 1.5 *Zimbabwe: batholith*

Q6. *As you read the following section, link it back to the diagram you produced in Question 3 to identify the context in which the weathering takes place. Re-draw the diagram incorporating the different processes mentioned here.*

Q7. *Try to identify any factors which have affected the type and effectiveness of the weathering processes.*

granite can expand and does so parallel to the surface. This is sometimes classified as a weathering process as the rock breaks up into fragments. The fractures are known as 'pressure release' and are shown in Figure 1.7. The photo in Figure 1.8 clearly shows the remaining blocks, and the curved nature of the joints. The rest of the blocks have moved by mass movement down to the flatter regions. On many Dwalas, detached slabs are found as remnants of this jointing (see Figure 1.7).

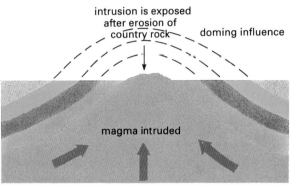

Figure 1.6 *Magma intrusion doming the surface*

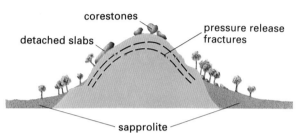

Figure 1.7 *A Dwala with pressure release jointing*

Figure 1.8 *Dwala with evidence of pressure release jointing*

Zimbabwe is underlain by a gigantic batholith. This is a huge granite **intrusion** (Figure 1.6), 600 km long by 350 km wide! Britain's equivalent is in the south west of England stretching from Dartmoor to the Scilly Isles. Magma from within the earth's mantle has risen to near the surface. It has cooled slowly and in the process, joints have opened up due to the magma contracting. Crystals were able to grow, especially those of quartz, mica and feldspar. The land above has domed upwards causing great fractures. The surrounding rocks were metamorphosed by the heat. Millions of years have elapsed and erosion processes, especially running water, have removed the material above the granite, to exhume (bring to the surface) the granite rocks. This then is the context in which weathering processes can take place.

The landforms of this area tend to be classified into two main groups, the *Dwala* or whaleback and the *Castle Kopje*. There are also intermediate types of landforms, showing characteristics of both, and in between these are areas of relatively flat land.

Dwala

Being an intrusion, the rock was formed under great pressures which surrounded and confined it. With the loss of this pressure due to exhumation, the

Castle Kopje

This landform (Figure 1.9) shows a different joint structure to Dwalas. The cooling magma produced contraction joints when the **igneous** rock reduced in size during the cooling process. These joints tend to be rectangular, as shown in Figures 1.9 and 1.10. As weathering occurs along these joints, with the

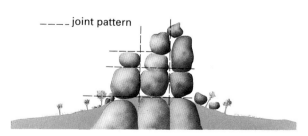

Figure 1.9 *Castle Kopje: evidence of rectangular joints*

increased surface area available for work, the end products are these piled up corestones. These can be equated to the tors found in Dartmoor today.

Figure 1.11 *Remnant mica and quartz particles and exfoliation flakes*

Chemical

The minerals of granite are highly significant to chemical weathering. With large crystals from the slow cooling, the feldspar is particularly susceptible to weathering. The processes of hydration and hydrolysis attack feldspar and break it down into potassium hydroxide and alumino silicia acid. The former is carbonated and removed in solution. The latter is broken down leaving the clay minerals. These processes are reflected in several ways. Firstly, in Figure 1.11 where the particles of mica and quartz have been left behind, the size of the particles reflects the size of the crystals. In other places standing water has collected in slight depressions called weathering pans or gnammas (see Figure 1.12). The water provides the catalyst for the chemical processes mentioned and materials from these processes are washed down, and clay in particular is deposited between the main outcrops (foreground Figure 1.10). This material is known as Sapprolite.

Figure 1.10 *Castle Kopje with balancing blocks*

The main processes of weathering

Physical

The climate has a major role to play here. Look in an atlas to find the latitude of Zimbabwe. This can be seen to be within the tropics with highly seasonal rain and high temperatures due to the angle of incidence of the sun. The long sunshine hours have important consequences for the weathering processes. The outside layers of rock heat up and expand during the day, yet cool rapidly during the night with few clouds to keep the heat in. As rock does not conduct heat easily, the stress due to quick expansion and contraction in the surface layer leads to flaking, a process often referred to as exfoliation (see Figure 1.11). The large diurnal (daily) range of temperatures can also affect minerals of different colours. The darker minerals expand and contract at faster rates than the surrounding minerals. This is caused by the low reflective ability of the colour. This also sets up great stresses in the mineral structure leading to the eventual breakdown of the rock. It is clear that in this region insolation weathering is most important.

Figure 1.12 *Weathering pans (gnammas): intensified chemical weathering*

Biological

The impact of biology cannot be underestimated. The pale colouring in Figure 1.13 shows up the lichens which grow in the water zones. These lichens produce acids which are concentrated within these hollows and intensify the chemical weathering. The roots of the fig tree (see Figure 1.14) break up the joints allowing more surface area for weathering as well as further breaking up the rock itself. A very unusual variation of biological weathering also occurs here. Scorpions hiding under the flakes of rock on the surface are food for baboons! The patches of lighter rock are where the flakes have been removed by the baboons in their hunt for the next meal. (See Figure 1.15.)

Figure 1.13 *Grey lichens in the weathering pan*

Figure 1.14 *Root structure developing along the joints*

Figure 1.15 *Exfoliation flakes on Imadzi Dwala*

Factors affecting weathering

Rock and its characteristics

The geology of the rock is a key factor in its weathering. Granite's weakest mineral is feldspar, whereas with limestone it is calcium. The process of carbonation changes the calcium carbonate to a bicarbonate which is soluble and easily removed. The effectiveness of this can be seen in the development of large pot holes such as Gaping Gill near Ingleborough in North Yorkshire. All major limestone areas in the world tend to have such features but they are often named in different ways.

The joints within the limestone are seen to be the key to the development of limestone pavements. The joints are caused because the limestone is a sedimentary rock and as it dries out it contracts. The joints become the focus of the carbonation. They are widened and form grykes whilst the remaining blocks stand out as clints (see Figure 1.16). Malham is one of the well-known pavements in Britain. Where the bedding planes are more easily weathered there is a more layered development as seen in Figure 1.17 with the Pancake Rocks on the West Coast, South Island in New Zealand.

Figure 1.17 *Bedding planes revealed in the Pancake Rocks*

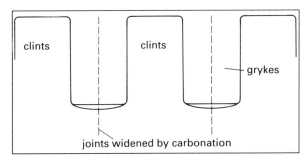

Figure 1.16 *Development of a limestone pavement*

The **lithology** is also important. Much weathering in **Karst** takes place underground. The water is able to infiltrate into the rock either through pore spaces between the minerals or down the joints. This leads to the development of underground caverns as joints and bedding planes of the sedimentary rocks are enlarged. The water then contains calcite and as the water drips from the roof tiny particles of lime are left to form stalactites. Similarly on the ground the drip splashes and part of it evaporates so that stalagmites develop upwards.

The climate

Variations in the climate affect the speed and effectiveness of weathering. Temperature, for example, affects the rate of reaction of the chemical processes. As the temperature increases, the rate of reaction increases and at a greater speed. Experiments have suggested that raising the temperature by 10°C more than doubles the reaction speed of the chemical processes.

It is not just the mean annual temperature but the diurnal range which is significant. As seen in the Matopas, insolation was effective due to the difference between day and night temperatures which affected the rapidity of expansion and contraction. Furthermore, if this diurnal range goes above and below freezing then this is likely to affect the freeze-thaw action. The **blockfield** on the summit of Glyder Fawr in Snowdonia (Figure 1.18, page 12) reflects this, as do the scree slopes in Figure 1.1.

Chemical weathering is strongly linked to the presence of water and so rainfall figures are an important factor. It is water which carries the acids along the fractures in the rock, enabling weathering to occur on the exposed surface area. Water itself has ions (oxygen and hydrogen) that react with the different minerals in the rock. The absorption of water (the opposite of dehydration) and the resultant swelling of the mineral breaks up the rock. Where water is present in large quantities, chemical processes tend to work faster.

It is not just the climate of today which is significant. Past climates play a key role in the formation of some landscapes, particularly periglacial conditions when freeze-thaw was

Figure 1.18 *Snowdonia: blockfield on the summit of Glyder Fawr*

speeded up and slope processes (see page 14) were very active. Scree slopes in places like the Lake District may owe their origin more to the conditions after the Ice Age in Britain rather than to present conditions.

Variations in temperature can also be brought about by aspect. South facing slopes in the Northern Hemisphere thaw more frequently during the day in comparison to north facing ones, allowing freeze-thaw to be more effective.

Biotic and soil

Vegetation can play a vital role in weathering. The presence of luxurious vegetation provides not only the root structure to break up rocks but also the rotting vegetation provides humic acids. The acids released from the **humus** attack the minerals in a process called chelation. The power of roots can often be seen in Britain where areas of tarmac and even hard tennis court surfaces are lifted and cracked as root structures develop. Often the luxuriant vegetation will support more fauna, from the smaller decomposers breaking down the humus, to the burrowing animals allowing the movement of water to the joints of the bedrock beneath. At the sea edge creatures by the glorious name of boring piddocks use their valves to rasp at the limestone, enabling them to burrow into the rocks.

Soil can have different influences. In some cases it protects landscapes against the effects of physical weathering. In others it provides CO_2 to further the process of carbonation. Acidity of water under soil is often greater than that of local rainwater and produces significant increases in chemical weathering.

Human impact

People influence weathering in different ways. Possibly the most significant in recent times has been the increase in acidity of rainwater. Although the link is questioned, the development of coal-fired power stations and heavy industry does seem to have affected the acidity of the rain. This has had important impacts in speeding up weathering processes and removing aluminium from rocks and soils. This is particularly relevant in areas such as Scandinavia where the acid rains are brought by the prevailing South Westerlies from Britain.

Quarrying certainly speeds up physical weathering with the development of more exposed rock surfaces and even by causing pressure release jointing. At the Poole's Cavern in Derbyshire, the faster growth and the different colour of stalagmites and stalactites is linked to the release of more lime as a result of quarrying and is an example of how chemical weathering has speeded up.

Another factor is our use of certain types of rock. Many fine buildings in the Central Business Districts in Britain's towns and cities have been built of Purbeck limestone and other sedimentary rocks, and are now being severely weathered. Statues and tombstones, depending on the rock type and siting, also reflect weathering (Figure 1.19). This gargoyle from New College, Oxford, shows what he thinks of the increasing rates of weathering!

Figure 1.19 *Gargoyle on New College, Oxford showing weathering*

Figure 1.20 *Granite weathering at different locations*

Q8. Plot the data in Figure 1.20 onto a scatter graph with mean annual temperature (MAT) °C on the y axis, and the percentage of clay in granite on the x axis.
  a) Describe the relationship between the clay content and the mean annual temperature.
  b) Which processes are likely to have been involved in producing clay?
  c) Explain why the graph should show this relationship.

Q9. Figure 1.21 shows a comparison in the depth of bedrock beneath the surface in two different parts of the same limestone area in Britain.
  a) Compare and contrast the two sets of data.
  b) Given the geology of the area, which landform feature is represented in both graphs? Draw a diagram of this feature.
  c) Now suggest what conditions could produce the differences between the areas A and B. Explain why this should be.

Q10. Study Figure 1.22.
  a) Fill in the vegetation zones.
  b) What could be represented by factor B?
  c) Describe the pattern of i) physical weathering and ii) chemical weathering, in relation to latitude and vegetational zones.
  d) Explain these patterns.
  e) Locate Britain on the x axis. Why is this a slightly unrealistic picture of the extent of weathering for this area?

SITE	1	2	3	4	5	6	7	8	9	10	11	12	13	14	15	16	17	18	19
% Clay Content	5	5	11	6	10	14	11	13	27	28	12	20	31	25	43	20	27	50	58
MAT °C	5	7	7.5	8	10	12	12.5	12.5	12.5	13	13	14	15	15.5	16	16	16	17	17

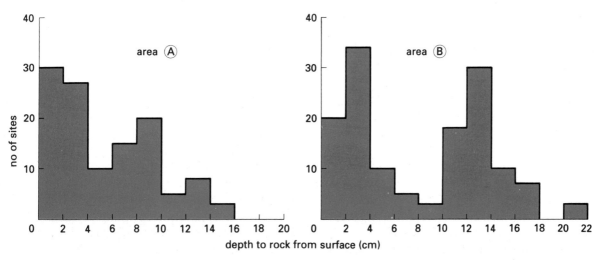

Figure 1.21 *Depth of limestone weathering*

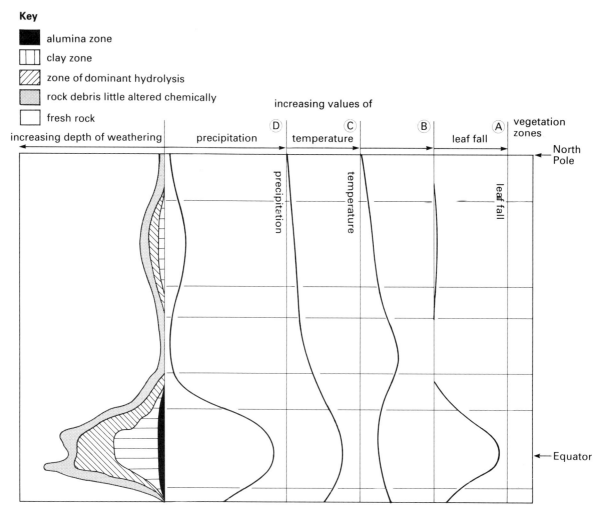

Figure 1.22 *Changing amounts of weathering with latitude & environmental factors*

ON THE WAY DOWN

The photograph of Mount Cook (Figure 1.1) establishes the link between the tectonic processes, weathering processes and mass movement. The rock avalanche is the process, as a result of gravity, working on the weakened rock and oversteepened slopes. The next section will look at the main processes operating on the slope and the impact that these have on the landscape, both natural and artificial. The shape or profile will be considered in relation to the processes operating on it and also in relation to how it affects the processes themselves.

Slope form

To measure and compare profiles (shape) the slope is divided into elements. An element shows a part of the slope where a constant shape is kept. The most common elements are shown in Figure 1.23. Slopes

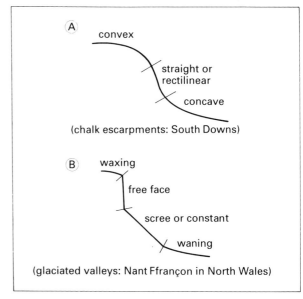

Figure 1.23 *Elements making up slopes*

are more complex than this diagram but it provides the opportunity to compare a slope with these set norms.

Another way to portray slopes is through maps. There are different ways of showing these but perhaps the most common is the morphological map. Figure 1.24 is a morphological map for the area beneath Mam Tor in Derbyshire. Each element has both the slope angle and type of form.

Processes at work

A process is the series of changes happening on the slope. It involves movement and simply shows what is going on. Each process tends to have certain characteristics and requirements and to also have different effects on the landscape.

Slide

This is when blocks of rock slide down a surface. For this to happen the main need is a slide plane – a surface over which sliding can take place. Lift up a flat surface (like a table at one end) and eventually friction will be overcome and objects will start to move over the surface. Planes are commonly found in sedimentary rocks which have **bedding planes**. These involve some impermeable layers which increase the water content, reducing friction. The slide planes of the Dwalas of the Matopos are due to the pressure release joints. Movement of slabs (see Figure 1.7, page 8) will occur downslope when friction is overcome.

Flow

A flow is the rapid movement of material with water down a slope. The three controlling factors are the abundance of water, the presence of unconsolidated material and enough of a slope angle to maintain momentum. The water not only adds weight to the debris but also acts as a lubricant and reduces internal friction.

The process of flow can take a variety of forms. Aberfan (the 1966 mud flow of South Wales) is a frequently quoted example due to the severity of the death toll. The extra water from excessive rainfall, the presence of a spring, the unconsolidated tailings from the coal workings all led to the very fast movement of mud down the steep glaciated valley slopes. A lahar is a type of flow where melting glacier ice or heavy rain combines with ash from composite cones and leads to fearsome flows from volcanoes (e.g. Mount Pinatuba in the Philippines). In periglacial climates, the top layers of ice will often thaw in the summer. The presence of water and the lack of friction due to the presence of ice below leads to the movement of the unconsolidated soil – a process called solifluction.

Slumps and rotational slips

The key characteristic here is that of rotational movement (at school, as a lesson progresses so the students slump down in their chairs!). This movement is often associated with clay. Clay is porous enabling it to hold moisture, and this leads to a propensity for it to collapse. The tell-tale signs are the lobe with convex slope at the base and the **graben** left at the top (see Figure 1.25).

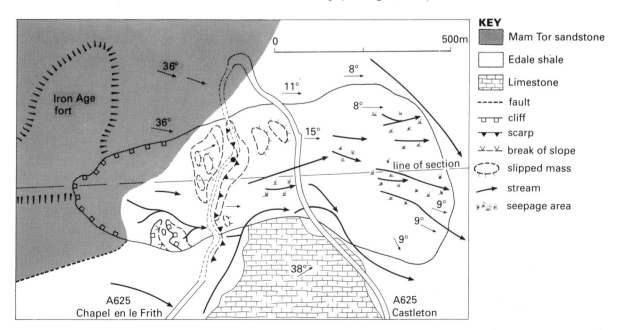

Figure 1.24 *Slope failure at Mam Tor*

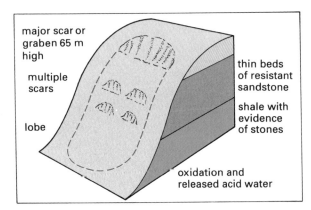

Figure 1.25 *Major slope features*

Fall

The determining characteristic of fall is the steepness of the slopes. Slopes with angles of 40° up to the vertical, are likely to be subject to fall as the main process. Often a free face will have little vegetation (Figures 1.1 and 1.26) and so the bare rock is exposed to weathering processes. Figure 1.26 reflects elements of this: zone 1 and zone 2 are a different colour to the surrounding zone 3 which is covered in green lichens. Zone 1 represents a rock avalanche similar to that in Figure 1.1 and zone 2 represents a chute down which material weathered by freeze-thaw or hydration falls. The end result is a pile of scree seen in zone 4 stretching down to the snout of the glacier. The collapse in zone 1 could have been caused partly by pressure release as rock has been removed by the glacier. In many cases the scree ends up as a scree slope often around the critical angle between 27–32° as shown in Figure 1.1. Where the scree is not removed a blockfield is produced (Figure 1.18).

Wash

Wash is close in definition to flow but in this case, the water tends to stay on the surface due to impermeable rocks or excessive rainfall. In desert areas this will often lead to sheetwash with the water spread over a large flat area. Intense convectional rainfall is unable to infiltrate fast enough and so washes off as a sheet. In some areas this water can be channelled and again the Dwalas show the development of **rills** (see Figure 1.27). Badland areas with plenty of weathered material produce gullying where the channels are more fully developed.

Figure 1.27 *Rill developments on the Dwalas*

Creep

This process involves the expansion and contraction of soil, either due to alternate wetting and drying, or due to freezing and thawing. As the soil expands, the only direction it can do so is parallel to the surface (shown as movement 1, Figure 1.28). In contraction however the main force is gravity, a vertical force (shown as movement 2).

The particle ends up at movement 3 because of the soil's tendency to have a cohesive force. Many grassy slopes in England have terracettes which look like animal tracks. Animals may in fact use them but their origin is from creep processes.

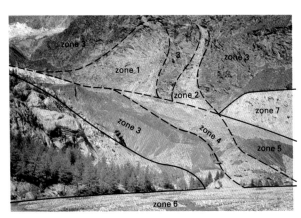

Figure 1.26 *Fall at the Fee Glacier*

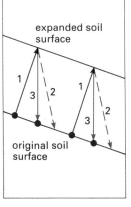

Figure 1.28 *Soil creep (heave)*

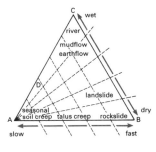

Figure 1.29 *Mass movement classification*

Raindrop impact (splash)

In areas of intensive rainfall, where there is little vegetation to provide protection via interception, the direct impact of a large raindrop can have important consequences. The splash of the drop often incorporates small particles of mud and splits with the majority heading downslope. The small particles of mud are left behind in the crack or pore spaces through which the droplet infiltrates. This blocks the crack or pore and allows less infiltration, and hence more wash, to take place. The impact of the drop also compacts the surface which again leads to less infiltration. This is particularly true in areas such as Nepal where people have removed vegetation. Without interception by the leaves, the slopes are exposed to the full impact of the splash.

> **Q11.** Using the headings of (i) rock and its characteristics, (ii) climate, (iii) biotic and (iv) human factors, re-read the previous section and list the ways in which these factors have played a role in the processes operating.
>
> **Q12.** Study Figure 1.29. Fill in the four parts which are missing (A–D) on the diagram.
>
> **Q13.** Compare and contrast the processes operating on slopes in a southern England chalk escarpment with those of a Zimbabwean Dwala.

Figure 1.30 Winnats Pass on the A625

DECISION MAKING EXERCISE

Mam Tor

Refer to Figures 1.24–25 and 1.30–1.
In 1979 the A625, the main link between Sheffield and Manchester, was closed due to severe slope failure. As the planning assistant for this district you have been asked to gather information to help decide whether this road should be repaired or whether an alternative route should be developed.

Your brief is as follows:

1 to decide what happened and why;
2 to decide whether the road is safe to repair and if so what should be done to ensure no further failure;
3 to assess another possible route that could be used indicating any difficulties that need to be assessed.

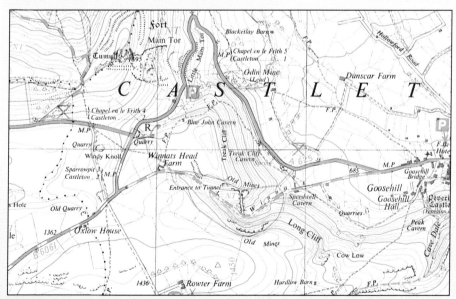

Figure 1.31 1:25 000 OS extract, Castleton © Crown Copyright

PROJECT SUGGESTIONS

1 Measuring slope angle and particle size (e.g. scree slope)

This will involve sampling, normally using a transect or linear sample with measuring facets (short lengths of similar angle) by one of a variety of methods (i.e. clinometer, gun clinometer or pantometer, using ranging poles). At various points along this transect the particle size can be measured. This is normally done using the 'A' axis, or longest axis, with callipers. A variation in location of slopes would be useful i.e. involving aspect, a stream at the bottom, a dry valley, an upland slope, a more gentle slope. Comparisons can then be made. Scree slopes, however, can be dangerous!

2 Weathering and gravestones

Here it is important to find the factors which seem to affect weathering within a local area. The gravestones provide the elements of time normally so difficult to include within the scope of A level projects. Using a variety of different churches to gain a range of environments, research can be undertaken to ascertain to what degree various factors have influenced the weathering. For example, age, aspect, geology and lichens can be considered. The difficulty is deciding how to measure or estimate the amount of weathering, and isolating each factor in the weathering process.
A variation in this theme is seen also in terms of buildings (see J Frew: *Advanced Geography Fieldwork*, Nelson).

GLOSSARY

Bedding planes the surface between the different layers of sedimentary rocks.

Blockfield an area of angular rock fragments found in a relatively flat region.

Denudation a cycle of the combined processes of weathering, mass movement on slopes and of erosion and deposition by an agent of transport, resulting in the wearing down of the landscape.

Erosion breakdown and removal of material by a variety of processes.

Fault crack in the rock or in the earth's crust along which movement can take place.

Graben valley caused by the ground sinking.

Granular disintegration rock is broken down into small particles or minerals.

Humus dark brown to black, decomposed, amorphous organic material within a soil profile.

Igneous rock formed when magma cools allowing minerals to crystallise.

Intrusion magma which has forced its way into surrounding rocks without reaching the surface.

Joint crack in rock not involving displacement.

Karst a region of limestone rock with underground drainage and surface hollows.

Lithology properties of the rock including resistance, porosity, permeability and strength.

Mass movement downward movement of regolith or rock under the force of gravity.

Metamorphic existing rocks changed in minerals and structure by heat, pressure and/or stress.

Rills small channels where water from sheetwash begins to concentrate.

Sedimentary sediment which is consolidated into rock, normally in a series of layers.

Tectonic forces uplifting and deforming force caused by the interaction of huge crustal plates which cover the earth's surface.

Weathering chemical alteration or physical disintegration of rock in interaction with atmosphere, biosphere and hydrosphere.

REFERENCES

R Collard, *Physical Geography of Landscape*, Unwin
R J Small, *Geomorphology and Hydrology*, Longman
M Clark & R J Small, *Slopes and Weathering*, Cambridge
A Goudie, *The Nature of the Environment*, Blackwell
Dalton, Fox, Jones (1990), *No 11 Classic Landforms of the Dark Peak*, Geographical Association 1990
J F Whitlow, *Granite Landform Features in Zimbabwe*

C Ollier, *Weathering and Landforms*, Macmillan Education
Geofile (Jan 1990), *Weathering*, Stanley Thomas
Geofile (Jan 1993), *Slopes & Slope Processes*, Stanley Thomas
Geographical Analysis (Feb 1989), *Quarries Issue 16*, (March 1989), *Scree Slopes Issue 18*, (June 1989), *Solifluction Issue 20*, Geographical Association
Geography Review (Sept 1991), *Crumbling Walls*, (Jan 1993), *Impact of Acid Rain on Building Stone*, Philip Allan Publishers

SECTION 2

All iced up

by Richard Walker

KEY IDEAS

- What and where is ice?
- Has the extent of ice varied in the past?
- How does ice move?
- What do the glacial landforms tell us about the operating processes?
- What are periglacial processes?

To study glaciation it is helpful to be able to see a glacier at first hand. Unfortunately in Britain none now exist but with the introduction of package tours and the relative ease of travel, many now have the opportunity to see glaciers during skiing trips. Even this can be misleading as so much of the glacier and the landscape is covered by snow.

This section draws heavily on one field study from the Saas Fee area of Switzerland (see Figure 2.1, page 20). The Saas Valley is found in the Valais Canton in southern Switzerland close to the Italian border and in the adjacent valley to Zermatt and the Matterhorn. Landforms found here are similar to those found in parts of Britain such as the Lake District and Snowdonia. These areas provide evidence of similar processes having operated in Britain in the past which produced their characteristic landscapes.

ICE AND ITS DISTRIBUTION

Ice forms where snow falls and accumulates. In some areas, like Britain, snow will fall but will almost entirely melt during the summer. As a result ice does not really form or last. Where snow does accumulate then the development of ice can take place. Study Figure 2.2 on page 20 which shows the start of the Trift Glacier above Saas Grund. Zone IIa shows snow. Contrast this with the opaque colour of zone V. What has taken place? The snow has been compacted. Closer inspection of zone IIa would show clear layers for each year's **accumulation**. As this builds up the weight squeezes out the air from the snowflakes and the density increases. Sometimes part of the snow melts and later refreezes adding to the ice. Gradually the compaction continues and ice is formed. Footprints in snow complete a similar process in a shorter time! Figure 2.3 on page 20 shows **névé** which has lasted through the summer and the opaque ice underneath. Notice how much of the rest of the snow has been removed, either by **ablation** (melting) or even by the wind. The ice has thus come from the sources up valley rather than snow accumulating here. Figure 2.2 shows the link between zones IIa and b, and zone IV, with the ice field above feeding the **corrie** below. Avalanches are common in summer and the ice curtains in zone I enable snow and debris to move easily into the corrie adding further weight and increasing the

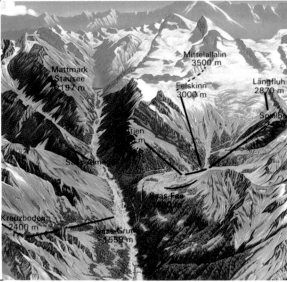

Figure 2.1 *Location of Saas Valley, Switzerland*

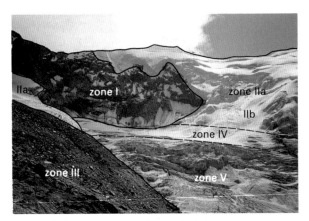

Figure 2.2 *Accumulation for the Trift Glacier*

Figure 2.3 *Contrast between névé and glacier ice*

compaction. This enables the ice to build up in what in fact is a hollow in zone IV. This is the start of the development of a corrie glacier.

The Saas Valley, towards which the Trift Glacier flows, is a tributary valley leading down to the Rhone Valley. The map in Figure 2.1 shows the area has several glaciers. So why does this area have glaciers? The answer must be linked to the altitude and latitude which causes the precipitation to fall as snow and to accumulate. Where else in the world are glaciers found? The answer is wherever there is sufficient snow surviving the summer melt. The two controlling factors are altitude and latitude. Areas of high latitude, such as the Antarctic, tend to be large collection zones for ice. Moving towards the Equator, altitude takes over as the key factor with most glacier development taking place in the Rockies, Andes and Himalayas. Nearer to the Equator, Mount Kilimanjaro is 5895 m high which enables ice to accumulate.

The scale of ice development is very different in different parts of the world. The Fee Glacier (Figure 2.1) is very small in comparison to some regions of ice. This allows for the classification according to size (Figure 2.4). The corrie glaciers such as Lagginhorn (Figure 2.6) are accumulations within the hollows. Valley glaciers move out into the valley below (Figure 2.14, page 25). When a valley glacier arrives at the low lying areas it spreads out to form a piedmont glacier such as Dobbin Bay (Ellesmere Island). The ice caps are linked to mountain ranges or extensive plateaux in areas of high latitude such as the four in Iceland. The ice sheets of today are limited to the Antarctic and Greenland.

Type	Example	Extent/Measurement
Corrie	Lagginhorn (Saas Valley)	1–2 km long
Valley	Trift and Fee (Saas Valley)	often 20–30 km long
Piedmont	Dobbin Bay (Ellesmere Island)	Variable
Ice Cap	Vatnajokull (Iceland)	Less than 50 000 km^2
Ice Sheet	Antarctic	More than 50 000 km^2

Figure 2.4 *Variation in size of glaciers*

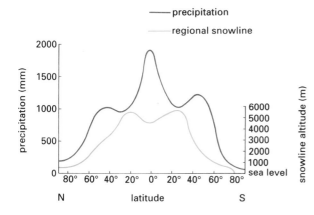

Figure 2.5 *Variations of the snowline with altitude*

Q1. *Study Figure 2.5.*
a) *Do the following places have a permanent snowline:*
Mount Fiji 35°N 3776 m;
Mount Cook 43°S 3764 m;
Mount Kenya 0° 5199 m;
Goldhöppugen 62°N 2469 m?
b) *Describe and explain the shape of the line representing the regional snowline.*

Figures 2.6 and 2.14 (page 25) show ice development on two different sides of the valley. Figure 2.14 shows the development on the Mischabel ridge with the Hohbalm Glacier on the right (western) side. Figure 2.6 has Lagginhorn central with the peak of Weissmies just to the south. The contrast is quite obvious with full corrie glacier and valley glacier developments on the east facing slopes and only partial corrie and some small glacier development on the west facing. How has this come about?

The Saas Valley runs approximately north–south (see Figure 2.1). The Fee Glacier is more protected from the winds and the sun. The prevailing winds from the south west bring snow. As the winds blow over the Mischabel Ridge, the snow is able to settle in the sheltered areas. On the eastern flank of Weissmies, in areas such as Lagginhorn, the area is more exposed so the snow does not settle. The sun's rays (insolation) are thought to play a role in the ablation of the snow. Yet remembering the **albedo** of snow (80–95 per cent) perhaps this is sometimes overrated. However, the higher afternoon temperatures will be significant to west facing slopes. In the Saas Valley the rain shadow effect may be more important with much of the moisture being deposited on the Mischabel Ridge on the west, leaving less moisture for the eastern side. Hence **aspect** also plays a vital role in glacier development.

Figure 2.6 *Weissmies Glaciers*

VARIATION IN AMOUNTS OF ICE

Knowledge about past climates has expanded greatly in recent times. Evidence has been gathered from deep cores taken from within ice sheets such as Greenland and the Antarctic, and from sea floor sediments. This involves measuring the isotopes O^{16} and O^{18} in the shells of plankton and also within the ice itself. Snowfall, linked to higher evaporation, will reduce the level of O^{16} in comparison to O^{18}. Using this evidence, as well as geological deposits and Carbon 14 dating, gives the opportunity to piece together a picture of the past temperatures of the earth.

There have been several **Ice Ages**, when up to 30 per cent of the land was covered in ice, compared with about ten per cent today. The most recent was during the Quaternary period about two million years ago to about 12 000 years ago (see Figure 2.7). This epoch is called the **Pleistocene** and has been followed by a generally warmer period known as the Holocene. Figure 2.7 shows the variations in temperature that have occurred in the past 800 000 years.

well. The presence of more material and gases in the atmosphere mainly due to volcanic eruptions, even a different arrangement of continents around the South Pole and earlier mountain building eras, could all lead to a decrease in temperature or an increase in the amount of accumulating snow.

For Britain, this meant times of colder periods (known as glacials) with warmer periods (interglacials) in between (see Figure 2.7). Britain is thought to have had at least four main glacials within the last Ice Age and maybe as many as 16. The four main colder periods have collaborative evidence from North America and the European Alps. Further evidence of change is found by examining the pollen from peat deposits where the pollen is able to survive. The pollen gives indications of the tree species alive during certain time periods, and with their known temperature tolerance levels, estimations as to warm or colder periods can be made. For example in Britain the presence of more pine and birch would suggest colder periods than that of today. Further to this, the growth of tree rings gives an indication of temperature, with faster growth coinciding with warmer temperatures.

Figure 2.8 shows the extent of the four main glacial periods in the UK. Some of the advances obliterated

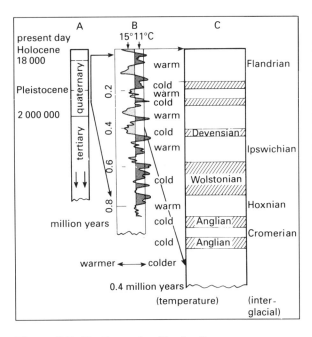

Figure 2.7 *The timescale of the Ice Age*

The causes of these Ice Ages are not fully known but the main ideas in vogue are those put forward by Milankovitch in the 1920s. These link the earth's orbit round the sun, its changes in tilt and even its wobbles around its axis to decreases in radiation reaching the earth. When the three effects coincide the decrease in radiation is thought to trigger the onslaught of an Ice Age. Other factors are thought to play a part as

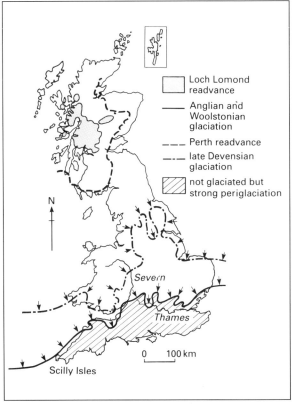

Figure 2.8 *Ice coverage of Britain during the Pleistocene*

22 ALL ICED UP

evidence from previous glaciation making interpretation very difficult.

Even since 12 000 years ago there have been colder times. The Little Ice Age is quite well chronicled. During the fifteenth century the glaciers in Europe started to advance and this lasted to around 1850. Historical reports show complaints from farmers in Norway and Iceland who lost fields to advancing ice. In Switzerland the homes of several villagers were destroyed. In England the Thames froze over for two months of the year and frost fairs were regularly held culminating in the biggest one in 1683–4. Paintings by Dutch painters from this period depict such events.

On an even shorter time-scale, fluctuations occur over decades and even on a yearly basis. Scientists in the 1970s suggested a new Ice Age was on its way, but now the Greenhouse effect is of greater concern! An understanding of the cycles of temperature change is crucial if a full picture of the greenhouse change in temperature is to be put in its proper perspective.

Signs posted on the Morteratsch Glacier near St Moritz, Switzerland, chart the retreat of the glacier back up the valley. The Fee Glacier charts its retreat by the location of the terminal moraine (see Figure 2.17, page 27) suggesting a retreat of 1 km per year since 1850. In some cases the seasonal change is crucial. In the Gulf of Bothnia, the seasonal freezing allows for use of the ice as a medium on which transport can take place. Often the frozen water stops transport. The building of the pipeline in Alaska from Prudhoe Bay to Valdez for transporting oil was linked to the frozen waters to the North of Alaska.

Q2.   **a)** *Plot the data in Figure 2.9 on a graph.*
   b) *Suggest what this information shows.*

Alpine	1965	1970	1975	1980
% of glaciers advancing	80	64	57	43
% of glaciers retreating	7	23	31	49
% of glaciers stationary	13	13	12	8

Figure 2.9 Glacier movement

MOVING ICE

It is difficult to imagine a solid actually flowing and yet this is exactly what ice does. A comparison can be seen in old churches where the bottom sections of old glass panes tend to be thicker than the top as the glass has slowly flowed downwards over the years. However, ice movements are quite complex and are best understood broken down into the component parts. Studies have shown that, in many glaciers, there is a slide component 'A' known as **basal slip** as depicted in Figure 2.10 on page 24. The pressure of the weight of the ice melts the part of the ice in contact with the bed. This water decreases the friction and allows movement in a series of small jerks. This cannot happen in what are termed polar glaciers as in this case the ice is frozen to its bed.

The second component is internal deformation (shown as B in Figure 2.10). As the term suggests, there is a change within the ice. As the ice builds up in the corrie, its weight puts pressure on the ice beneath it. Two different changes can take place. Firstly the ice crystals themselves can alter shape and move past each other. Or secondly, if the pressure is too great, a slip plane is set up. In this situation, movement occurs along this plane. This is reflected on the surface by a series of transverse **crevasses** due to the differential movement. These can be seen in zone V of Figure 2.2 (page 20).

These components produce differing effects. In the hollow called the corrie, a form of rotational movement takes place (see Figure 2.11 (I), page 24). With ice and debris added on the upper section, there is a strong downward pressure at A. With more melting at B there is also an important outward component. The overall resultant movement is rotational seen in Figure 2.11 (II). As the ice moves away from the backwall, a crevasse opens up called a **Bergschrund** (see Figure 2.11).

Further down the glacier, particularly where there is a major change in gradient, other flow patterns are set up. Where there is an increase in gradient, the bottom section can be moving faster than the top. The slip planes are downward in direction (see Figure 2.12, page 24). The evidence for this is seen in the crevasses opening across the glacier (see Figure 2.2, page 20) just below zone IV. In other areas the gradient decreases. Here the top section moves faster than the bottom with the resulting slip planes moving upward. The evidence is often seen in the bulging surface and further crevasses.

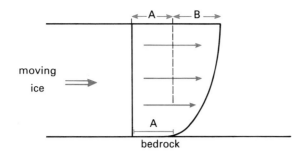

Figure 2.10 *The two components of glacial movement*

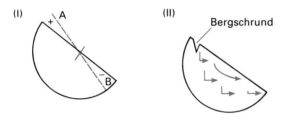

Figure 2.11 *Rotational movement within a corrie*

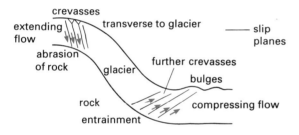

Figure 2.12 *Extending and compressing flow*

An interesting thought is captured in the question: what goes forward as it retreats? The answer of course is a glacier! The ice will still be moving forward as its snout retreats.

Q3. *Study Figure 2.13 which shows the mass balance for a glacier.*
 a) *Work out the total accumulation, total ablation and balance for each month;*
 b) *represent this data on a graph;*
 c) *annotate the seasonal changes;*
 d) *work out the total mass balance for the year. If this overall trend continued, what effect would it have on the glacier?*

Mid-1980s	Direct snowfall	Avalanches	Windblown snow	Meltwater	Evaporation
Oct	+22	0	-2	-10	-1
Nov	+94	0	+2	-5	0
Dec	+85	+3	+1	-1	0
Jan	+210	+3	0	-2	0
Feb	+55	+14	+3	-3	-1
Mar	+3	+12	+1	-15	-6
Apr	0	+8	-2	-43	-8
May	+1	+8	-3	-192	-20
June	0	+1	+1	-100	-14
July	0	0	0	-33	-15
Aug	0	0	0	-20	-16
Sept	+1	0	0	-10	-3
Total	+471	+49	+1	-434	-84

+ accumulation   – ablation

Figure 2.13 *Mass balance for a glacier*

LANDFORMS AND ICE EROSION

Landforms left in areas of ice erosion give vital evidence about the processes forming them. Firstly landforms in the study area will be described.

Corries, arêtes and pyramidal peaks

The hollow in which the ice forms (seen in Figures 2.6 and 2.14) is called a **corrie**. Figure 2.14 shows the corrie development under Nadelhorn with the Hohbalm Glacier flowing out. It shows the characteristic shape with steep walls surrounding the ice, giving rise to a bowl-shaped feature. In between these features steep ridges known as **arêtes** are often found separating one corrie from another, either at their side or even backing on to each other. Where several corries are found back to back then **pyramidal peaks** or **horns** are formed. Figure 2.14 shows the ridges leading up to the peaks (in the clouds). The Matterhorn is the most famous near this area.

Overdeepened valleys

The main Saas Valley has steep slopes leading down to a much more gentle base and this can be seen in Figures 2.1 (page 20), and 2.15 for the Rhone Valley. This shape is often referred to as U-shaped but some authors link it more to a parabola. This is in sharp contrast to the typical V-shaped river valley in mountainous areas.

Figure 2.14 *Corrie development under Nadelhorn*

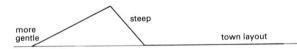

Figure 2.15 *The Rhone Valley*

Features of glaciated valleys

There are several features found in association with glaciated valleys. Firstly there is one called a **roche moutonnée** (rock sheep!). It is the variations in the resistance of rock that are the key here. Figure 2.15 shows two roche moutonnées in Sion in the Rhone Valley. These features have resisted erosion, while glaciation has removed the original less resistant surrounding rock, and there is a gradual slope up to the highest point and a much steeper slope down towards the town. In the background, the bare rock reveals another feature. The spur from the old ridge has been removed by the ice action to form a truncated spur shown by the steep, craggy rocks.

The settlement of Saas Fee is perched in a tributory valley with a clear drop of 240 m down into the Saas Valley. The photo in Figure 2.1 shows the tortuous route of the road from Saas Grund to Saas Fee in response to the steepness. The same feature can be found on the eastern side up to Kreuzboden. These steep sides in glaciated valleys are referred to as hanging valleys. They are the valleys of tributaries eroded and cut back by the main glacier. Often they can be seen with waterfalls cascading down into the main valley.

Other features which are present but less apparent are ribbon lakes. The long thin lakes in the valley bottom (for example Mattmark) are found in many valleys. The one in Figure 2.1 happens to be partly artificial.

Leading up from the ribbon lake is a steep slope into the Alps as seen at Monte Moro in the Saas Valley. This is referred to as a trough end.

What do these features suggest about erosion?

Effectiveness

The distinctiveness of the features found in all areas of glaciation suggests that the erosion process has characteristic aspects and so must be extremely effective to produce these results.

Smoothing

Much of the rock surface of a glaciated area is smoothed. The process responsible for this is called **abrasion**. As the ice moves along the bed, rock fragments are crushed between the ice and the bedrock. This has a sandpapering effect and the crushed rock, known as rock flour, can be seen in what is called glacial milk, the white coloured water flowing out of glaciers.

Grooves

Within the smoothing, grooves are found and these are explained by the fact that pointed rocks will exert a great pressure with the weight of ice above. Rather like the way in which high-heeled shoes will mark a floor, whereas a flat-soled shoe spreads the pressure. These grooves are called **striations**.

Broken nature (plucking)

This is seen on the back wall above the corries (Figure 2.6) or on the downward side of the roche moutonnées. There is difficulty in interpreting this as much of it could be caused by freeze-thaw or hydration. It is debated as to whether ice actually erodes here. In some ways the term plucking gives a wrong impression. Unbroken rock is stronger than ice so the ice will deform around the rock and will not be able to pluck it away – the roche moutonnée remains. However if the rock is already broken or fractured by weathering, then the rock can be entrained. The ice partly melts due to the pressure and the meltwater flows round the rock. It freezes again with the pressure being released, so gradually enveloping the rock in ice. This process is known as **regelation**. Hence the rock is incorporated in the glacier. This process would tend to leave jagged surfaces. How much is weathering and how much is entrainment is still open to debate.

Steep slopes

The ability of the glacier to produce these distinctive valley slopes is not doubted. The presence of the hanging valley (Saas Fee, Figure 2.1) is also due to the power of the main glacier relative to its tributary. Overdeepening is the term used here and it is seen to be a positive feedback situation. In particular, where a group of corrie glaciers meet and coalesce, the extra ability of the ice to erode is shown by the presence of a trough end where there is a sudden gradient change. Ice erodes and increases the gradient, and therefore it moves faster and so erodes more, and so increases the gradient, and so on. This may be linked to **compressive flow** where the material at the base of the slope is removed effectively in the upward movement of the slip planes. This develops the slope further and maintains the same steep gradient. Often in the valley bottom a lake builds up and is known as a finger or ribbon lake as found in the Lake District. The old interlocking spurs are removed to leave what are known as truncated spurs (Figure 2.15).

Corrie shape

The characteristic shape of a corrie must be linked to the processes mentioned earlier and to the rotational movement of the ice. This would explain the jagged nature of the back wall and the smoothed lip or threshold at the exit. The change in gradient beyond the lip is shown in the extending flow out of the Trift Corne in Figure 2.2.

Issue

To what extent is ice an erosional process? This may seem an unusual question but it is argued that ice is mainly a transporting agent. Figure 1.26 (page 16) shows where the rock fall has been in the Fee Glacier and also a chute for weathered material. This chute extends onto the glacier and the scree material can be seen at the snout. Under present conditions, ice in these Alpine areas may be more effective carrying the material which has fallen on to it rather than carving out much itself. However the rock flour and presence of the features mentioned indicate that erosion and removal are both highly effective.

GLACIAL DEPOSITION

Study Figures 2.16 to 2.23. These illustrate many of the depositional features associated with valley glaciers. These can be grouped, according to the amount of sorting, in terms of size and also shape.

Unsorted

This type of deposition is mostly called **till** and consists of a great variety of blocks of different sizes, mainly with jagged shapes and often with very fine material interspersed. Till links to the source of the material either by weathering (freeze-thaw and hydration) and mass movement (fall) from the valley sides or by the erosional processes of abrasion (fine) and entrainment (variety). This material is dumped in a variety of places. The **terminal moraine** is found at the snout of the glacier. Figure 2.16 shows the position of the snout before the glacier retreated. The **lateral moraine** is at the side. Figure 2.18 shows the composition of the lateral moraine as gulleying has taken place. The glacier is in retreat and a small **recessional moraine** seems to have built at the snout with the material from the rock fall (see Figure 1.26). In some cases where two glaciers meet the two lateral moraine joins to form one **medial moraine** within the main stream of the glacier.

Figure 2.16 Deposition features beyond the Fee Glacier

Sorted

Much of the material associated with glaciers is affected by meltwater and so its origin is referred to as fluvio-glacial. Depending on the time in which the material has been in the meltwater, it will show elements of sorting as the flow slows with larger material deposited first. The meltwater will tend to remove the jagged edges to produce rounder particles through the process of **attrition**. Meltwater under the ice often forms **eskers** which are long depositional ridges following the meandering path of the stream. Beyond the snout the flow tends to spread the load of material out via a series of distributaries to form an **outwash** plain (see Figure 2.16). This is graded in its deposits until finally only the smallest material is deposited in the proglacial lake. Due to the seasonal nature of meltwater, the summer melts will produce a much thicker layer of deposits compared to the winter freeze. This leads to the development of **varves** under such lakes (see Figure 2.17).

Figure 2.18 *Lateral moraine with gullying*

Combined

In some cases the deposited material shows only some sorting. For example, crevasses near the snout fill up with material from the glacier either by mass movement, ablation or by meltwater. Due to a partial meltwater origin, mounds known as **kames**, when left by the retreating glacier, will show some elements of sorting. These sorts of crevasses can be seen in Figure 2.19.

Figure 2.19 *Snout with crevasses*

Transportation

Figures 1.26 and 2.19 also show the different ways a glacier can transport. Material is carried on the surface (supraglacial) as well as contained within the bands of ice (englacial). Part of the englacial will come from the erosion process and part from mass movements which have been covered later by snow and ice. Rock material can also be carried underneath (subglacial) and will tend to be used in the process of abrasion.

Q4. Study the data given in Figure 2.20 on page 28. B, C and D give data obtained from the four sites shown in A.
  a) Set up hypotheses to link the sorting, size and orientation of particles to the four sites.
  b) How would you collect this information?
  c) Present the information in C and D in diagrammatic form.
  d) Analyse the data in relation to your hypotheses.
  e) Suggest what the landform at site 3 might be and give your reasons.

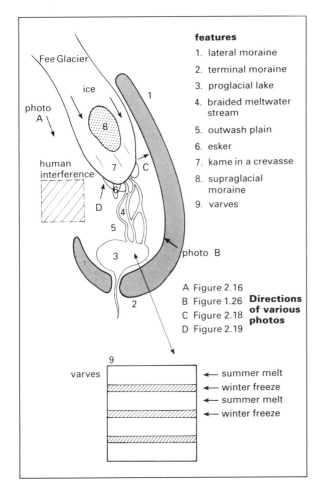

Figure 2.17 *Depositional features of the Fee Glacier*

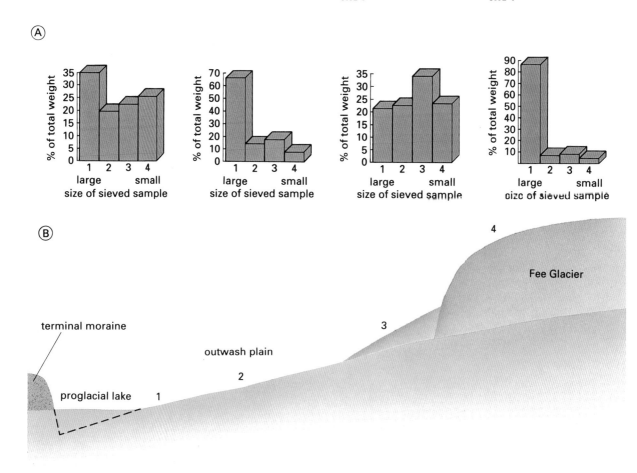

Figure 2.20 *Data from the Fee Glacier*

PERIGLACIAL PROCESSES

Peri in Greek means 'around' or 'about'. **Periglacial** covers the regions which flank areas of ice and glaciers. This is estimated to be about 25 per cent of the world's land surface, a not inconsiderable amount. The periglacial zone is associated with relatively treeless tundra in barren high latitudes as well as in mountain ranges stretching into lower latitudes. Figure 2.21 shows these areas for the Northern Hemisphere. The map shows the influence of mountain ranges such as the Rockies on this distribution, as well as the continental climate stretching the zones south. The key element to these areas is the low temperatures for most of the year which lead to frozen ground. **Permafrost** is the name given to this perennially frozen ground. In some cases, there is a summer thaw as temperatures reach up to 15°C or more in continental areas. This creates an **active** layer where surface zones can thaw down to 6 m. In some areas the variation in soil and vegetation produce localised thawing which is given the name of **discontinuous permafrost**. Where the soil can hold moisture, such as in clay, permafrost is near the surface; on the other hand sandy soils with good drainage only lead to permafrost at greater depths. In the Northern Hemisphere, south facing slopes receive more insolation which increases localised thaw.

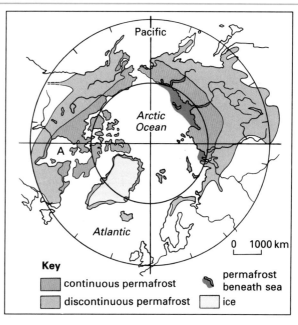

Figure 2.21 *Distribution of permafrost*

Figure 2.22 *Development of ice wedges*

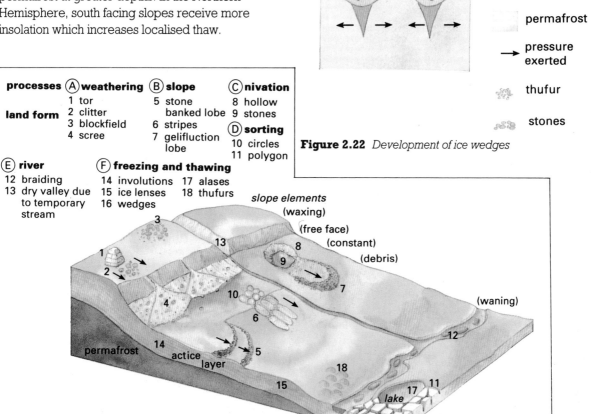

Figure 2.23 *Landform features produced by periglacial processes*

To grasp the development of periglacial landforms, two important points need to be understood. Firstly, the actual process of freezing and thawing is crucial to the formation of the landforms and secondly the nature of the soil and sediment varies considerably. Together these points help to explain the development of virtually all the key features dealt with here.

The process of severe winter freezing, for example, involves contraction. As a result, cracks appear in the ground, rather like drying mud. During summer, meltwater and sediment will flow into these cracks. As freezing takes place again the **ice wedges** (see Figure 2.22, page 29) will grow and widen the cracks still further.

The presence of water in different quantities in parts of the active layer is important in the formation of **ice lenses**. As solid ice grows in the soil in winter, it draws water in from the surrounding pores. Therefore silts provide the best medium for lenses. The lense expands and domes up the surface of the land, a process known as **frost heave**. This may be responsible for features known as thurfurs (see Figure 2.23, page 29). Because the ice forms in separate zones within the soil it is known as 'segregated' ice.

Regolith with pebbles and rocks within it, produces the next effect. The stones are able to release heat quickly during the night and as a result any water underneath the stones is more likely to freeze. With smaller pebbles **pipkrakes** or thin ice needles form which lift the pebbles. Larger stones attract more water, as in the ice lense, and as the water expands, the stones are forced upwards towards the surface. In many cases these stones tend to move into the zones above the wedges and so form **stone polygons** and **circles**. Once this is established the larger stones tend to freeze more easily than the finer ones in the centre and so exert pressure which domes up the central section of the ground (see Figure 2.22).

The next step is to realise that both freezing and thawing will take place on the surface first. As the active layer freezes downwards it exerts a pressure on the more flexible but still thawed material below. Due to the uneven freezing downwards, the mobile layers become squeezed and twisted into what are known as involutions. This is particularly true if the layers are made up of fine material with a high water content.

As pressure is exerted downwards by freezing, and from the sides by encroaching permafrost, so water is trapped on all sides, known as **talik**. This happens particularly under lakes. The water eventually freezes, and in doing so expands, doming up the surface. Over a number of years this can produce features of up to 60 m high which are called pingos.

During times of thaw, ice lenses and wedges can melt and collapse, which leads to hollows where marsh collects in the bottom. If there is less melting, smaller rounder rises remain. The scenery here is referred to as thermokarst because of the comparison with the solution hollows of limestone scenery. Melting where wedges are prevalent can create much deeper hollows known as alases and often lakes will form. The insulating effect of the water may prevent the permafrost from re-establishing (see Figure 2.23).

Slope development and processes

The processes involved on the slopes are also linked to freezing and thawing. For example frost creep occurs with expansion parallel to the existing surface (see soil creep in Figure 1.28, page 16) and with vertical drop. This process is more effective in periglacial conditions than in temperate climates and pipkrakes can help the process.

The thawing on the surface leads to a wet unconsolidated mass lying on top of an impermeable frozen layer. With little vegetation to stabilise it, this active layer is able to flow, a process known as gelifluction. This is a term for solifluction within specifically periglacial environments. The material moving down the slope often produces a lobe with a convex slope (see Figure 2.23). The sorting process mentioned earlier can lead to stones around these lobes producing stone banked lobes.

The significance of aspect is seen here. The south facing slopes experience more thaw which leads to asymmetrical valley profiles as gelifluction causes movement and creates gentler slopes. The north facing slopes are more sheltered and therefore have steeper slopes.

The effectiveness of slope processes can be seen in the relict features in areas in Britain where periglacial action was significant. For example coombe deposits found at the bottom of chalk escarpments are said to have moved during these times. Clitter or the scatter of rocks spreading down slope from tors shows further evidence of this movement. A large gelifluction lobe has been formed beyond the mouth of Cheddar Gorge as material has flowed out over the surrounding clay.

Thawing can be so effective that slope wash occurs with rapid movement, like sheet wash, over a wide area. Otherwise the water can be concentrated in rivers over the impermeable surface, with the permafrost preventing the water soaking through the chalk or limestone, but allowing dissection by meltwater channels. It is this process that is one of the reasons given for the development of the now dry valleys in the permeable limestone areas such as the Cotswolds.

Nivation

The combination of weathering and slope processes produces an important effect which links back to glaciation. Where snow drifts occur, weathering occurs underneath. Gelifluction can then remove this weathered material, which expands the hollow. This is especially true where the material is fine because coarser material will tend to remain in the hollow. If later ice is able to form then this can start the development of a corrie. This effect of snow on the landscape is known as nivation.

DECISION MAKING EXERCISE

A new UK ski resort

Obtain an OS map extract for an upland area of the UK which has been glaciated (e.g. Snowdonia, Lake District, Scottish Highlands).

1. Try to account for the glacial history of the area. Use tracing paper and draw in the upland areas, working down to the main valleys and any low lying area. Using the maps, identify and name any glacial features referred to in this chapter.
2. Imagine that there is a plan to develop this region into a ski resort. As a planner for your area you are required to produce a plan showing how this development should take place. The requirements are as follows.

Communications – road and/or rail.
Mountain restaurant.
Ski lifts and cable cars.
Information centre.
Mountain rescue post.
Camping site.
Ski schools.
Hotel development.

3. Present a case for a public hearing which follows one of these two views:
   a) the development should go ahead;
   b) this is just the sort of development which should not be considered for this area.

PROJECT SUGGESTIONS

Projects are mainly going to be associated with till fabric analysis and changes in slope angles.

1 Distinguishing between different deposits

Deposits which have been laid down in different environments will show varied characteristics. From these characteristics, the possible origin and method of deposition can be established. There are four aspects to be measured. The orientation of the 'A' or long axis is measured using a compass; the angle of dip of the 'A' axis using a clinometer; the size of the 'A' axis using callipers; and the shape either by the Cailleau Roundness Index or Powers scale of Roundness. By using graphs for orientation and dip, standard deviation for size, and comparison of the indices, the four characteristics can be compared.

Mass movement, ice and meltwater will produce deposits of a very different nature. The use of random sampling with a minimum of 30 stones will give a representative sample. (See Figure 2.20).

2 Slope angles

Due to the scale involved, the best angles to measure are those to do with depositional features. Certain depositional features have recognisable shapes, for example moraines will tend to be shaped asymmetrically. A careful choice of depositional feature is important. Once established, the main needs will be two ranging poles and a clinometer. The better the clinometer the more accurate the results.

GLOSSARY

Ablation melting of snow or ice from the area.

Abrasion sandpaper effect crushing rocks into flour underneath the glacier.

Accumulation collection of snow turning to ice.

Active layer layer of soil subject to periodic thawing created above the permafrost.

Albedo the reflective nature of the surface.

Arête steep ridge between two glacial areas.

Aspect direction a valley side or slope faces.

Attrition wearing away of particles of debris by contact with other particles.

Basal slip movement of glacier over bedrock.

Bergschrund deep tensional crevasse formed around the glacier head, as ice moves downslope.

Compressive flow movement of ice, via slip planes, towards the surface, caused by a decrease in gradient.

Corrie hollow in which the glacier develops.

Crevasse a crack in the ice opened due to stress.

Esker a sinuous ridge of material laid down by a river under a glacier.

Frost heave upward dislocation of soil and rocks by soil water freezing and expanding.

Hanging valley the tributary valley is left much higher than the main valley due to the glacier cutting it back.

Ice Ages time periods during which ice covered much more extensive areas than today.

Ice wedge a near vertical sheet of ice tapering downwards.

Ice lens area of ice often with convex surfaces, existing underground in periglacial conditions.

Kame a relatively unsorted depositional mound left from a crevasse in the ice.

Lateral moraine collection of unsorted material at the side of the glacier.

Medial moraine a ridge of debris formed on the surface of a valley glacier where two lateral moraines have joined.

Névé (firn) snow surviving more than a year.

Outwash material carried and deposited beyond the snout of the glacier by meltwater.

Periglacial climatic conditions beyond the edge of glaciers with particular processes.

Permafrost ground with a temperature below 0°C over more than one year. The permafrost zones are of two types: continuous and discontinuous.

Pipkrake ice needle.

Pleistocene epoch during which the recent ice age took place.

Pyramidal peak (horn) mountain peak created by several corries backing onto one another.

Recessional moraine formed at the snout as the glacier retreats.

Regelation process where ice melts under rock pressure and refreezes as pressure is released.

Regolith broken rock debris at the base of a soil, overlying the bedrock.

Roche moutonnée asymmetrical hill with gentle slope up one side and steep slope on the other.

Stone polygons or circles areas of ground separated by cracks, forming appropriate shape, caused by frost in periglacial surfaces.

Striations grooves cut in the surface of rock by rocks carried in the ice base.

Talik layer of unfrozen ground between permafrost and seasonally thawed active layer.

Terminal moraine pile of unsorted material dropped at the snout of the glacier.

Till unsorted material comprising of a combination of fine and coarse material.

Varve layered deposit under proglacial lake with bands of different thickness relating to summer and winter deposition.

REFERENCES

R Collard (1988), *Physical Geography of Landscape*, Unwin

Clowes & Comfort (1982), *Process and Landform*, Oliver & Boyd

R J Small (1989), *Geomorphology & Hydrology*, Longman

Knapp, Ross, McCrae (1989), *Challenge of the Natural Environment*, Longman

Time Life Books (1983), *Glacier*

Geofile, (April 1992), *No 192 A case study of Iceland*, (Sept 1990), No 158 Glaciers form and process, (April 1990), *No 150 Fluvioglacial processes*, (April 1988), *No 109 The Pleistocene in Great Britain*

SECTION 3

Water on the move

by Ian Selmes

> **KEY IDEAS**
>
> - The hydrosphere is a dynamic environmental and human system.
> - How is water measured and controlled within drainage basins?
> - What is the political and economic significance of this control?
> - How is river load transported, used in erosion and deposited?
> - What is the nature of drainage channels and river valleys?
> - How are rivers monitored and managed?
> - Why do flood hazards occur, how might they be predicted and restricted?
> - Just how important are clean and plentiful supplies of water?

THE HYDROSPHERE

Water...

Water, sunlight and air are the fundamental resources of our planet. Together they enable life to exist. Without any of them the earth would be barren. Where one or more of them becomes scarce the nature of the place is altered. That there is water and that it is cycled and recycled through the environment leads to some of the most commonplace and recognisable features of the physical environment. Yet because it is unevenly distributed it is a major factor in influencing the human environment – where people live and work and in what density, their health and development, their wealth and power.

Water in the hydrosphere, whether as a liquid, solid (ice) or gas (water vapour), on the ground,

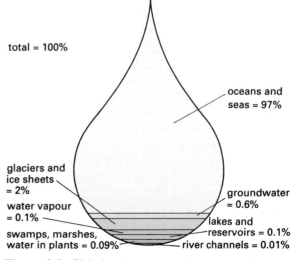

Figure 3.1 *Global water resources*

total = 100%
oceans and seas = 97%
glaciers and ice sheets = 2%
water vapour = 0.1%
groundwater = 0.6%
lakes and reservoirs = 0.1%
swamps, marshes, water in plants = 0.09%
river channels = 0.01%

WATER ON THE MOVE **33**

underground or in the atmosphere, amounts to an estimated 1.4 billion/km^3; a very large amount. Figure 3.1 on page 33 shows how it is distributed.

> **Q1.** *What proportion of the earth's water is freshwater and where is this concentrated?*
> **Q2.** *Why might this distribution of water present difficulties for people wanting to make use of this resource?*

Though there is a total quantity of water in the earth's environment, the state of that water (as a gas, liquid or solid) does not remain constant. Water molecules change from one state to another in a constantly balancing equation in which there is no net loss or gain to the environment. This continuous circulation of water molecules between ocean, atmosphere and ground is the **hydrological cycle** (Figure 3.2).

channels. In terms of a quantitative study of the hydrological cycle, **drainage basin** output (**discharge** in a river) is the product of input as precipitation minus evapotranspiration plus or minus water in storage.

> **Q3.** *Which of the ocean, atmosphere and ground is the most dynamic and the least dynamic part of the hydrological cycle?*
> **Q4.** *If more water was stored in glaciers and ice sheets what effects would this be likely to have on sea levels and on evaporation rates? Why?*

Water everywhere...

One of the variables in the hydrological cycle is time. The balance of water held in different states and in the various stores of the hydrological cycle varies from time to time. For the most part river channels are adjusted to contain the amount of water that flows into them from other stores. Perhaps the most common hazard presented by hydrology is short-term and usually localised changes taking place in the stores and flows. Unusually large and intense precipitation is occasionally experienced in most places. The result is flood, even in the most developed parts of the world.

In June and July 1993 the mid-west prairie lands of the USA were deluged with rain for more than two months. On some days as much as 125 mm of rain fell. The world's third longest river drains a third of the country. The River Mississippi and its main

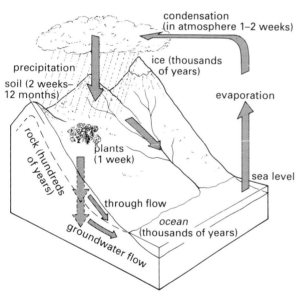

Figure 3.2 *The hydrological cycle (storage times)*

Within this cycle water is stored for different lengths of time in the ocean, the atmosphere, as interception by plants and animals, in glaciers and ice sheets, in river channels, and in lakes and reservoirs. It may also be stored in the ground as soil moisture or as groundwater in permeable rocks. Between these stores water moves or flows in various ways; by evaporation from open water and transpiration from vegetation; by condensation and precipitation; as runoff in channels; by infiltration into the soil and by seepage deeper into the rock. Precipitation may also flow overland before reaching river channels as throughflow in soil, or as groundwater flow in rock and then into river

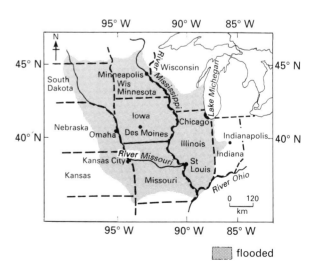

Figure 3.3 *The Great Mississippi flood 1993*

tributary the Missouri could not cope, rising by mid-July to almost 15 m above normal flood level. Twenty million hectares of land in eight States were inundated with floodwater (Figure 3.3), despite massive efforts to build high banks, or levées, of sandbags at key points along the rivers. Crops and homes were destroyed, the course of the rivers changed considerably and thick mud was left across the entire landscape. The river's commercial transport of grain, oil and coal was prevented from reaching its destination. Not since 1927 had there been such a flood on the Mississippi. Since then the US Army Corps of Engineers had built 300 dams and reservoirs to make sure it did not happen again. The power of nature is such that water gets everywhere and there is little we can do to prevent massive, though rare and short-term, imbalances in the hydrological cycle.

But not a drop to drink!

Some areas of the world are arid and tend to support low population densities although most areas experience moderate rainfall, averaging around 700 mm a year, and more people live in these areas. Just as there can be extremes of excess input into the hydrological cycle, so there can be droughts. As with floods these are localised and relatively short-term, but their effects are perhaps even more devastating to the communities involved.

In 1992 southern Africa was parched of rain for half the year. From Angola on the Atlantic coast to Mozambique off the Indian Ocean, and from Zaire to South Africa, in an area that includes some of the most fertile land in Africa, more than 100 million people faced starvation. In a normal year the Inter-Tropical Convergence Zone (ITCZ, see Section 4) with its thick band of cloud would move south bringing seasonal rain to the savanna. Without the rain, grain crops failed across the entire region necessitating vast and expensive imports for countries that are struggling to develop. Zimbabwe had to import two million tonnes of maize at a cost of US$ 450 million and Zaire spent $300 million. Rivers and reservoirs ran dry; irrigation and hydroelectric schemes failed; water rationing was instituted in towns and cities and the water available became even more polluted. Bankruptcy on the Transvaal of South Africa resulted in suicides amongst close-knit Africaner families. Millions of farm labourers were left without work or income. Only the expensive imports of food saved the region.

The hydrological cycle is not then a constant or simple model of water and its movement within the environment. Nonetheless it is to a large degree predictable. This has led to monitoring of many river systems and, from the understanding gained, control of these rivers to benefit the people who live nearby.

INVESTIGATING DRAINAGE BASINS

In order to measure and monitor the environment a suitable area has to be identified; one that is a sub-section of the total and is repeated across the globe both in terms of its characteristics and in the processes operating in it. In hydrology the area for study is a drainage basin.

What is a drainage basin?

All rivers, no matter what their size, flow within valleys whose slopes face inward on that river. These slopes rise to high points before descending into the valley of another river. The ridges that divide slopes facing into different valleys and therefore separate water draining into different rivers, are known as watersheds. The area of the valley surrounding a particular river and within the watershed is the drainage basin. It may be small, as for a single stream, or larger as in the area drained by a main stream and its tributaries. Whether it be the River Amazon in Brazil or the Gwash in Rutland, England, a drainage basin exists and the measurements that are made of it and the dynamic relationships within it are comparable.

A world scale view of water in the environment would be a broad generalisation, taking no account of the dynamic nature of hydrology both spatially and temporally. If a national study of hydrology were to be made, the unit of study would generally be a subjective human creation. Isolated investigations of parts of a river channel are of local importance but can only be interpreted in terms of the drainage network and the general physical processes affecting it. It is therefore a drainage basin that provides the most satisfactory unit for a hydrological study. It is an integrated functioning unit in dynamic equilibrium (where a change in one component causes a change in another).

River regimes

The quantity of water that flows in a river channel is known as the runoff or discharge of a river and it is measured in cumecs (cubic metres per second). The discharge is likely to vary in the short term, both within and between drainage basins, for a wide variety of reasons. These concern the nature and amount of precipitation and the physical characteristics of drainage basins. However there is an average pattern of discharge for any particular drainage basin when measured over a number of years, called the river's **regime**. This average discharge, expressed as an annual **hydrograph**, suggests the general response of a river to climate and to certain physical characteristics of the drainage basin that are relatively stable factors. Consequently there are similarities in the regimes of rivers where areas experience similar climates or types of environment.

Climatic factors that affect the regime concern the annual input from precipitation and its seasonal variation, and also the loss from evapotranspiration. This may be influenced by the height of the drainage basin through relief, rainfall and temperature changes. The larger the area of a drainage basin the greater the total water input is likely to be. Steep slopes in a drainage basin with thin soils and impermeable rock are likely to lead to higher discharge levels (Figure 3.4). The seasonal changes in precipitation and temperature cause variation in the amount of water in storage and in the speed with which it moves into a river.

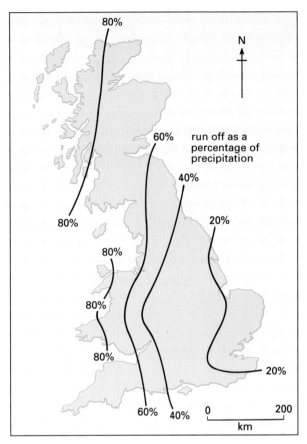

Figure 3.4 Runoff as percentage of precipitation

Q5. Select an OS map of an area known to you. Choose a river on the map, trace its length and use the contours and/or spot height clues to draw in its watershed. How large an area does the drainage basin cover?

Q6. Study Figure 3.4 and using an atlas suggest reasons for the pattern of runoff rates across Britain.

In Britain river regimes tend to exhibit a simple pattern of one high discharge and one low discharge in the year owing to the rivers' relatively short length. Yet there is a contrast between the regimes seen on lowland and upland rivers. The drainage basin of the River Thames in south-east England covers almost 10 000 km² of gently sloping chalk, sandstone and clay with thick soil and precipitation of around 600 mm a year (Figure 3.5a). The River Ystwyth (Figure 3.5b) flows in a smaller drainage basin in central Wales. In its 41 km route to the west coast, down a steep gradient on bedrock of Silurian grits and shales, the Ystwyth passes through areas of hill sheep farming, forestry and some lowland pasture. This is supported by upwards of 1200 mm a year of rainfall.

In Europe, regimes are often more complex with two peaks a year being common. On the River Rhone there is a high level of discharge in spring from snowmelt. Another peak occurs in early winter from seasonally heavy rainfall (Figure 3.5c). The timing of the peaks is determined by the climate at different places along the river. A later peak near its mouth arises from Mediterranean rainfall. Closer to the source, the alpine conditions cause higher overall discharge and less seasonality in the regime.

The larger the drainage basin the more numerous the climatic conditions found within the area. The rivers of West Africa are dominated by the River Niger whose drainage basin is 100 times bigger than that of the River Thames. Its main tributary, the River Benue, drains an area equivalent to that drained by the River Rhine. Over its 4400 km length the Niger passes through tropical savanna areas characterised by a distinct wet and dry season. However, the total annual precipitation varies from 260 mm at Gao in the Sahel of Mali to 2500 mm at Port Harcourt near its mouth in Nigeria, while the River Benue experiences over 1200 mm a year. In the wet season the discharges are high but recede rapidly with the dry season and reach a minimum as the wet season starts (Figure 3.5d). The Niger exhibits less variation in discharge than the Benue because it has vast floodplains of alluvium into which water drains and then emerges as groundwater in the dry season.

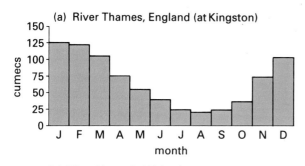
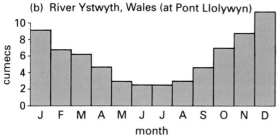
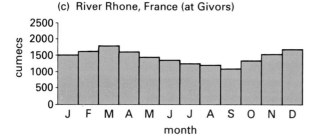
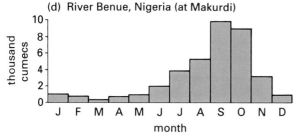

Figure 3.5 *River regimes: a) River Thames b) River Ystwyth c) River Rhone d) River Benve (tributary of River Niger)*

Q7. Describe and suggest reasons for the differences in the regimes of named upland and lowland rivers in Britain.

Q8. In what ways are the British river regimes different to those of rivers in the Mediterranean and savanna regions? Suggest difficulties that might arise for local people from the regimes of rivers in these environments.

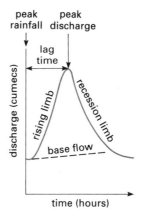
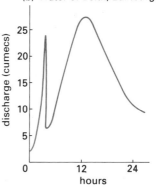
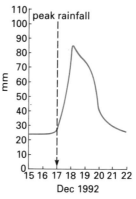
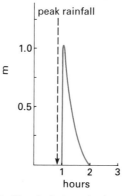

Figure 3.6 *Classical storm hydrographs: a) general characteristics b) Water of Leith, Edinburgh c) River Avon on clay (Bathford) d) Nahel Yael, Negev Desert, Israel*

Storm hydrographs

Short-term changes in a river's discharge arise from factors that are more variable and less predictable than those responsible for its regime. These variable factors include the intensity and duration of the precipitation, together with the relief, geology, vegetation and land use in the drainage basin.

In studying the reaction of a river to a rainstorm over a number of hours, a certain pattern of discharge can be expected (Figure 3.6a). From its base flow before the rainfall (the water then gets to the river by groundwater flow) there is a lag time whilst the rainwater travels from the ground on which it fell to the river channel and then flows to the gauging station. The rising limb of the **storm hydrograph** is mostly the product of overland flow. The proportion of precipitation that flows overland depends upon the intensity of the rainfall, the type of vegetation and its ability to intercept the rain, the permeability and infiltration capacity of the soil or rock or artificial surface, and the preceding rainfall history. Thus drizzle falling on wet ground in a British winter is likely to cause a steep rising limb compared to that from a thunderstorm in summer over highly permeable chalklands. Similarly, rain falling on a moorland landscape of heather or grass will not be intercepted as much as on an area of forestry and rivers, and will therefore exhibit a shorter lag time and a higher peak discharge.

After the precipitation has ceased and overland flow has also finished there is a recession limb to the hydrograph, as water enters the channel from throughflow and moves down the channel to the gauging station. According to the drainage basin characteristics, the discharge will continue to decline until it returns to base flow. Should there be a period without further precipitation the water table may fall and so will the base flow. On permeable ground the water table may fall below the valley bottom, in which case the river runs dry until the groundwater is restored by subsequent rain. Each of these events will alter the precise nature of a hydrograph on a river and between rivers in different drainage basins (see Figure 3.6, page 37).

> **Q9.** *Figure 3.6 shows storm hydrographs produced under different circumstances. For each hydrograph describe its main characteristics and suggest causes of the patterns described (you may need to refer to an atlas).*

> **Q10.** *The urban storm hydrograph for the Water of Leith that flows through Edinburgh in Scotland exhibits an initial rapid and sharp peak discharge followed by a second and slightly greater peak. Why might this be?*

The politics of water control in drainage basins

Some drainage basins may extend over little more than a church parish while others cover entire countries. Since water is a vital resource, everyone wants to make use of the water that is available to them. Where the ground is permeable there may be underground stores of water called artesian basins. Places where rivers flow have more readily available resources. Yet the demands on water supplies are often such that there is not enough to meet everyone's needs, especially where the annual hydrograph is very variable or where the quality of the water has been reduced. Dams and reservoirs may be built to store the water for use in drier times and such engineering also allows the discharge in a river to be controlled. However when a river flows through a number of countries the control of the water supplies can present huge geopolitical problems with the stronger hand being held by the countries in the upper reaches of a river.

The first agrarian civilisation developed in Mesopotamia, using the waters of the rivers Euphrates and Tigris to irrigate parts of modern day Syria and Iraq. Together these countries control most of the length of the rivers (1875 km of the Euphrates' 2320 km length, and 1385 km of the 1658 km long Tigris). Yet it is Turkey that holds the key to the hydrology of the region, controlling the headwaters and the main rainfall areas of both drainage basins.

Until the 1990s none of the three countries were facing a water shortage, although the pollution levels did rise downstream. Problems have occurred with new irrigation and hydroelectricity projects under development in Turkey and Syria. The first Turkish project, the Keban Dam on the Euphrates, was completed in 1973. The Grand Anatolian Project (GAP) will eventually consist of 22 dams and 19 HEP stations across both the drainage basins. It will provide half the country's electricity needs and irrigate almost ten per cent of Turkey. Almost as grand is Syria's Ath-Thawrah HEP and irrigation project on the Euphrates. The net result of full implementation of these projects would be a reduction of Iraq's water supply from the Euphrates to less than a third of what it was and to less than the minimal needs of the country.

In January 1990 Turkey temporarily stopped the flow of the Euphrates to cause a reservoir to build up behind the Ataturk Dam. Headlines in Syrian and Iraqi newspapers predicted war with Turkey. The Gulf War of 1991 intervened but the hydropolitical problem has not gone away. All that is preventing completion of the projects is the economic weakness of both Turkey and Syria.

> **Q11.** *Why might the quality of water be expected to decline downstream in a river? How might the climate of Iraq exacerbate this situation?*
>
> **Q12.** *Identify other drainage basins that include a number of countries and discover how the water is managed and what problems are caused for identified groups.*

Drainage patterns

A river is rarely a single channel. Almost all but the most recently formed of rivers, such as on glaciers, consist of a network of channels in a drainage basin with the smallest channels feeding into larger and these in turn into yet larger channels. Tributaries of increasing size flow together and eventually into the main river in a hierarchical manner. The spatial layout of these networks of interconnected channels is known as the **drainage pattern**.

Drainage patterns are multitudinous, for their shape is determined by the underlying rock type(s) and structure. Nonetheless the most common patterns are dendritic, trellis and radial. A dendritic pattern consists of a branching tree-like set of channels where the confluences are at an acute angle. Such patterns are found where the geological structure does not influence the development of the drainage pattern, usually because the river flows over a single type of rock and there is a regional slope.

In contrast, a trellis pattern of drainage contains channels that join at right angles. Such a pattern occurs where the river network has developed on a regional slope (followed by the main or consequent channel) that crosses a number of gently dipping sedimentary rocks. Tributaries (or subsequent rivers) develop at right angles along less resistant rocks such as clay. These subsequents in turn acquire their own tributaries flowing straight down their valley sides either in the same or the opposite direction of flow to that of the consequent. Classical trellis patterns developed in south east England on the dome of chalk that once linked the North and South Downs. As the rivers eroded the chalk the other rocks were exposed and etched out to create the modern landscape and drainage pattern.

A radial pattern of drainage is a simple pattern of separate rivers flowing out from a central high point like spokes from the hub of a wheel. This pattern develops on a dome where sedimentary rocks have been folded upwards or on a volcanic cone. In Britain the Lake District exhibits a radial drainage pattern but not because there is a dome there today. There was once a dome of carboniferous limestone whose remnants are found to the east and west of the modern Lake District. The radial drainage that developed on this dome was reinforced by valley glaciation, superimposing the radial pattern on the modern landscape.

In areas that do not contain a general slope but rather have a series of hollows on a plain or plateau the drainage is known as deranged. The pattern is unco-ordinated and disconnected with short channels heading in various directions, often into lakes or marshes. Conversely in areas of pervious rock, such as the carboniferous limestone of the Pennines of England and the Massif Central of France, the drainage is interrupted. Channels may start near hilltops on impermeable rock but then disappear down swallow or pot holes on reaching the limestone. The vertical joints in limestone are chemically weathered and eroded as are the horizontal bedding planes. In this way the river flows in an underground channel following the rock structure of the area and reappears on the surface lower down the slope as a spring. This happens where the river meets the water table or where the limestone meets a less permeable rock type.

> **Q13.** *Choose OS and Geological Survey maps that cover the same area. Select a river, identify its watershed, drainage basin and drainage pattern. Use the Geological map to explain why that drainage pattern may have developed.*
>
> **Q14.** *The drainage pattern within a drainage basin may also determine the shape of a storm hydrograph. What differences could be predicted between hydrographs of a dendritic drainage network and that of an evenly spaced trellis one?*

Measuring river networks

Both rivers and the drainage basins from which they derive their water vary considerably in size. Nonetheless the processes that operate in any drainage basin are common and the characteristics of a drainage network are of hydrological significance. In the 1940s and 1950s R E Horton and

A N Strahler were among the first geographers to quantify these characteristics and to demonstrate their significance, and thereby enable drainage basins to be compared.

In comparing the spatial characteristics of a channel network the first technique is called stream ordering. The standard method employed is Strahler's stream ordering. In this a hierarchy of tributaries is recognised from the fingertip channels with no tributaries of their own (1st order channels) through progressively larger channels with the order of the main channel signifying that of the drainage network as a whole. Downstream of the confluence of two 1st order channels there is a 2nd order channel. Downstream of where two 2nd order channels meet is a 3rd order channel and so on. However the accuracy with which a network of river channels is ordered will depend upon the scale of the map employed and the ruggedness of the relief (maps with a scale smaller than 1:25 000 omit some headwaters and the degree of this inaccuracy increases with more rugged relief). Moreover, channels are dynamic and yet a map shows their position at one moment in time.

According to Horton's Law of stream numbers the higher the channel order the fewer the number of channels, and these numbers exhibit a geometric relationship. Analysis of the drainage network therefore continues with the bifurcation ratio where the number of channels in an order is summed and divided by the frequency of the next order. The resulting ratio suggests the average number of channels of an order needed to produce a channel of the next order. For a drainage network the individual sums may be added and divided by the number of orders, minus one. For most natural drainage networks the bifurcation ratio falls between three and five. Furthermore the average channel length is directly related to the order of the channel (Horton's Law of stream lengths), higher order channels being longer. Yet the precise ratios are dependent on the size and shape of the drainage network. A dendritic drainage pattern will tend to show a smaller bifurcation ratio than a trellis pattern in a similar sized drainage basin. With the drainage pattern influencing the shape of a storm hydrograph the bifurcation ratio has a secondary function of suggesting the relative efficiency with which a network will cope with water that falls in a drainage basin. Shorter tributaries spread along the main channel in a trellis pattern will tend to cause a slight rise in the discharge which is maintained for a relatively long time and so does not cause flooding. A dendritic pattern with its long tributaries tends to join the main channel in a more concentrated manner and generally leads to a longer lag time with a sharp peak discharge that may cause flooding (an inefficient network).

Another numerical description of drainage basins concerns Horton's Law of stream areas in which the higher the order of the main channel the greater the area of the drainage basin. To compare drainage basins, drainage density can be calculated which involves measuring the length of channel in each square kilometre of the drainage basin and calculating the mean length of channel. The value obtained reflects the climate, relief and geology of an area. The finer the texture of sedimentary rock particles, from chalk through carboniferous limestone to loess, the less permeable the rock and the higher the channel density on the surface. Volcanic rocks and ground wet from precipitation, anchored by dense vegetation or on steep slopes will also tend to support high drainage densities.

Q15. *For the river studied in Q13 trace the drainage network, order the channels, calculate the bifurcation ratio and the drainage density. Use the OS and Geological map evidence to suggest reasons for the figures obtained.*

Q16. *Compare the drainage network characteristics in Q15 with those of the river in Figure 3.7. The River Gwash in Rutland, England, drains into the Rutland Water reservoir, flows over clays and rests on Jurassic limestone. Use Figure 3.7 and explain the differences with the river studied in Q13 and Q14.*

Drainage basins and their economic significance

A drainage basin is not only a relief and hydrological unit, it is also important for the people and economy of a particular river valley. With the river network of tributaries from the extremities of the area joining to make larger rivers as they flow downhill into the lowlands and generally to the coast, there is a natural collection and distribution channel. Raw materials and goods produced upstream can be carried down to a port at the mouth of the river. Materials and goods imported may also be carried to all parts of the area by the same transport system. Canals, railways and roads tend to follow valleys, and their networks have a similar function. With transport in a particular direction, i.e. downstream, there is often a sequence of related economic functions, such as the forestry and fishery industries of the River Tyne in north east England (Figure 3.8).

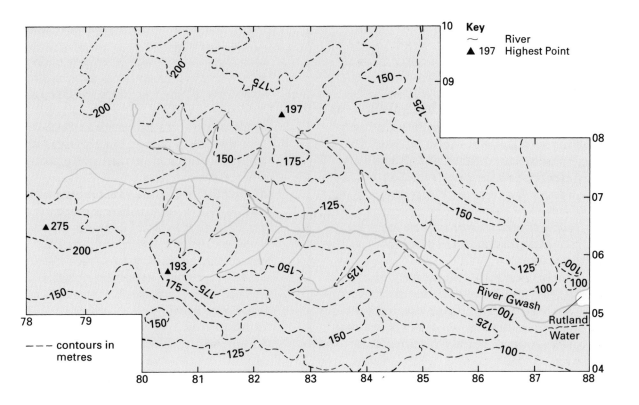

Figure 3.7 *Drainage basin of River Gwash, Rutland*

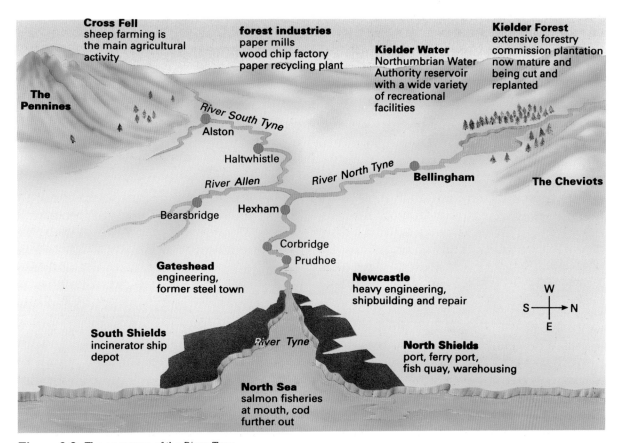

Figure 3.8 *The economy of the River Tyne*

WATER ON THE MOVE 41

Q17. *What are the main land uses of north east England and how do they relate to the rivers? The false colour Landsat image on the inside back cover (Figure 3.35) shows north east England from the border with Scotland to Flamborough Head. The dark green areas are the upland moors of the Pennines (left) and the North York Moors (near the coast). The deep blue sinuous lines running towards the east coast are the rivers Tyne, Wear and Tees. The red and pale blue areas represent farmland. The mid-blue are urban areas. The white streaks running across the image are condensation trails left by a high-flying aircraft.*

a) *Place a sheet of tracing paper over the image. Trace the coastline, the upland moors, the three rivers, the outline of the urban areas and of the farmland. Create a key. Consult an atlas to identify and label the names of the rivers, main urban areas and uplands. Work out an approximate scale.*

b) *Study a land use map in an atlas together with other relevant clues and suggest the type of farmland shown on the image as red and as pale blue.*

c) *Create a matrix with columns for distance from one of the three main rivers of north east England, and rows for land uses (moor, two types of farmland, urban).*

d) *Over the image place a transparent 1 cm grid. Take a regular point sample over north east England. Record the occurrence of each of the land uses in the appropriate distance cell of the matrix. Illustrate the pattern of land use with appropriate statistical diagrams, perhaps using a spreadsheet and related graphics package.*

e) *Describe and suggest reasons for the pattern of land use discovered.*

DYNAMIC DRAINAGE CHANNELS

River channels, their dimensions and internal characteristics are adjusted to their function of transporting water and sediment. As described above, the discharge of a river will vary both in the short and long term and from place to place within the drainage basin. In consequence the processes of transport, the nature of the load and the characteristics of the channel will also change. Nonetheless there are relationships between characteristics of the channel that are common to all rivers.

Figure 3.9 *River bedload*

Figure 3.10 *Suspended river load and bedload*

River load

The sediment that a river transports is known as the **load** of a river. This material is derived from the denudation processes of weathering, mass movement of the valley slopes and from erosion of the channel. The load may be of three types; dissolved, suspended or bedload (Figures 3.9 and 3.10). Dissolved load consists of minerals that are soluble such as calcium and iron. Suspended load is solid particles like clay that are held in the water by its turbulence. Bedload consists of material that is coarser and cannot be lifted by the moving water. The actual size and weight of particles that form suspended load or bedload depends upon the velocity of the river (Figure 3.11).

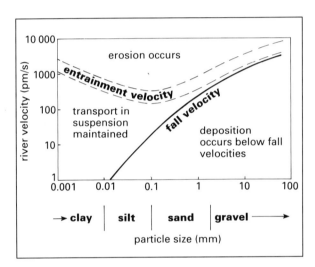

Figure 3.11 Hjulstrom's curve

Q18. Study the river and particle movement relationships suggested in the Hjulstrom Curve of Figure 3.11.
a) The entrainment velocity is the velocity needed for a river to pick up sediment. Describe the pattern of entrainment velocity of load from clay to pebbles. Why do the very small clay particles tend to have a higher entrainment velocity than sand particles?
b) The competence of a river is demonstrated by the largest size of particle that it can transport. What does the Hjulstrom Curve suggest about a river's competence?
c) Within a channel the velocity of a river will vary at different times. If a river's velocity were to decrease from 1 m/sec to 0.1 m/sec what effect would this have on the river's load?

The average velocity found at a point in a river will depend upon the slope, the shape of the channel and the roughness of the bed and banks of the channel. To find the average velocity a current meter ought to be set to read at 0.6 of the depth as measured from the surface. Alternatively, if the velocity is measured on the surface the average velocity in the channel may be found by multiplying the result by 0.8. These adjustments take account of velocity variation with depth and in the cross-profile, together with the impact of friction on water flow. With channel roughness decreasing and channels becoming both wider and deeper downstream, the velocity of a river also tends to increase downstream.

The different types of load are transported by different processes. The bedload is too large for a river to entrain. Instead it may be rolled or slid along the bed by the force of flow of the water, a process called traction. In this process large particles are moved over smaller particles that act like ball bearings. Slightly smaller particles may be lifted by turbulence, but being too large to entrain, fall back to the bed and then maintain motion by bouncing, known as saltation. The falling of such particles may cause other particles to move in a similar manner through impact. Particles that can be entrained are then carried along with the flow of water in suspension. This load often gives the river its colour, such as the milky green of glacial rivers or the browns from clay. The dissolved load similarly moves with the flow.

A function of transported load is to cause the wearing away or erosion of the channel. Figure 3.11 shows that velocities higher than those necessary for entrainment are required to cause erosion. Without load the river velocities in normal river flow are not high enough to erode solid rock, though unconsolidated sediments such as clay or sand may be eroded. The frictional drag of bedload moved by traction is known as hydraulic action. Where saltation or suspension occurs the load may be thrown or rubbed against the channel causing abrasion. When the transported load hits against the channel and other particles, the particles are likely to decrease in size and become more rounded. The impact of this attrition tends to increase in a downstream direction. Finally, according to the acidity of the water different minerals in the bed and bank will be dissolved and at different rates. So in all rivers there are clear relationships between velocity, the type of load transported and the processes of erosion that operate.

Q19. *The photograph in Figure 3.12 shows a pothole in the bed of a river. Describe and explain the process(es) that are likely to have caused this feature. In the future what is likely to happen to the pothole shown and others nearby? In what ways will this change the channel?*

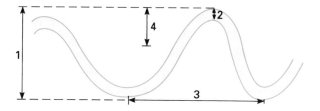

Figure 3.12 Pothole in a river bed

1 = meander width
2 = channel width
3 = meander wavelength
4 = meander amplitude

sinuosity = $\dfrac{\text{real length of channel}}{\text{straight line distance}}$

NB: normal ratios for
sinuosity = 1.5 to 4
channel – width: 3 m – wavelength = 1:10 to 1:14

Figure 3.13 *The geometry of a river channel*

Channel geometry

The shape of river channels, both in their cross-section and in plan, varies downstream according to changes in discharge and in load. The cross-section near the source will tend to show a wide and relatively shallow channel measured at **bankfull discharge**. Often the channel will not be full of water but will contain various streams of water within a boulderfield. Such a large width-depth ratio involves an extensive wetted perimeter, much friction in the water (particularly with the large size of the bedload) and low velocities. Downstream the width and depth increase in response to the greater quantity of water in the river but the width-depth ratio will be smaller as the channels efficiency for transport increases (the most efficient channel shape being semi-circular) and so the velocity rises. However the cross-section shape also varies according to the flow of water.

In plan a river channel can be seen to meander, a term named after the River Menderes in western Turkey, as much of the energy of rivers is used in lateral erosion of the channel. Only artificial channels are straight for any distance. Figure 3.13 shows some of the geometric characteristics used to describe river channels. With accurate information about river plans available from OS maps, some consistency in the ratio between these characteristics has been discovered. Wavelength of meanders also exhibits a strong positive correlation with mean annual discharge.

Q20. *Select an OS map and a section of a river. Calculate the geometric characteristics shown in Figure 3.13. Select another section of the same river and of the same length. Measure the characteristics again. What changes are apparent downstream?*

Certain downstream changes are common in the channel plan of most rivers. The channel tends to become more sinuous, the detour the river takes from the straight line becoming greater through increasing meander width and decreasing meander wavelength. Where the energy of the river is focused laterally, erosion occurs on the bank hit by the main current.

Along a channel a sequence of shallow water flowing over bedload (riffles) and deeper water sections (pools) may be observed. At low flow conditions, as the river water flows over a riffle, the area of highest velocity is caused to swing to one side of the channel. As a meander is approached this faster water is forced to collide with the outer bank, undercutting it and occasionally causing bank material to topple or to slump into the water, leaving a steep concave bank (a river cliff) and deep water. The rate at which this erosion occurs depends on the bank material; the larger the proportion of clay the more resistance offered. The river bank deflects the fast water round the meander producing a helicoidal flow comprising a downstream surface movement and a river bed flow at right angles back across the channel. This pattern of flow carries the material eroded from the outer bank of a meander slightly downstream and across the channel onto the shallow convex point bar where it is deposited in the slower flowing water. However it is infrequent floods that cause most rapid change in the channel shape.

Slowly the meander changes shape, becoming more sinuous (Figure 3.14), and migrates downstream. Where a channel is extremely sinuous, during a flood a river may erode through the narrow neck and straighten the channel, the abandoned meander becoming cut off from the new channel. While it contains water it goes by the name of an oxbow lake, but these silt up rapidly into marshland (Figure 3.15).

Some channels, such as glacial meltwaters, flow through outwash material where large volumes of bedload gravel is transported by traction. The wide and shallow bankfull channels that result, cause islands of sediment to build up and the water to separate into many braided channels. Rapid daily change in the pattern of braiding is common.

A similar phenomenon occurs at the mouth of a river if it flows into an almost tideless sea or a lake. The sudden decrease in velocity causes rapid deposition of huge quantities of silt that may divide the main channel into a number of distributaries, known as a delta. The flocculation or clotting of clay particles in salt water also contributes to this rapid deposition at the mouth of the river channel.

Figure 3.14 *Sinuous meanders on the River Manifold, Peak District*

Q21. *With reference to changes in the channel shape downstream along a river channel, construct an annotated diagram from source to mouth that illustrates these characteristics and the processes responsible.*

Figure 3.15 *Remnants of a cut off meander, River Manifold*

CREATING RIVER VALLEYS

A river flows downhill from source to mouth. As it flows the energy of the river is used to erode the channel and to transport load. Where there is inadequate energy to transport the sediment then deposition occurs. The net result of these processes is a river valley of a certain shape both in the long and cross-profile. As the river flows downstream the balance of the processes alters and so does the shape of the valley in which it flows.

The graded long-profile

Where all the energy of the river is used in transport and there is no net gain or loss of sediment then the river is said to be **graded**. In the long-profile of a river the graded profile is a smooth concave shape. If there is any excess energy it is used to erode the channel bed, the result being a decrease in gradient and so of velocity. If there is inadequate energy to transport the load it is deposited, steepening the gradient and increasing the velocity so that more sediment may be transported.

A river channel in the floor of its valley is therefore in a state of dynamic equilibrium with negative feedback. A change in one variable leads to an opposite change in another variable causing grade to be maintained in the long term. However, if sea level were to fall then the steeper gradient created near the coast would lead to an increase in erosion and a new long-profile would be formed within the old one. Erosion would cause the new long-profile to gradually extend up valley as the river adjusts to the new conditions. Such a sequence of events is called **rejuvenation**. A polycyclic long-profile with more than one concave section is evidence of the sea level change, each graded section relating to one sea level position.

> **Q22.** *Figure 3.16 gives measurements for the valley of the River Tees in north-east England. Use this data to draw graphs of the river's long-profile. Draw one graph on normal graph paper, another using semi-log paper (distance on the x-axis), and a third graph on log-log paper. Describe the shape of the Tees' long-profile? What does this shape suggest about the history of the Tees? What advantages and disadvantages for illustrating long-profiles are highlighted by the three types of graph paper?*

Distance from source (km)	Height (m)
0	754
2	600
4	560
8	520
10	505
16	420
20	370
25	300
30	235
35	204
40	180
45	150
50	137
55	120
60	100
65	88
70	64
80	41
90	28
100	21
118	10

Figure 3.16 *River Tees long-profile data*

Cross-profiles

The cross-profile of a river valley also adjusts dynamically. Near the source, on a relatively steep gradient, the river possesses excess energy (particularly during bankfull discharge). Erosion is rapid and focused on the bed of the river channel. Nonetheless the upper reach of a river valley in Britain (a humid, temperate environment) is not usually a gorge. The denudation processes of weathering and mass movement operate on the exposed valley sides. Weathering creates a mantle of loose material on the surface which moves down the slopes under the influence of gravity. On reaching the river this material is transported away, especially at times of peak flow. In Britain, where the rates of river erosion and slope destruction are balanced, the valley sides are opened up to form a fairly steep sided, narrow and relatively deep valley (Figure 3.17). With erosion concentrated on the river bed a river will wind its way around the base of the hills (spurs) creating the landform called interlocking spurs.

As a river flows downhill its gradient decreases and so does its erosion, though as shown in the

Figure 3.17 *River valley cross-profile in Peak District, temperate humid environment*

earlier section on channel geometry, erosion still occurs where the water flows faster, such as towards the outside of meanders. Deposition also tends to increase towards the mouth as energy levels fall. This means the river valley tends to become shallower (from less vertical erosion in the channel), wider (from lateral erosion and migration of the meanders, wearing back the valley sides further towards the mouth) and the valley sides tend to become lower (from the accumulation of material at the base of the slopes).

With the wider valley its floor also tends to be made flatter by alluvial deposits. This floodplain arises from the point bars made as the meanders migrate and also from deposition during floods. In Britain the rate of floodplain build-up is slow, at just a few centimetres a century with only occasional floods. In areas of frequent flooding and very shallow gradients, such as on the Mississippi, the floodplain may fill in at a rate of two metres a century.

Q23. *Consider the evidence found on the slopes of valleys that suggests the following processes of mass movement are operating: soil creep; mudflow; landslide (see Section 1).*

Q24. *Figure 3.17 is of a valley in the limestone area of the Peak District in England. The photograph in Figure 3.18 shows a river valley cross-profile in a tropical arid environment. Describe the shape of each cross-profile. Suggest reasons for these shapes with use of the processes described above.*

Figure 3.18 *Tropical arid valley, Bazman in South Iran*

WATER ON THE MOVE **47**

Q25. *Select an OS map that includes the valley of a river along much of its length. Find a location near the river's source and others further downstream. Draw a cross-profile of the valley (from watershed to watershed) at each position. How does the shape change. Why does it change in these ways?*

In an arid or semi-arid environment with heavy episodic or seasonal precipitation the discharge of the rivers is highly variable (see Figure 3.5d, page 37). The consequence of this is that for a short period the rivers have enormous energy to transport load. Rapid bed erosion in the source area outstrips the rate of weathering and mass movement, and leads to gorge formation. In the long-profile, headwaters tend to possess a steep gradient followed by an extensive low gradient floodplain. Vast spreads of alluvium occur across the floodplains that, under natural conditions, tend to be flooded regularly. Indeed the flash floods common after a downpour in arid areas (Figure 3.6d, page 37) present a major hazard. The huge volume of load swept along valleys that also contain lines of communication is a hazard both to travellers and to maintaining long-term access.

When a negative change of sea level is experienced rejuvenation occurs. As suggested earlier, this leads to the formation of a polycyclic long-profile. In the cross-profile the effect of increased rates of bed erosion is for the channel to become incised into the floodplain. The former floodplain left above the new level of the river is then fossilised and called a terrace. Every time the sea level falls another set of terraces is created. However, should sea level remain stable for enough time the river may become graded. Lateral erosion of meanders then starts to wear back the terraces and the new low ground around the river becomes the new floodplain.

Q26. *The sketches in Figure 3.19 show the valley cross-profiles of the River Thames in England and the River Benue in Niger, West Africa. Describe the profiles and explain what they suggest about sea level stability in each area.*

Evidence that a flattish area near to but now above the level of a present river is a river terrace may be gained by sediment analysis. The size of particles deposited by a river will depend upon the velocity of water. Nonetheless river deposits possess different characteristics to deposits left by the sea, glaciers and meltwater streams. By measuring the roundness of particles in a sample of the deposit and creating a frequency distribution graph of categories of particle roundness (see Figure 3.20) the characteristic of that deposit may be seen. Comparison with similar graphs from deposits known to have been laid down in various environments can confirm whether the sampled deposit was laid down by a river in the past.

Q27. *What are the characteristics of the riverine deposits shown in Figure 3.20 and how do these compare with those formed by other processes?*

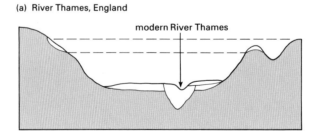

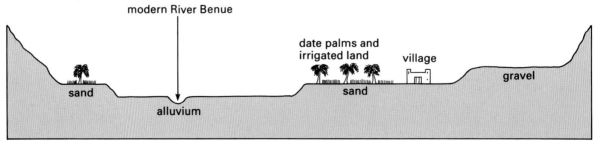

Figure 3.19 *River valley cross-profiles: a) River Thames b) River Benue*

48 WATER ON THE MOVE

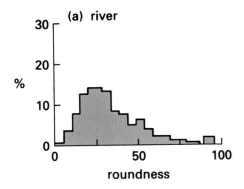
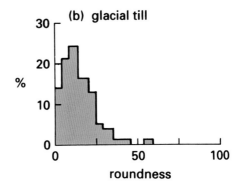
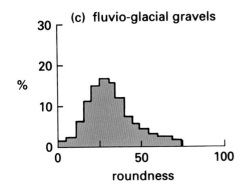

Figure 3.20 *Particle shape in three environments of Wensleydale, North Yorkshire: a) river b) glacial till c) fluvio-glacial gravels*

MONITORING AND MANAGING RIVERS

With rivers acting as a system in which a variety of factors interrelate to create channel and valley characteristics in dynamic equilibrium, the monitoring of these variables in a river allows effective management of its water resources. Such management can help to ensure reliable water supplies and to prevent or control floods.

Monitoring the drainage basin

In the first part of this section, concerning the hydrological cycle, it was suggested that quantitative studies may be made of the water within a drainage basin. There is a water balance of inputs and outputs illustrated by this formula:

precipitation (input) = river discharge (runoff) + evapotranspiration (output) +/– storage

It is possible to measure each variable, to varying levels of accuracy, allowing the dynamics of the drainage basin to be understood and predictions to be made about future behaviour.

With input for a drainage basin deriving from precipitation, an accurate knowledge of how much water enters the area and with what intensity, is of fundamental importance to the hydrologist. Measurement involves a rain gauge of a known diameter that records the depth of water collected over a certain period of time. In Britain a 13 cm diameter rain gauge is standard. Its rim needs to be high enough above the ground to avoid splashing in of water from beyond the collection area. Also, to minimise turbulence and wind speed created by the obstacle of the rain gauge, the gauge is best set in a pit with an open grid roof. The rain gauge funnel minimises evaporation and automatic rain gauges with tipping bucket mechanisms (Figure 3.21, page 50) avoid the necessity for daily emptying of the standard rain gauge.

A difficulty in measuring precipitation arises from the area involved as rainfall will vary across the drainage basin. The pattern of this variation can be monitored using a set of randomly placed rain gauges spread around the drainage basin. An average for the drainage basin may be calculated by plotting lines of equal precipitation (isohyets) and measuring the area between the isohyets. The average precipitation is the proportion of the drainage basin between two isohyets multiplied by the mean precipitation between the isohyets, averaging the figures for the parts of the drainage basin and expressing the amount in millimetres.

Figure 3.21 *Automatic tipping bucket raingauge*

The discharge for a drainage basin is monitored using a gauge at or near the river's exit from the drainage basin and on a straight section of channel. However the geomorphological factors (slope, rock permeability, etc) and human factors (urbanisation, land use) that influence discharge vary within a drainage basin and so tributary gauging is useful too. Discharge equals the cross-sectional area of the channel multiplied by the velocity of the water, expressed in cumecs. Velocity is measured with a current meter as explained on page 43. Where frequent discharge measures are to be made it is more convenient to employ a standard cross-section as provided by a rectangular weir or a v-shaped flume (Figure 3.22). With a known cross-section, only the depth of water requires monitoring, this being transformed into discharge using a rating equation.

Figure 3.22 *Flume to measure river discharge*

The other variables in the water balance (evapotranspiration and storage) are not so easily measured. The potential evaporation may be found using an open pan of water (lysimeter). Potential transpiration may also be estimated based on sunshine, temperature and windspeed levels (Penman equation). These measures, however, overestimate actual evapotranspiration. The rate at which water infiltrates the ground can be measured, but the amount of water stored in the ground cannot be measured directly without destroying that ground. For groundwater it is possible to monitor the level of the water table in a well. Consequently a full and accurate measurement of the passage of water through a drainage basin is impossible. Nonetheless since input and the main output of discharge can be monitored accurately and regularly it is valuable to do so. The figures obtained will show a relationship reflecting the particular geomorphological and human characteristics of the drainage basin. It should be remembered that these relationships will vary between drainage basins, the measurements from one only being representative of that drainage basin. Each drainage basin's precipitation and discharge needs monitoring if the management of water resources is to be informed.

Predicting floods

The major reason for monitoring and understanding water movements in a drainage basin is to predict and give warning of discharges that are likely to cause flooding. The human and economic hazards from flooding are obvious from the description of the 1993 Mississippi flood in the USA earlier in this section. The causes and consequences of floods in Britain and in Bangladesh are considered below.

In the sub-section on investigating drainage basins the nature of a storm hydrograph was considered. A steep rising limb and short lag time are generally linked to intensive rainfall. Should the discharge rise above bankfull then a flood occurs (Figure 3.23). In attempting to predict how frequently this event is likely to take place the historical record of discharge levels for a river is used. This leads to unreliability because the length of time over which discharge records have been made for a particular river is usually short. In Britain records of over a century are rare. Since factors responsible for discharge vary between drainage basins no general rule applies.

To obtain a recurrence interval for a certain discharge on a river, a number of techniques are used. One of the most commonly employed uses the following formula.

recurrence interval for discharge Y = number of years of discharge record / frequency of discharge Y in that time

It is usual to employ maximum annual or monthly discharges in this calculation.

Figure 3.23 *Flooding at Bath*

Once recurrence intervals for a number of discharges on a river have been worked out a graph may be drawn of discharge against recurrence interval on semi-log graph paper. A straight line through the plotted points may be extrapolated to allow prediction of recurrence for any discharge on that river. Knowledge of the average expected interval between floods of a certain size allows informed decisions to be made about the extent and type of flood defences needed on a river. The 100-year recurrence interval is often used in designing defences.

Storm hydrograph records for a river are vital for producing a flood warning so that measures may be taken to limit the hazards to people and property. The knowledge that on a certain river a rainfall of X tends to lead to a peak discharge of Y in a lag time of Z hours is the basis for a flood warning. Once rainfall X has been recorded the likely future behaviour of the river is known. If this discharge is likely to exceed bankfull a flood warning may be issued to civil and emergency services and appropriate action taken. Unfortunately the relationship of the variables that affect discharge in a drainage basin are not constant and flood prediction is not completely reliable. This accounts for the inaction of individuals and authorities which is sometimes seen in the face of predictions.

West of England, December 1992

After four years of scanty rainfall that left reservoirs half empty and hosepipe bans common across England, the last week of November 1992 saw almost 170 mm of rainfall. In each of the first days of December there was another 25 mm – the heaviest downpour for 50 years. The cause was a series of deep depressions. Rivers such as the Avon in Wiltshire burst their banks, flooding meadows, disrupting railways and roads and inundating low-lying pubs and houses (Figure 3.31, page 58). Rivers throughout the west of England and in South Wales were on flood alert, with sandbags barricading many properties. Fallen trees and landslips were further hazards. Animal stock had been moved to higher ground and the flood protection built through major towns prevented them experiencing similar devastation. Nonetheless the hard packed soil stopped much of the water reaching aquifers and the English water shortage continued.

India, Nepal and Bangladesh, July 1993

South Asia depends on the monsoon to bring rain, without which there would be droughts. 1993 saw the heaviest rains in four decades. Water from the Himalayas brought flooding in the lowlands. Most of Bangladesh lies within a few feet of sea level and the land used by 21 million people was covered in water. Landslides set off by the floods washed away crops, 100 000 people lost their homes and at least 3000 people lost their lives in Nepal. In the Punjab and other northern states of India millions fled their homes as the water rose, destroying their granaries. Epidemics of a new strain of cholera and of diarrhoea spread among the refugees in makeshift camps where water supplies were contaminated with sewage and animal carcasses and they were isolated from food and medicine.

Q28. *Why are floods more common around the River Ganges than the River Avon?*

Q29. *Suggest why the South Asia floods of 1993 were more damaging than the west of England flood in 1992? (Consult an atlas and other resources.)*

A VITAL RESOURCE

Water is vital for life, also for many of the physical processes that operate on the earth and for many of its landforms. The quantity of water available influences domestic, industrial, recreational and agricultural use, population distribution, diet and lifestyles. Water quality has similar implications as well as affecting health and the state of the environment. A lack of water quality or quantity presents a hazard that can occasionally develop into a human disaster.

Water in Britain

It is estimated that in 1830 the average daily demand for water in Britain was 18 litres per person. By 1990 around 140 litres were used each day in the home with a further 190 litres a day used for other purposes; a total of 13 410 million litres per day being consumed in England and Wales. The difficulty of supplying water in Britain is increased by the unfortunate distribution of precipitation, evaporation and population (see Figure 3.24). In the early 1990s the problem was compounded by drought. Its intensity increased towards the south east of England where there was as little as a third of normal rainfall; this had dire environmental consequences (Figure 3.25). In 1974 a national water plan was devised

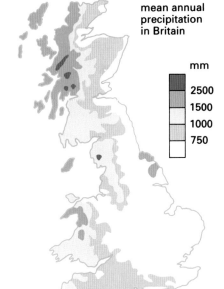

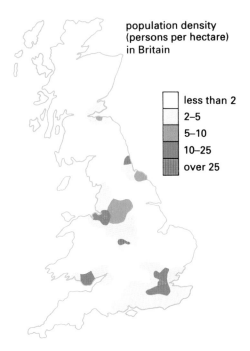

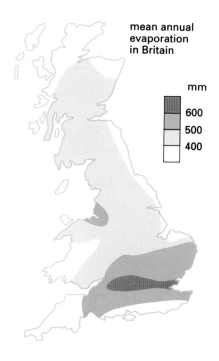

Figure 3.24 *Water supply problem in Britain*

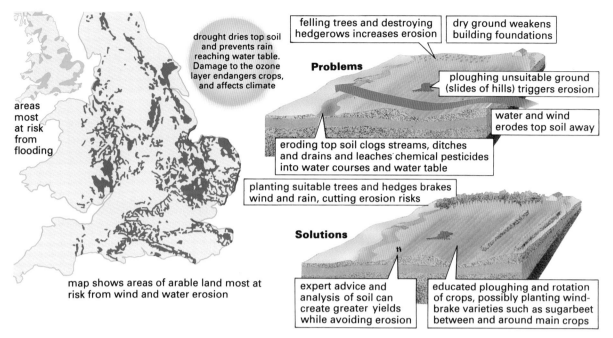

Figure 3.25 *The impact of drought in England*

involving the transfer of water from areas of surplus to areas of deficit (Figure 3.26). The Kielder Water Plan in Northumbria is one part of the national plan that was completed in 1982. A large reservoir was created at Kielder to regulate supply to Tyneside and a pipeline was added to transfer water southwards to the Rivers Wear and Tees (Figure 3.27). However in 1989 The Water Act created ten privatised regional water service companies preventing further progress on an integrated national water plan. Water shortages are likely to remain a feature of life in eastern England for many years to come.

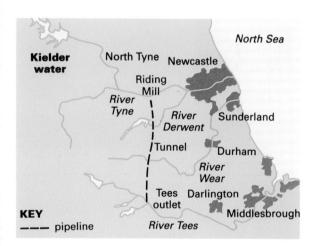

Figure 3.27 *Kielder water scheme*

Q30. Why is there a physical problem of water supply in England?

Q31. Summarise the various techniques that could be used to alleviate this problem together with some of the consequences of a lack of water.

Q32. Suggest why the privatisation of water supply may have hindered fulfilling the national water plan.

Figure 3.26 *A national water plan*

WATER ON THE MOVE 53

Water in Kazakhstan

The area around the Aral Sea has a continental arid climate. The sea is fed by two large rivers, the Amudar'ya and Syrdar'ya. In summer mean monthly temperatures rise above 23°C, which is ideal for growing cotton – a vital raw material for the Russian textile industry. Because of the summer drought about half of the river flow was diverted during the 1960s for irrigation. As a consequence the Aral Sea has shrunk with devastating consequences for the physical and social environment.

The shoreline of the Aral Sea has retreated to a distance of up to 60 km, isolating former fishing settlements and destroying livelihoods. Without the sea to slowly absorb insolation, summers are hotter, shorter and drier. Humidity levels have fallen. Winters are longer and colder. With less water to disperse the pesticides and fertilisers used to maximise cotton yields, the Aral Sea is more saline and polluted. Wildlife has been decimated. The excess heat has reduced the vitamin content of locally grown vegetables and reduced the pasture land for cattle. Malnutrition has become widespread and tap-water undrinkable unless boiled. Anaemia, rickets, toxic hepatitis, kidney stones, joint problems and cancers of the oesophagus, stomach and skin have all become commonplace. The environment has been reduced to a state in which it is virtually uninhabitable because its water resources have been abused.

Q33. *Is the price of this irrigation worth paying for industrial development? Justify your answer.*

Q34. *Referring to an atlas and using other ideas, suggest alternative sources of water and means of growing crops in such an area that would not involve an ecological disaster.*

Unhealthy water

The United Nations made the 1980s the International Drinking Water Supply and Sanitation Decade. The aim was to make clean water and sanitation available to everyone by 1990. By the end of the decade the World Health Organisation (WHO) reported that improved community water supplies had reached another 18.4 million people. Nevertheless at least 25 million people die every year from water-borne diseases such as bilharzia, guinea worm, cholera and diarrhoea. Millions more are affected by the insects that breed in freshwater and cause malaria and river blindness.

Bilharzia (schistosomiasis) is one of the world's main health problems. As Figure 3.28 shows, the disease is caused by a parasitic worm which develops in a freshwater snail. Once mature it drills through the skin of people in the water, damaging the kidneys, bladder, liver and spinal chord. Lethargy, weakness and paralysis ensue. The cycle of the disease continues when infected people defecate in the water. Where irrigation schemes have been built in Africa, bilharzia has also spread. Expensive drugs

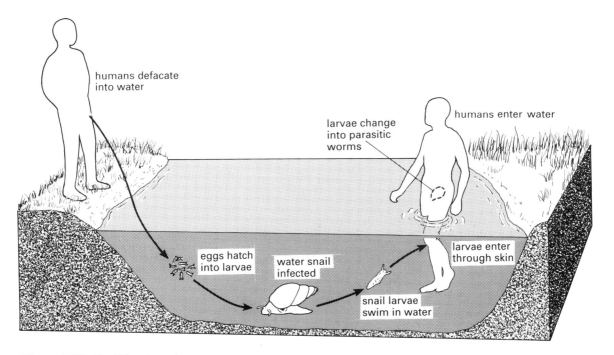

Figure 3.28 *The Bilharzia cycle*

can be used to curb the disease but the cheaper and favoured approach is to change public health habits. By centralising villages away from irrigation canals and by providing communal latrines and clean water supplies there is less contact between people and the worms. Rapid urbanisation can however lead to overuse of these facilities and urban sprawl up to the canals. Over 200 million people in at least 70 countries are still affected by bilharzia.

Cholera was rampant in Britain until the mid-nineteenth century. A Dr John Snow working in Soho, London, discovered that if a certain stand-pipe for communal water supply was turned off then the incidence of cholera declined. Poverty, causing poor sanitation especially in shanty towns, is the underlying cause. After natural disasters such as cyclones, earthquakes, and in war zones, rotting carcasses, water supplies or food contaminated with sewage often leads to cholera epidemics. In Britain tourists may catch cholera whilst abroad but our clean water supplies and high levels of sanitation mean domestic cholera is no longer a danger to public health.

> **Q35.** *Find out about the other water-related health hazards mentioned above, noting their cause(s), the vector involved and how they may be controlled.*

River pollution and its control

The quality of river water can be measured in a number of ways. Common measures include its pH (a measure of acidity), its dissolved oxygen content (the amount of oxygen dissolved in the water expressed as a percentage of milligrams/litre), and the biological oxygen demand (the amount of oxygen in mg/l needed to fully oxidise all the organic pollutants; the smaller the BOD the purer the water). It is important to monitor oxygen levels because fish and water plants require dissolved oxygen to survive. Another significant measure is a Biotic Index (Figure 3.29); the presence of different creatures signifies the water quality. The concentration of ammonia compounds, nitrates and suspended solids are also important to the health of a river.

In Britain the 1989 Water Act created the National Rivers Authority (NRA). One of its responsibilities is for water pollution control. Reporting in 1990, the NRA warned that many rivers had long been abused and that there had been a significant deterioration in the quality of nearly 6300 km of freshwater since 1985. Wide regional variations in this decline were

pollution levels		animal life	dissolved oxygen* example figures	
			5°C	20°C
↓	clean	stonefly nymph, mayfly nymph. Unpolluted water is full of fish and insect life. Waterbirds such as herons and ducks are present.	14.0	10.0
	low	freshwater shrimp, caddisfly larva. Species at the end of the food chain such as large fish and waterbirds disappear first.	11.5	9.5
	high	bloodworm, water louse. Snails and freshwater mussels disappear. Very few fish survive.	5.5	3.5
	very high	rat tailed maggot, sludge worm. Heavily polluted water often looks clear but supports little or no life.	2.5	1.5

dissolved oxygen in milligrams per litre (mg/l)

Figure 3.29 *Biotic index*

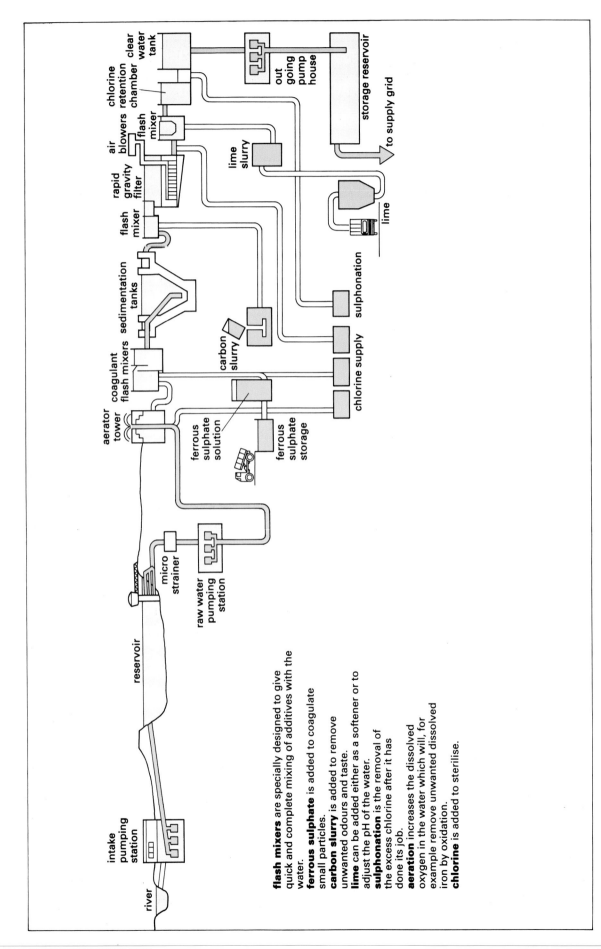

Figure 3.30 *Water treatment works*

apparent, with a third of rivers in Thames and the south west affected, to just five per cent in Northumbria. The finger was pointed at legal and illegal discharges from agriculture and industry, as well as drought, acid rain, runoff from contaminated land and reduced government expenditure on sewage treatment in the run up to water privatisation. Over the same period improvements in water quality were found along 4600 km of rivers, canals and estuaries.

The closure of mines in the early 1990s has been a major cause of river pollution. As tin and coal mines have been shut their pumps have been switched off and the mines have flooded. When the Wheel Jane tin mine near Truro in Cornwall closed in February 1990, millions of gallons of an orange cocktail laden with heavy metals (iron, zinc and cadmium) escaped into the River Fal. In addition to the river's strange colour, there was a threat to shellfish, especially oysters, and to the water drawn from private wells in the area, a major tourist haunt. Similarly the River Pelena near Port Talbot, South Wales, was turned bright orange by the iron oxide and sulphur pollution from an abandoned coal mine. The River Wear near Sunderland also contains high levels of sulphuric acid, iron, manganese and aluminium washed out of closed coal mines in the Durham area. This threatens the water supplies of 150 000 consumers.

River pollution from cattle slurry and silage effluent, particularly from dairy farms, has been a significant contributor to the decline in oxygen in rivers. Nitrates draining into rivers from artifical fertilisers are most common in the east of England.

They can set off a chain reaction in the human body leading to stomach cancer. Lead in the drinking water of an estimated 7.5 million people can damage the central nervous system, particularly in children. It derives from acidic water which dissolves the lead pipes and storage tanks in cities such as Glasgow, Manchester and Birmingham. Aluminium in the water of cities across the north of England may cause senile dementia. Sewerage smells and is unsightly, and causes diarrhoea in Britain, but can lead to cholera elsewhere. The list of hazards from polluted freshwater is long and growing as the affects of particular pollutants are becoming clearer.

This rising tide of river pollution in Britain can be limited by natural purification processes. Storage in lakes and reservoirs causes suspended matter to settle at the bottom, organic impurities to oxidise in the upper water layers and colour to be reduced through the bleaching affects of sunlight. Also as water filtrates through the soil, suspended matter and bacteria are removed. Nonetheless pollution levels are rising and water treatment before consumer use is becoming all the more important (Figure 3.30). This involves storage of the water, screening, aeration, coagulation, flocculation, sedimentation, filtration, pH adjustment, disinfection and water softening. All this is costly. In 1991 the ten water supply companies spent £2 billion on improvements. The final cost of implementing European Union standards on sewage and drinking water may exceed £10 billion; costs that will be reflected in the water bills of consumers.

DECISION MAKING EXERCISE

Bath floods

The River Avon in Bristol rises in the Cotswolds and flows through the City of Bath on its way to Bristol and the Severn Estuary. For England it is a large river, its drainage basin covering 1552 km². It used to flood parts of the city on a regular and extensive basis (Figure 3.23, page 51). Contributory factors included the river's shallow and very sinuous channel. However it is a river with good hydrological records on which to base an analysis of flooding and flood protection.

1 Read the article in Figure 3.31 and summarise the impact of flooding in Avon and Somerset during December 1992. Using Figures 3.31 and 3.32 (page 59) suggest why it has become particularly important to protect the centre of Bath from such floods.

2 In the section on predicting floods (page 50) a statistical technique was introduced to calculate the recurrence interval of particular discharges. Use the Bathford discharge record from 1971–91 (Figure 3.33, page 60) to calculate the recurrence interval for discharges of 30, 50, 100, 150 and 200 cumecs (+/–10) and plot these on semi-log graph paper. Before improvement the bankfull discharge of the Avon was 170 cumecs. How often could one have expected the River Avon to have flooded? The 100-year recurrence interval (event) is often chosen for flood

protection schemes. What discharge would this protect against?

Following the severe floods of 1960 calculations were made and a model constructed to show how changes to the river between Bathampton and Twerton could allow the water to flow more quickly and thereby alleviate most of the flood hazard.

3 Consider Figure 3.34 on page 60 and the aspects of channel geometry described earlier. Identify and justify specific engineering works that might have had the desired effect.

In December 1979 and 1985 the River Avon contained 254 and 218 cumecs respectively and the centre of Bath was not flooded, due to £2 million having been spent on protection. The peak discharge in late 1992 was 187 cumecs.

4 Using the above information and the rest of Section 3, present a report for the National Rivers Authority that might help convince the Department of the Environment that flood protection is both viable and economically justified in British cities.

It Never Rains . . . but it pours

FAMILIES, farmers and business people were definitely not singing in the rain this week as they braved the culmination of the wettest November in 50 years.

The torrential rain began on Monday with rivers bursting their banks all over the region after more than an inch fell in 12 hours.

As the rivers overflowed, rail services were disrupted and roads were blocked.

From Chew Valley to Bath people were left with dead cars, mud on the carpets and mini rivers in their car parks.

In Bathampton the River Avon burst its banks, flooding the meadows and reaching low-lying pubs and homes in the area.

Farming couple Anthony and Sarah Creed of Meadow Farm had taken the necessary steps at the weekend to move their stock out of the meadows, fearing that floods were inevitable.

Meanwhile, staff at the Beefeater pub and restaurant in Bathampton were forced to bale out their dungeon area of the restaurant when flood waters reached the front door and car park.

The flood waters receded on Wednesday, leaving the staff hard at work to clear up.

Staff at the Old Mill Inn in Bathampton had crated up the cellars but the nearby boathouse was affected by flooding.

The worst-hit areas were in towns around the River Avon in Wiltshire and in the Chew Valley in Avon.

Villagers in the Chew Valley area had plenty of headaches as the rain poured down.

In a low-lying area laced with streams and close to the reservoir, parish councillors kept a constant check on local problems.

Chew Magna's parish clerk, Douglas Wooding, reported yesterday all was clear after he had done his rounds of the village.

He said: "All the rivers are down, and there's a very little run-off in places, but I don't think either the rivers or the streams will be flooding."

But the village had come close to big problems, said the National Rivers Authority.

Mr Phil Hewett, spokesman for the NRA, said it had been "touch and go" around Chew Magna.

He said: "Properties were very nearly flooded. People living in homes near places where there were floods were very nearly evacuated.

"We then had a day of respite on Tuesday. On Wednesday we had more rain and that brought the River Chew up to its previous levels, but it has now dropped back again."

Drivers were forced to abandon their cars and 157 children were evacuated from a school in Chew Stoke because of flooding outside, while the route between Saltford and Newton St Loe was shut on Monday.

Keynsham also stood by with bated breath as the roads came under threat from a full-to-bursting River Avon.

Weathermen said November had been the wettest for 50 years with three inches of rain falling in three days and they are warning that there is still more to come.

A further ten millimetres of rain is expected tonight.

Public relations manager for the NRA Annabel Lillycroft said: "We can say that the worst is over for the time being, but it is winter now and the river levels are more than usually high.

"We only need to have a small amount of rainfall and we could have some problems again."

The NRA is also warning, incredibly, that despite the record rainfall this week even more needs to fall to ensure the Wessex region remains drought free.

After four years of scanty rainfall, which has seriously depleted reserves, the Chew Valley Lake, made to hold around 4,500 million gallons of water, has taken in around 180 million gallons of water in just a few days.

But the vast reservoir, which supplies a large area of Avon and parts of North Somerset, is still only 80 per cent full.

Bath Evening Chronicle

Figure 3.31 *It never rains . . . but it pours*

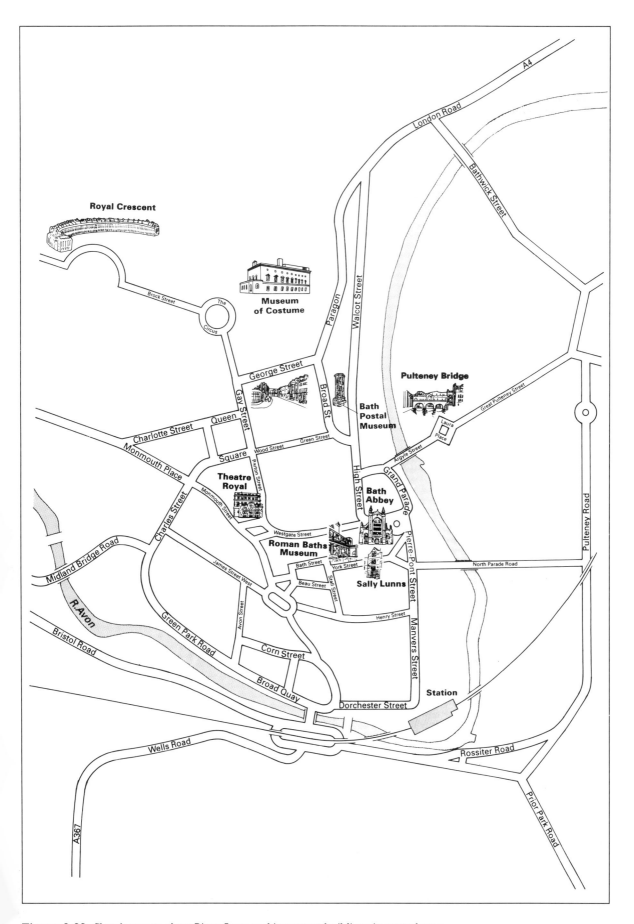

Figure 3.32 *Sketch map to show River Avon and important buildings in central area*

	Maximum Monthly Discharge (cumecs)											
Year	Jan	Feb	Mar	Apr	May	June	July	Aug	Sep	Oct	Nov	Dec
1971	140	170	75	119	22	165	14	12	7	55	78	41
1972	104	204	120	51	33	18	11	7	8	8	38	208
1973	37	66	15	34	29	53	47	43	11	12	13	22
1974	97	226	43	12	9	13	8	8	191	40	129	129
1975	146	60	113	41	13	8	10	8	23	7	23	53
1976	13	32	32	11	7	6	3	7	74	52	163	161
1977	116	135	68	36	48	68	10	64	11	11	76	131
1978	140	69	53	43	85	11	55	55	5	4	13	61
1979	60	157	102	42	227	100	10	15	7	8	16	301
1980	76	135	73	85	13	14	7	9	11	89	105	55
1981	32	33	171	39	42	33	12	14	23	63	49	149
1982	94	53	193	30	11	19	46	13	11	45	108	153
1983	159	156	58	69	60	54	15	10	29	53	44	103
1984	166	79	47	16	21	10	5	9	22	28	118	64
1985	153	129	41	47	17	39	13	44	17	22	10	250
1986	191	82	53	43	61	12	7	31	19	20	117	111
1987	97	61	84	92	15	17	9	5	5	45	101	41
1988	110	137	61	17	11	10	14	10	41	87	48	42
1989	35	130	124	56	14	6	9	5	6	15	33	233
1990	102	209	38	13	7	6	7	8	6	13	12	38
1991	116	49	72	28	14	14	12	5	12	17	58	22

Figure 3.33 *River Avon discharge records (Bathford) 1971–1991*

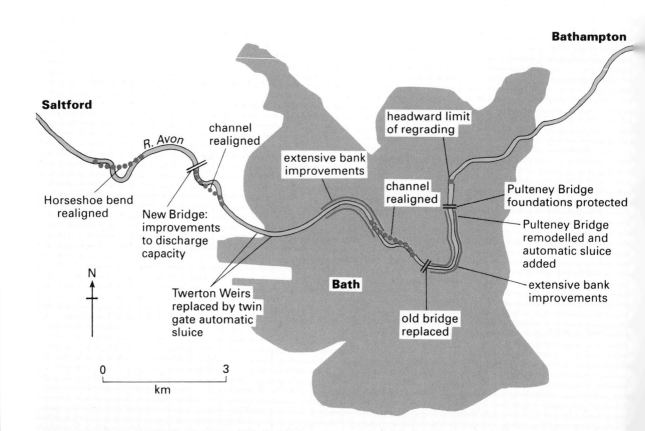

Figure 3.34 *Bath's flood protection scheme*

PROJECT SUGGESTIONS

1 Changes in the long-profile of a river

Having gained permission for access, identify suitable sites along a river at which common measurements can be taken. Measure a cross-sectional area at bankfull using a tape and ranging poles; a velocity profile using a flow meter at regular depths and positions in the cross-profile; particle roundness based on a Cailleaux Index. Analyse with scale diagrams of the first two measures and a histogram for roundness the characteristics of the sample. Compare the findings down the long-profile. Suggest explanations.

2 River water quality

Select a stretch of river that includes agricultural land, industrial and urban development. Map the land use on a 1:25 000 scale OS map. At regular intervals along your sample reach, take water samples and measure pH, dissolved oxygen and water temperature. Record measurements on your map. Construct graphs of the reach and changes in these environmental characteristics, adding the land uses. Focus on any changes in these characteristics and apparent relationships with land use. Follow up the relationships by investigating the highlighted land uses further.

3 Drainage basin dynamics

Identify two local rivers and contact the regional office of the NRA, requesting daily/hourly data on precipitation and discharge for a particular period of time. Create a hydrograph for each river to investigate the relationships. Contrast and consult OS and Geological Survey maps to help explain the differences.

Alternatively, obtain from NRA mean monthly maximum discharge figures over as long a period as possible. Calculate recurrence periods for different discharges. Graph your calculations. Also obtain information about past floods and flood protection on your river. Assess and comment on your findings.

GLOSSARY

Denudation a cycle of the combined processes of weathering, mass movement on slopes and of erosion and deposition by an agent of transport such as a river resulting in the wearing down of the natural landscape.

Discharge the amount of water passing a certain point in a river in a second, measured in cubic metres per second or cumecs. **Bankfull discharge** is when the channel is full to the top of its banks. Further discharge results in a flood.

Drainage basin an area of infacing slopes drained by a single network of streams and enclosed by a watershed.

Drainage pattern the spatial arrangement of streams within a drainage network; generally either dendritic, trellised or radial.

Graded long-profile the concave shape resulting from a no net gain or loss of sediment, in which all the energy of the river is used in the transport of load.

Hydrograph a graph showing the temporal relation between precipitation intensity and discharge, either annually or over the few days of a storm event.

Hydrological cycle the movement and storage of water molecules between land, sea and air.

Load the material carried by a river; either dissolved, suspended or bedload. This material may be transported in solution, or by the processes of suspension, saltation or traction. During transport the load may erode (or wear away) the channel bed and/or banks by hydraulic action, abrasion and corrosion. Also during transport the load may become smaller in size and more rounded through attrition.

Rejuvenation a relative fall in sea level causes renewed vertical erosion in a river bed, resulting in floodplains forming terraces and in polycyclic long-profile shapes.

River regime the average annual pattern of discharge in a river.

REFERENCES

Department of the Environment, Room A238, Romney House, 43 Marsham Street, London SW1 3PY

Friends of the Earth, 26–8 Underwood Street, London N1 8XE

National Rivers Authority regional headquarters: Anglian Water, Kingfisher House, Goldhay Way, Orton Goldhay, Peterborough PE2 5ZR

Thames Region, Kings Meadow House, Kings Meadow Road, Reading RG1 8DQ

Severn Trent Region, Sapphire East, 550 Streetsbrook Road, Solihull, B91 1QT)

Nature Conservancy Council, Northminster House, Peterborough PE1 1UA

Meteorological Office, London Road, Bracknell, Berkshire, RG12 2SZ

Water Aid, 1 Queen Anne's Gate, London SW1H 9BT

A Doherty & M McDonald (1993), *River Basin Management*, Hodder & Stoughton

C Kirby (1987), *Water in Great Britain*, Pelican

B J Knapp (1979), *Elements of Geographical Hydrology*, Allen & Unwin

M D Newson (1975), *Flooding and Flood Hazard in the United Kingdom*, OUP

M D Newson (1982), *Hydrology: measurement and application*, Macmillan

K Richards (1982), *Rivers – form and process in alluvial channels*, Methuen

R C Ward (1975), *Principles of Hydrology*, McGraw-Hill

SECTION 4

In the air

by Ken Grocott

> **KEY IDEAS**
>
> - The atmosphere stores both heat and water vapour and transfers it horizontally and vertically to approach an equilibrium state.
> - The atmosphere is a filter for incoming solar radiation as well as a barrier for outgoing terrestrial radiation, which helps regulate surface temperatures.
> - The storage and transfer of carbon dioxide between the atmosphere, ocean, and earth has important implications for global warming.
> - Human activity in the last two centuries has released large amounts of heat and pollution into the atmosphere and this has had an impact on our climate.
> - The weather of an area can be understood on three main scales; the global macro level; the regional meso level; and the local micro level.
> - The atmosphere is both a resource for human activity, and a hazard to it.
> - Improved monitoring of the atmosphere by weather satellites and stations enables humans to forecast the weather with increasing accuracy.

THE ATMOSPHERE – A STRANGE BREW

Without our atmosphere, life on earth would not exist. As well as being the air that we, other animals and plants breathe, our atmosphere behaves like a blanket, allowing **short wave solar radiation** to pass through it, filtering out the harmful ultra-violet rays, then trapping the **long wave terrestrial radiation** that the earth gives off on being warmed by the sun. This is the **Greenhouse Effect**. Without it, earth would be about 32°C cooler than it is today.

The term Greenhouse Effect has been used in recent years to describe the warming of our atmosphere by human activity. For example the release of the Greenhouse gas carbon dioxide by burning coal in power stations or trees in our rain forests. However, it is important to realise that the Greenhouse Effect is a natural phenomena, which has been exaggerated by industrial pollutants and deforestation in the last two centuries.

What is the atmosphere made of?

The answer to this question depends on where you are because the three ingredients of gas, liquid and solid vary slightly from place to place.

Gases

The gaseous components of the atmosphere shown in Figure 4.1 do not change a great deal.

% Ground level dry air composition	
Nitrogen	78.09
Oxygen	20.95
Argon	0.93
Carbon dioxide	0.03
Ozone	trace

Figure 4.1 Dry atmospheric components

However, in a polluted industrial area, such as Baia Mare in Romania (*Geography Review*, September 1993) the atmosphere may contain more than a dozen components such as carbon monoxide, nitrogen dioxide, and sulphur dioxide. These are measured in parts per billion, and although typical figures for these pollutants may only be a few hundred parts per billion, they have an important role in the atmosphere (see page 87).

On average, water vapour forms four per cent of the atmosphere at the surface, but it is variable both temporally and spatially. Water vapour is virtually absent above 10 km. Over the 70 per cent of our planet which is ocean, and over our forests, water vapour will evaporate into the lower atmosphere, but over our deserts, water vapour amounts are negligible.

Liquids

Water will form in the atmosphere droplet by droplet through the condensation process, when vapour is cooled into a liquid. We see these droplets of liquid as clouds, fogs and mists, which unlike vapour are visible to the eye. Clouds, because they are able to absorb solar radiation directly, and reflect it back to space, as well as reflecting terrestrial radiation back to the surface of the earth, play an important role in the **heat budget** of the atmosphere.

The change of state of water in the atmosphere from vapour to liquid to solid and vice-versa also causes heat exchange in the atmosphere, cooling or warming it. When water evaporates from ocean water into the atmosphere, it stores latent heat energy; when the water condenses back into liquid in clouds, this latent heat is released back into the air to warm it.

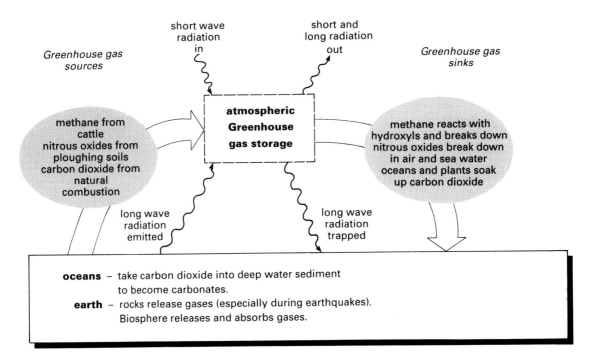

Figure 4.2 *Flow diagram of Greenhouse gas equilibrium and disequilibrium*

64 IN THE AIR

Solids

Apart from ice, solids occur most frequently in our atmosphere as aerosols. These are minuscule particles of dust from deserts, smoke and soot from industry, salt from sea spray, or volcanic ash, which float around in the air. They are important because they provide microscopic surfaces on to which water vapour can cool and condense; such particles are often called **condensation nuclei** or **freezing nuclei**, because they form the nucleus to water droplets and ice particles in the air. Aerosols are also important because as solids they can also directly absorb solar radiation on route to the earth's surface, and scatter some of it back to space.

When present in large amounts, for example in the ash cloud of a major volcanic eruption, aerosols can reduce the amount of radiation reaching the surface, and therefore reduce temperatures in a region temporarily. Such a screening effect is limited to a few months, because as the dust also acts as condensation nuclei for raindrop formation, rainfall will tend to increase and scour the aerosols out of the atmosphere. The largest volcanic eruption to date this century, that of Mount Pinatubo in the Philippines in June 1991, was observed to lower mean global temperatures by a few tenths of 1°C some six to 18 months after it occurred. Such a change appears quite small globally, but is impressive considering that it was triggered by one volcano. What could several eruptions in one year do?

Q1. *Draw a second diagram like Figure 4.2 which shows Greenhouse gas disequilibrium.*

Changes in atmospheric composition caused by human activity

In recent years, climatologists, scientists and environmentalists have become increasingly concerned about the effects of artificial or **anthropogenic pollution** in the atmosphere. The increase of Greenhouse gases in the troposphere comes from greater releases of natural gases such as carbon dioxide and methane, but also increasing emissions of artificial gases such as chlorofluorocarbons (CFCs), which are used as refrigerants and propellants. Both types of emission are thought to be disturbing the **atmospheric equilibrium**. The assumption behind equilibrium in the atmosphere is that providing the sources or inputs of gases are balanced with the sinks or outputs of gases, then the composition of the atmosphere will remain stable.

For example, the Greenhouse gas carbon dioxide has a source in forest fire and volcanic eruptions, but a sink in the ocean waters, in which it gets incorporated. There are other sources and sinks of carbon dioxide, but providing they remain in balance the amount of carbon dioxide in the atmosphere will remain roughly constant (Figure 4.2). Anthropogenic changes such as burning the rain forests both increases the source of carbon dioxide by releasing the carbon in green leaves into the atmosphere, and reduces the sink of it, by destroying the green plants that assimilate carbon dioxide in photosynthesis.

Burning coal and oil is another source of carbon dioxide which was stored as a sink in the carboniferous rocks laid down 250 million years ago. If the inputs of any gas into the atmosphere exceed the outputs, that gas will build up in concentration; if that gas is a Greenhouse gas, a warming of the atmosphere will occur.

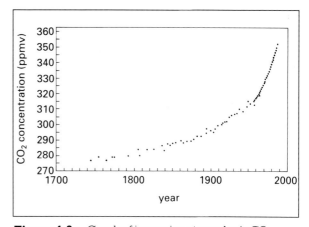

Figure 4.3a *Graph of increasing atmospheric CO_2*

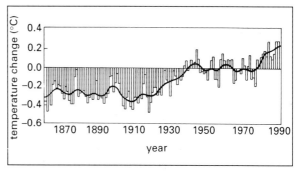

Figure 4.3b *Global warming graph*

Q2. a) *Describe the change in global temperatures shown in Figure 4.3b.*
b) *How is this trend different from that of carbon dioxide in Figure 4.3a?*
c) *What conclusions can you draw from a comparison of Figures 4.3a and 4.3b? What else would you need to know before suggesting a connection between the two graphs?*

Is global warming likely to continue in future?

This is a difficult question to answer, because so much depends upon how the effects of warming are absorbed by the oceans and the ecosystems around the world. Figure 4.5 indicates some of the possibilities: with **positive feedback**, the warming is likely to snowball and be exaggerated; whilst with **negative feedback** the warming is likely to be damped down or reduced in importance.

Both positive and negative feedback are operating simultaneously in the atmosphere, which makes forecasting long-term trends difficult. Nevertheless, by using complex models of the atmosphere known as General Circulation Models (GCMs) climatologists try to quantify by how much temperatures and rainfall are likely to change if present-day trends in the atmosphere continue.

Even then, the pattern of global warming is not easy to forecast, because different models make different assumptions about the future. Some models assume that some Greenhouse gases will decline in future because of agreements made by the developed economies with the Montreal Protocol of 1987, some assume the same level of gases as exists today, whilst others assume a steady increase in Greenhouse gases as the developing economies of the world increase their industrial output. The evidence from readings taken of carbon dioxide concentrations in the atmosphere is that it is increasing (Figure 4.3a) and that global temperatures are also on the increase (Figure 4.3b).

Apart from the difficulties of estimating the feedback effects of a warmer atmosphere, there are also problems in working out the importance of each Greenhouse gas in the future. Each Greenhouse gas has a different lifespan in the atmosphere before it is broken down; further, each gas is increasing in atmospheric concentration at different rates. As a result, Greenhouse gases vary in their contribution to global warming, so their presence in the atmosphere must be monitored carefully in future.

A further difficulty in forecasting the trend in global warming is that however sophisticated the GCM and the computer on which it is run, its accuracy can only ever be as good as the information which is fed into it. This information comes from thousands of weather stations worldwide, and yet there are still large areas of the world for which little long-term climatic information exists. We need to know about the past if we are to predict the future.

Scientists are filling the gaps in our knowledge of past climatic behaviour by using a variety of methods. One way of finding out how carbon dioxide affected the climates of the past is to measure how much of it existed in bubbles of air trapped centuries ago in our polar ice sheets.

The evidence from this suggests that a colder Ice Age earth had lower atmospheric carbon dioxide levels; but whether a colder planet has the same atmospheric circulation is doubtful. Such research in the most inhospitable parts of our planet should help us to look into the future with more accuracy. Most scientists seem to believe that global warming will continue into the next century; the main debate now seems to be how quickly it will happen.

Greenhouse gas	Chemical symbol	% annual rate of increase	Lifetime in atmosphere (yrs)	Atmospheric concentration in 1800 (parts per million by volume)	Atmospheric concentration in 1990 (parts per million by volume)	% Estimated effect on global warming at 1990 values
Carbon dioxide	CO_2	0.5	50–200	280	350	61
Methane	CH_4	0.9	10	0.8	1.72	15
Nitrous Oxide	N_2O	0.25	150	0.28	0.31	4
Chlorofluorocarbon 11	CFC 11	4	65	0	0.0028	2
Chlorofluorocarbon 12	CFC 12	4	130	0	0.0048	7

Figure 4.4 *Greenhouse gases*

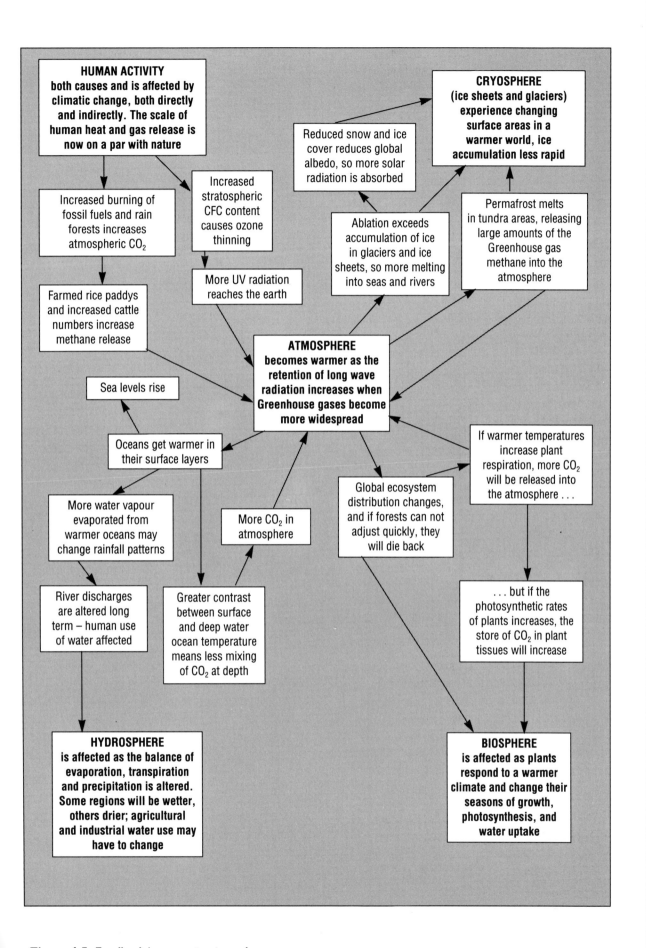

Figure 4.5 *Feedback in a warmer atmosphere*

Q3. *If asked as a climatologist to advise the government on controlling emissions, which Greenhouse gas would you say needs our most urgent attention? Use all the information on lifetimes and rates of increase in Figure 4.4 to justify your answer.*

Q4. *Look at Figure 4.5. What are the implications of atmospheric warming on human activity?*

Sandwiches of air

Research by French meteorologists in the last century revealed that the atmosphere is not one but four distinct layers around the earth. We live in the bottom of this pile of air, so the air pressure on us is greatest; in fact, half of the weight of the atmosphere is in the bottom 5 km of the atmosphere. Pressure is the weight exerted by the air on the surface of the earth, and is measured in millibars (mb). The average surface pressure is about 1000 mb, but this decreases with altitude.

Each layer has a different combination of gases. In some layers, the dominance of one gas in particular causes a change in temperature which is sufficiently different from the atmosphere above and below to be seen as a distinct layer.

The atmosphere is therefore layered according to temperature differences with altitude. The temperature of the atmosphere changes considerably with altitude. There are two layers where the temperature decreases with height. Such a phenomenon is called a temperature lapse rate; the temperature lapses or falls with height. This happens in the **troposphere** (0 to 15 km) and in the **mesosphere** (50 to 80 km).

There are also two layers where the temperature increases with altitude. This latter situation is unusual, and is called an **inversion**, i.e. the usual situation of cooling with height is inverted or reversed. This happens in the **stratosphere** (15 to 50 km) and the **ionosphere** (80 km upwards). In the stratosphere, the increase in temperature is caused by **ozone** (O_3) which absorbs short wave radiation from the sun. Finally, there are layers of the atmosphere where temperature varies little with height. These are **isothermal layers**, and are known as pauses because the trend below them halts before being reversed.

Q5. *Plot the data in Figure 4.6 on graph paper to establish the four thermal layers of the atmosphere, and the changes in pressure with height. Use two horizontal axes for temperature and pressure, and plot height on the vertical axis.*
a) *On your graph, label the four layers of the atmosphere.*
b) *Now mark on the three pauses; tropopause, stratopause, and mesopause.*
c) *To give a human dimension to the scale, mark on the height at which modern jet aircraft fly. Label on the height of Mount Everest too.*
d) *Why does temperature decrease with altitude in the troposphere and mesosphere?*
e) *Atomic oxygen (O) is present in the ionosphere in some concentration; as temperatures rise in this layer, what must it have the capacity to do?*

Height (km)	Temperature (°C)	Pressure (mb)
0	10	1000
15	−50	250
35	−60	8
50	80	2
60	80	1
80	−40	0
85	−40	0
140	140	0

Figure 4.6 *Height, temperature and pressure*

Ozone – help or hazard?

Much has been written in recent years about the thinning of the ozone layer, which lies at a height of 20–40 km in the stratosphere. In 1985, satellite images showed that the thinning of ozone in the stratosphere over the poles was so great that the term ozone hole was coined. This implies that there are parts of the stratosphere where ozone is absent entirely, which is not the case. However, scientists are increasingly concerned that up to 70 per cent thinning of ozone by pollutants such as CFCs will have long-term implications for life on earth (Figure 4.7); ozone in this instance is a beneficial screen.

Ozone forms when oxygen molecules absorb solar ultra-violet radiation. Oxygen then splits and combines with other oxygen molecules to form O_3, which further absorbs ultra-violet radiation.

Ozone thinning occurs when pollutants take up the spare oxygen molecules which ozone needs to form, so preventing ozone production. It is also attacked by chlorine based chemicals such as chlorine monoxide, which stem from CFCs. Normally, chlorine monoxide is destroyed by nitrogen oxides

Hole in ozone layer expands to endanger humans for first time

THE hole in the ozone layer over Antarctica, which this year is the earliest, biggest and deepest yet, last week covered inhabited land for the first time when it extended to the edge of South America and the Falkland Islands.

The World Meteorological Organisation in Geneva and the US National Aeronautics and Space Administration both registered the northern tip of the hole, at present a vast ellipse nearly the size of North America, touching Tierra del Fuego last Sunday, Monday and Tuesday, and the Falklands last Wednesday, before rotating eastwards out into the Atlantic.

People underneath the hole, from Argentinian and Chilean shepherds to British troops on the Falklands, were subject to a reduction of nearly 50 per cent in the protection from the sun's harmful ultraviolet light, UVB, that the ozone layer in the stratosphere around the earth normally provides.

Excessive UVB is known to cause skin cancer and eye cataracts and may affect the human immune system, as well as causing damage to plants and animals.

Although scientists said that the brevity of last week's episode meant that it was unlikely that anyone had suffered harm, Rumen Bojkov, chief of the WMO's environmental programme, said it was "a significant and very unfortunate development for the world".

Dr Bojkov called on all countries to accelerate further the phase-out of chlorofluorocarbons (CFCs) and other man-made chemicals responsible for destroying ozone, when they met to discuss the issue in Copenhagen next month. He said yesterday: "Until now the ozone hole was in general affecting only penguins but it is now clear that in certain circumstances it can reach South America.

"Ozone destruction is also getting worse in the northern hemisphere, and because of the long lifetimes of CFCs in the atmosphere, it will continue to get worse for years to come, whatever we decide to do. It is essential for the world to act now – not later, now."

Jonathan Shanklin, one of the scientists of the British Antarctic Survey who revealed the existence of the ozone hole in 1985, said that populated areas had last week suffered the highest and potentially most harmful incidence of UVB ever recorded.

"This was totally unexpected, and it should be a warning to us once again that we are playing with fire in altering the chemistry of the atmosphere," he said.

The ozone hole, an area in which the ozone layer has been severely depleted by up to 70 per cent of its normal thickness, suddenly appeared over Antarctica in the early 1980s; its principal cause was proved to be the chlorine contained in CFCs, the chemicals widely used in aerosols, refrigerators, foams and solvents.

The hole appears in September and October, inside the polar vortex, the high-speed winds that circle Antarctica, when the sunlight of the south polar spring causes the chlorine to react with ozone molecules and break them down.

Although the process of phasing-out CFCs world-wide was begun by the Montreal Protocol of 1987, and accelerated in London in 1990, ozone depletion has continued to get worse because the commonest CFCs remain in the atmosphere for 100 years or more after being released. In April, scientists announced that the ozone layer over Europe last winter had thinned by up to 18 per cent.

The hole over Antarctica has grown in the past three years and this year began to form earlier than before.

By September 23 it covered 8.9 million square miles, nearly the size of the entire North American continent, a 15 per cent increase on 1991. Last week British Antarctic Survey scientists at the Halley and Faraday bases recorded their lowest readings of the ozone in the atmosphere directly above them.

The hole is amoeba-like, its edges constantly changing with the polar wind system and rotating in a clockwise direction, and last week it was elliptical in shape: on October 4, its northern tip touched Tierra del Fuego, and remained there for two more days before moving eastwards to cover the Falklands.

Both the Argentinian city of Ushuaia (population 10,000) and the Chilean city of Punta Arenas (population 100,000) were covered by the hole, Dr Bojkov said.

On the days in question the American Nimbus-7 satellite operated by NASA reported ozone readings of about 170 dobson units over Tierra del Fuego, and 220 over The Falklands; the normal ozone reading is about 300.

"These are by far the lowest ozone values ever observed at these inhabited latitudes," Dr Bojkov said.

Douglas Parr, air pollution campaigner for Friends of the Earth, said last night: "This is a development whose potential is frightening and shows that the atmosphere is capable of springing unpleasant surprises.

"People should realise that we are still pumping out thousands of tons of CFCs and other ozone-depleting chemicals into the air each year. At Copenhagen next month we shall be looking to the British government to take a lead in accelerating the phase-out process."

Ministers from the 83 countries who are signatories to the Montreal Protocol, including Britain, meet in Copenhagen on November 23 to discuss bringing forward the target for total CFC phase-out from its current date of 2000 to 1996 or possibly earlier.

Helping The Earth Begins at Home week starts on Saturday and aims to show householders how to cut carbon dioxide emissions and save money.

The message will be that whenever we turn on the heating, switch on a light, cook a meal or run a washing machine, we use energy which costs money and may be damaging the planet.

Michael Howard, the environment secretary, was due to launch the week in London today as part of the government's long-term campaign to persuade consumers to be less wasteful.

The Times, 4 October 1992

Figure 4.7 *The ozone hazard*

in the stratosphere, but in the long winter nights at the poles, nitrogen oxide freezes in ice clouds, and so is not available to break down chlorine monoxide. This gas then accumulates to thin ozone, which in turn means more ultra-violet radiation reaches the earth's surface, damaging living tissues.

Ozone also occurs in the lower troposphere, where it acts as a poison. At the surface ozone is made when sunlight causes a reaction in hydrocarbons, and carbon monoxide, both common ingredients in vehicle exhaust emissions. Large cities with heavy traffic such as Los Angeles and London therefore have high ozone levels (see page 87) and **photochemical smogs**. As well as having health implications, such smogs have an impact on the urban microclimate; the local weather affecting towns and cities (page 88).

> **Q6.** *Using the information in Figure 4.7 on page 69 summarise the hazards of ozone thinning.*

FURNACES AND FREEZERS

How is the atmosphere heated? Most of the troposphere is heated indirectly from below by long wave radiation given out by the earth, once it has been heated by the short wave radiation from the sun.

The radiation reaching the surface at any given point depends upon several things, such as seasonal variations in the angle of incidence of the sun's rays, the reflectivity of the surface, known as the **albedo**, and the amount of interception of incoming radiation by water vapour, cloud type and dust. Albedos vary from surface to surface; a white layer of fresh snow will reflect as much as 85 per cent of incoming radiation, whilst a dark coniferous forest may reflect only ten per cent.

> **Q7. a)** *Using Figures 4.8a and 4.8b, calculate the percentage of radiation reaching the surface on a clear day compared to a cloudy day.*
> **b)** *If a cover of snow existed beneath the cloud cover in Figure 4.8b, what percentage of radiation would reach the surface then?*

Different cloud types will also reflect different amounts of radiation; thinner cirrostratus will reflect about 40 per cent whilst thicker cumulonimbus will reflect up to 90 per cent.

Not all latitudes receive the same amount of radiation over the course of a year, which means that the amount absorbed and emitted by the land and oceans will also vary. The impact of cloud layers is known to be very important in affecting radiation reaching the surface.

Cloudy parts of the globe, such as the equatorial rain forests and mountainous regions, will get less radiation because cloud both absorbs and reflects radiation back to space. Radiation inputs to the same location on a cloudy day compared to a clear day are quite different (Figures 4.8a and 4.8b).

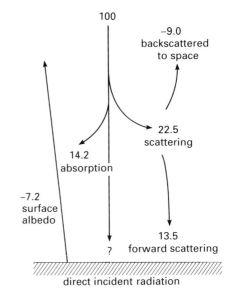

Figure 4.8a *Incident radiation on a clear day*

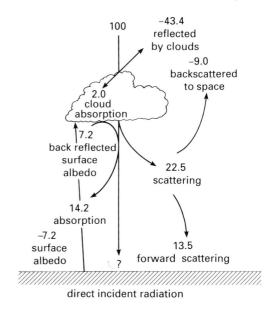

Figure 4.8b *Incident radiation on a cloudy day*

What happens to the radiation once it reaches the surface?

Once radiation has reached the surface, what happens next depends upon whether that surface is land or sea. On land, the radiation will warm the ground and this heat will be passed sideways and downwards by conduction through the earth, and upwards by radiation and then convection into the atmosphere. Over the seas, which cover some two-thirds of the earth's surface, heat is passed sideways and downwards mainly by convection and upwards into the atmosphere by conduction. Convection moves radiation in two ways; by direct transfer of sensible heat where warm water or air takes it to a cooler place, e.g. the upper atmosphere or deeper oceans, and indirectly by latent heat. The latter is moved when water changes state from water or ice to vapour in the atmosphere, locking this energy in its molecules, and releasing it when the reverse change of state, such as vapour to water, occurs.

The difference between land and sea is very important for heat distribution around the globe. Differences in the way rock, soil and water heat up and cool down can be explained by their **specific heat**. This is the amount of heat required to raise a mass of a substance through 1°C. The specific heat of sea water is much greater than that of sand, rock or dry soil. It takes five times the amount of heat to raise the temperature of sea water by the same amount as a comparable mass of dry soil, which means that the seas warm up more slowly than the land. Conversely, they cool down more slowly than the land, releasing their heat more gradually.

The oceans are therefore great reservoirs of heat energy; it has been calculated that the cooling of a metre layer of sea water by 0.1°C releases enough heat to raise a 30 m layer of air above it by 10°C.

Consequently, the ocean has a regulating effect on climate; it takes longer to warm up in summer, but then longer to cool down in winter, so mean temperatures in places near the oceans do not fluctuate greatly. Inland, the effect of **continentality**, or distance from the oceans, gives greater temperature ranges.

> **Q8. a)** The figures below show mean summer and winter temperatures in degrees Celsius for the two hemispheres of the earth. Identify which is north and which is south on the basis of continentality.
>
> | | Winter temperature | Summer temperature |
> |---|---|---|
> | A | 8.1 | 22.4 |
> | B | 9.7 | 17.1 |
>
> **b)** Calculate each temperature range.
> **c)** Temperature readings throughout the year at similar latitudes will reflect continentality. The temperature data in Figure 4.9 shows this. Plot the temperature data as four temperature curves on one graph (months across the horizontal axis, temperature on the vertical axis).
> **d)** Using an atlas to help you, explain the variations in temperature shown. Why does Port Stanley have a lower mean temperature than Southampton when both are coastal? What other factors must influence temperature?

Why does the atmosphere move heat?

This question can be answered by considering the data in Figure 4.10 on page 72, which shows the radiation budget (radiation coming in compared to radiation going out) for the globe by latitude. Radiation is shown here in units of 100 000 calories per cm^2 per year.

Location	Latitude	Distance from sea (km)	J	F	M	A	M	J	J	A	S	O	N	D	Mean
Winnipeg	50 N	2000	−20	−17	−9	3	11	17	19	17	12	5	−6	−14	1.5
Southampton	50.55 N	0	5	5	7	9	12	15	17	17	14	11	7	6	10.4
Kiev	50.30 N	700	−6	−4	0	7	15	19	20	19	14	8	1	−4	7.4
Port Stanley	51 S	0	9	9	8	6	4	2	2	2	4	5	7	9	5.6

Figure 4.9 *Temperature data in degrees Celsius for four locations circa 50 degrees from the equator*

Latitude	Radiation in	Radiation out
0	2.20	1.90
10	2.20	1.90
20	2.10	1.90
30	2.00	1.90
40	1.80	1.85
50	1.50	1.80
60	1.10	1.55
70	0.70	1.40
80	0.50	1.30
90	0.30	1.25

Figure 4.10 *Radiation movement*

> **Q9. a)** *Graph the data in Figure 4.10 on one grid, with separate lines for radiation in and out.*
> **b)** *Where the graph shows greater radiation in than radiation out shade it in and label it as radiation surplus.*
> **c)** *Where the graph shows lower radiation in than radiation out shade it in and label it as radiation deficit.*
> **d)** *Mark on the latitude where the change from surplus to deficit occurs.*
> **e)** *Mark on the latitude of Winnipeg, Southampton, Kiev and Port Stanley (Q8)*
> **f)** *What does the graph suggest for the radiation budget for each of the four locations? How does the temperature data in Q8 conflict with this?*

What becomes clear from Figure 4.10 is that low latitudes on average absorb more radiation than they emit, whilst higher latitudes emit more than they absorb. If this situation persisted, the equator would be a furnace, at least 15°C hotter than now, and the poles would be freezers, some 25°C cooler than at present.

Clearly, this does not occur on earth today. The reason we do not have furnaces and freezers at the extremes of our planet is that the surplus radiation from lower latitudes is transferred to higher latitudes and higher altitudes; put simply, the heat is transferred polewards and upwards. This is why the atmosphere moves; it is an attempt to balance the differences in temperature between the equator and the poles.

In fact, the atmosphere moves 80 per cent of the surplus radiation polewards; the remaining 20 per cent is transferred by warm ocean currents such as the North Atlantic Drift (Figure 4.11).

Almost every movement in the atmosphere, from violent hurricanes sweeping warm moist air away from the tropics, to depressions in middle latitudes mixing warm tropical and cold polar air, can be explained by this transfer of heat polewards and upwards.

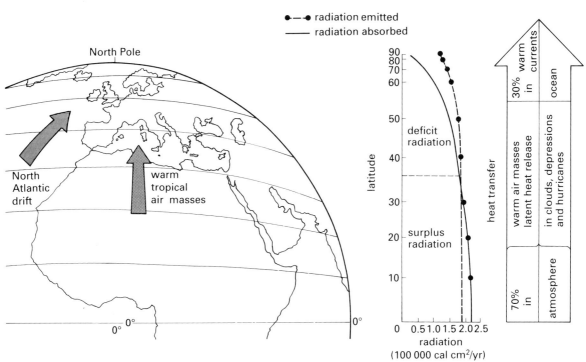

Figure 4.11 *The transfer of heat*

How does the atmosphere circulate?

The transfer of heat polewards by air masses and ocean currents indicated in Figure 4.11 occurs because of movement initiated in the atmosphere and oceans by the difference in temperatures between poles and equator. This movement stems from the fact that warm air rises rapidly over equatorial latitudes to form a low pressure zone; this uplift both instigates compensatory movement of nearby air and helps to move some of the surplus heat away.

Air on either side of this low pressure zone is drawn in to replace the rising air, and is in turn uplifted by the strong upward convection currents. The area where this occurs is called the **Inter Tropical Convergence Zone** or **ITCZ** (Figure 4.12). On global satellite images the ITCZ is usually easy to pick out as a discontinuous band of cloud, some of it tall cumulonimbus, which gives the equator high and regular rainfall.

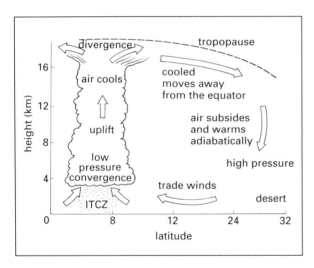

Figure 4.12 *Hadley cell section*

The air which has been lifted into the upper troposphere over the ITCZ will cool with altitude, and eventually be ready to sink, as it is now denser. This subsidence of air occurs on either side of the equator at a latitude of about 30°. As this air has lost most of its moisture content over the equator and is sinking, it is both dry and warm. It becomes warmer for two reasons; firstly, it is returning to the source of heat, i.e. the surface, and secondly, it is returning to where the pressure on it is greatest. When air is compressed it warms up, (an **adiabatic** warming).

This can be experienced when air in a bicycle pump is compressed in the tube. Such a warming does not require an external addition of heat, merely a pressure increase.

The subsidence of warm dry air at latitudes 30° north and south explains the existence of deserts in these locations. When air sinks back to the surface to form zones of high pressure, a dome of air called an **anticyclone** is created. As air accumulates in an anticyclone it tends to slide off sideways to zones of low pressure.

Cells and streams

This sideways movement of air is either back towards the equator to the ITCZ or poleward. When air returns to the equator it has completed a cycle (Figure 4.12); this rotation is called the **Hadley cell** (after George Hadley who in 1735 first suggested this rotation existed). The return of air from high pressure over the deserts to low pressure at the ITCZ creates constant winds, which are known as the Trade Winds, on each side of the equator. The rotation of the earth, which is anticlockwise as seen from the north pole, causes a deflection of these winds to the right in the Northern Hemisphere and to the left in the Southern Hemisphere, which pushes the Trades towards the west. This is the Coriolis force, which deflects all surface winds from a straight line path.

Historically, these Trade Winds have been important in transporting ships from Europe to the Americas on voyages of exploration and later voyages of trade and barter. Columbus was helped by these winds in his journeys to the Caribbean from Spain.

Q10. a) *Explain why the ITCZ between the Hadley cells will not always be over the equator.*
b) *What will the seasonal movement of the ITCZ mean for the areas between it and the deserts?*

When the air subsiding over the deserts moves polewards and not towards the equator, it is again fulfilling the role of dispersing heat away from the equator. In the past, the circulation of air in the middle latitudes was thought to be similar to the Hadley cell; this time air would move polewards in the lower troposphere, meet colder polar air at about 60° north and south, rise over the polar air creating the **Polar front**, and then return as cooler air in the upper troposphere towards the deserts.

This cell was called the Ferrel cell (after George Ferrel who published a paper in 1856 suggesting this cell existed).

More recent research into atmospheric movement in the middle latitudes suggests that the Ferrel cell does not exist in the way that was first thought. Instead, air moves in waves which meander between the deserts and the Polar front in a westerly direction in the Northern Hemisphere, and climb from low altitude to high altitude and back again as they move clockwise around high pressure systems and anticlockwise around low ones. These are **Rossby waves** named after Carl Gustav Rossby, who modified Ferrel's ideas in 1941 (see Figure 4.13).

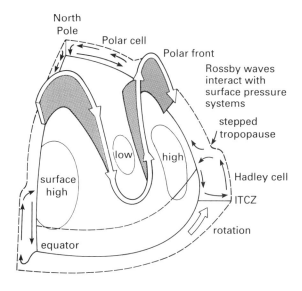

Figure 4.13 *Rossby wave rotation*

Q11. *State the main differences between the Ferrel cell model and the Rossby wave model for air movement in the middle latitudes.*

Polewards of the Polar front lies another circulation cell, the Polar cell. Here, air subsides away from the high pressure exerted by cold air sinking at the poles, and moves towards the Polar front low pressure area, where it is warmed on contact with the tropical air moving from the other direction. On warming, it rises into the upper troposphere, and returns back to the poles in the upper troposphere.

Originally then, the three cell model of Hadley, Ferrel, and Polar cells was used to explain how winds and **air masses** moved around the atmosphere to transfer their heat and moisture. Today, we are left with a two cell model (Hadley and Polar) with a middle latitude dominated by Rossby waves and **jet streams**. The latter are high velocity streams of air, flowing at around 300 km per hour, at the core of the Rossby waves. Jet streams are about 400 km wide, and at a typical height of 10 km from the surface they speed up west to east air flights considerably. They result from the sharp change in pressure between the different zones of air circulation at the tropopause. Each zone of circulation gets progressively lower towards the poles, so the tropopause is stepped; the Hadley cell extends up to 16 km at the tropical tropopause, the Rossby wave zone 12 km, but the Polar cell is flattest, extending up to 9 km at the polar tropopause.

This stepping of the atmosphere causes sharp pressure changes in the upper atmosphere. This in turn creates a **pressure gradient** which the wind will blow down from higher pressure to lower pressure. The steeper the pressure gradient is, the stronger the wind will be, so a jet stream is created when this gradient is very steep, e.g. where the polar front exists between the Polar cell and the mid-latitude rotation. The Coriolis deflection in the Northern Hemisphere deflects the jet streams from west to east.

The circulation of the atmosphere both near the surface and in the upper troposphere and stratosphere has a bearing on almost every aspect of human activity. Temperature differences create pressure differences; pressure differences cause winds to blow; winds blow air masses and ocean currents; air masses and ocean currents affect both day to day weather and long-term climate characteristics of an area. Winds can wreak havoc or be harnessed to generate energy, and the rains that they bring can cause flooding and landsliding or welcome relief from drought.

How do ocean currents affect human activity?

Although the oceans would move because of temperature and density differences in their waters, global winds which blow across their surfaces transfer their wind energy into the water to create ocean currents, which in turn transfer warmth and coldness to the places they flow past. The pattern of the ocean currents in the northern Atlantic and Pacific is broadly that of a clockwise rotation, i.e. moving water north along the western side of the ocean, and south along the eastern side. Such large circulations in the oceans are called **gyres**.

The north Atlantic gyre generated by the Trade Winds and the Westerlies has been historically important for trade. In previous centuries, when international trade was dependent on sailing ships, the constant wind patterns and ocean currents around the Atlantic governed trade there. Ships from

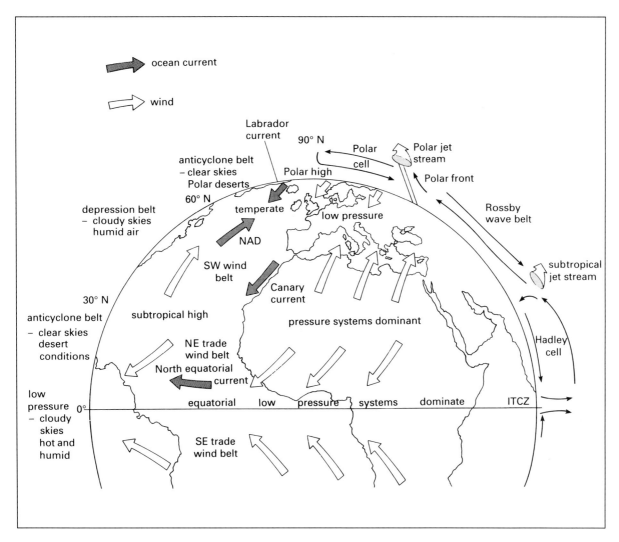

Figure 4.14 *Simplified atmospheric and oceanic circulation (Northern Hemisphere)*

European colonial powers such as Britain, France, Germany and Spain used the Canary current and north east Trades to sail to the west coast of Africa, where raw materials such as rubber, cocoa, coffee and ground nuts were produced by plantation agriculture.

This route became part of the trade triangle in the Atlantic; it also became known as the slave triangle, as West African slaves were taken to the plantations in the Americas as labour. The second leg of the triangle was across the Atlantic to the Americas using the Equatorial current. Ships were frequently becalmed in the Doldrums under the ITCZ, where lateral air movement is limited; were it not for the Equatorial current, many would not have made their destination.

On exchanging slaves for merchandise such as cotton and tobacco in the Americas, seafarers completed their circum-Atlantic route by returning to Europe on the Westerlies and the North Atlantic Drift (NAD). The NAD is responsible for keeping the shores of Britain ice-free during winter, ensuring that trade can continue all year round.

Contrast this with the St Lawrence seaway in eastern Canada, which freezes in the first three months of the year because of the cold polar air flowing south on the Polar cell, and the cold Labrador current offshore. Icebreakers and ice control booms enable this seaway to remain open for longer today, but up to 90 days of freight transport can still be lost each year.

ELEVATOR GOING UP – MID-LATITUDE DEPRESSIONS

Once the global movement of air masses has been simplified using models of circulation, an analysis can be made of what happens to individual air masses when they move away from their source area, taking their moisture and heat with them. Air masses are classified according to their temperature, humidity, and stability, and these three properties may be quite uniform across the hundreds of square miles of area and hundreds of metres of altitude that form an air mass.

Air mass characteristics are derived from where the air mass originated. Air pushing north away from the Saharan high pressure over Europe would be warm and dry, with a tendency to rise, or be unstable, until it has been cooled by mixing with colder air to the north.

Air masses over the Northern Hemisphere are classified as Arctic, Polar, or Tropical in terms of their temperature, and maritime (humid) or continental (dry) in terms of their moisture content. So another way of classifying the Saharan air mass would be Tropical continental (Tc).

Q12. *Suggest the type of source area where Tm, Pm, Pc and Am air masses may originate.*

Between the months of November and May, British weather is dominated by the passage of low pressure systems called **depressions**. Such systems occur when Tropical maritime (Tm) air moving away from the sub-tropical high pressure zone mixes with Polar maritime (Pm) air moving away from the Polar high pressure cell. The mixing of air masses occurs along the broad zone called the Polar front (Figure 4.11), and because the Tm air is warmer and less stable, it begins to slide over the cooler Pm air causing an indentation along this front.

This indentation reveals itself as a wedge of cloud which may grow to become a fully fledged depression. The cloud is forming as the humid tropical air is cooled with altitude, causing the water vapour in it to condense at its dew point. The polar air will also be humid but less so because of limited evaporation over colder seas.

The depressions that track across the Atlantic into Europe can now be seen by weather satellites many days before they arrive, so knowledge of how they develop is improving; the wedge of cloud is the first sign that tropical and polar air are mixing in the west Atlantic, and pushing eastwards on the westerly winds as an 'embryo' system. What happens after this stage is vital if an accurate forecast of the weather associated with the depression is to be made.

What meteorologists believe happens next in a depression is that the Tm air continues to slide along and over the Pm air, creating a low pressure centre, and then rotates clockwise out of the system in the upper troposphere (Figure 4.15).

At the same time, the Pm air is drawn into this ascent, curling around until it too spirals clockwise into the upper troposphere. As this creates a deficit of air at the surface, cold air from the upper troposphere descends in an anticlockwise fashion; some of this air sinks to form a cold air front at ground level, behind the front formed by the ascent of warmer air, whilst the rest flows over the top of the depression. This model of depression development is called the Conveyor Belt Model and is quite complex because it shows that simple ideas of Pm and Tm air mixing to form warm and cold fronts at the surface can not fully explain what happens; some connection with the upper troposphere is needed. We now know that the jet stream on the Polar front provides this connection.

Depression weather

Although ideas on depression formation are still being refined, the sequence of weather associated with them is reasonably well understood. Depressions have a warm front which is a band of layered cloud and rain at an angle of about 3° to the ground, at their leading edge. The stratus cloud at the lower end of this front gives steady prolonged precipitation. Behind this is the warm sector; a zone of warm ascending Tm air with lower level stratus or broken up cumulus cloud. Following on behind this is the **cold front** which is a steeper front at an angle of around 6° from the ground, which has denser cloud such as cumulonimbus rising up to the tropopause. This gives shorter bursts of intense precipitation.

The two fronts are shown on a weather map as bands of cloud and precipitation, but radar and satellite images reveal that each front has a hierarchy of precipitation cells. Within each front are meso scale cloud cells, up to 30 km^2 in area, and within these are individual micro cells, up to 3 km^2 in area. Micro cells usually carry the greatest amount of rainfall as individual downpours which may last from a few minutes to half an hour.

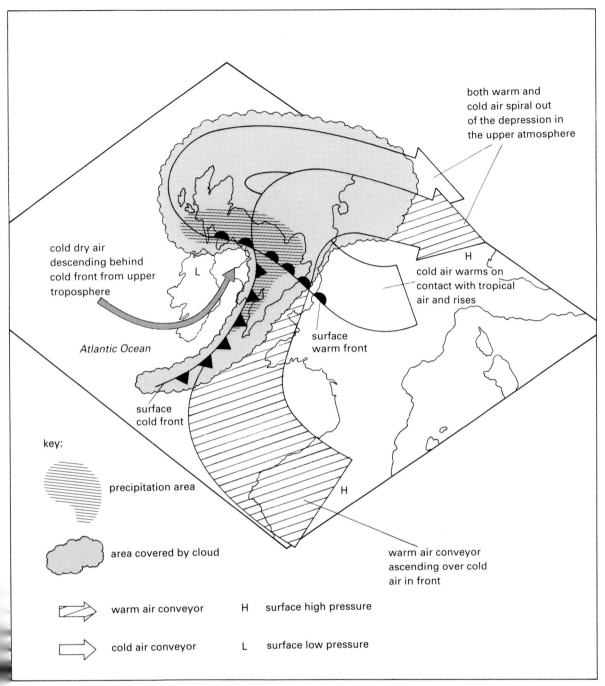

Figure 4.15 *The depression conveyor belt*

Depressions as hazards

In the early morning of 16 October 1987, the worst storm for three centuries struck the south east of Britain. This exceptionally strong depression wreaked havoc on the region; transport was brought to a standstill (Figures 4.18, 4.19), the worst power failure since the 1939–45 war occurred, and many buildings were demolished or severely damaged (Figure 4.20). 3000 miles of telephone lines were brought down, 15 million trees were felled, and insurance claims for the damage totalled over £1.1 billion. It took several days before most of the transport services of the south east recovered. Nineteen people were killed.

Over two years later on 25 January 1990, a second storm swept across Britain from the west.

Although this was forecast well in advance by the Meteorological Office, it too brought devastation to most of southern Britain. This storm killed 47 people, felled five million trees and caused power failure in over one million homes. Insurance claims were estimated at £750 million. Share prices of insurance companies such as Commercial Union, General Accident and the Guardian Royal Exchange fell in anticipation of more claims.

Two such extreme events occurring so close together focused a lot of attention on the ability to forecast weather well in advance. The weather sequence that was monitored at Oxford for the 1990 storm is shown in Figure 4.16. Figure 4.17 shows the weather map at mid-day when the storm was near its worst.

Q13. a) Using the weather information provided in Figures 4.16 and 4.17, say when the warm front, warm sector and cold front passed over the city. Justify your answer.
b) From which direction was the wind blowing when it was at its strongest?
c) Which aspects of the weather in the depression appear to be linked? Why is this?

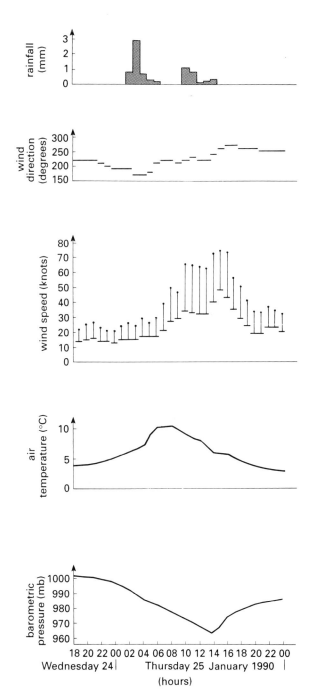

Figure 4.16 *Depression weather sequence chart*

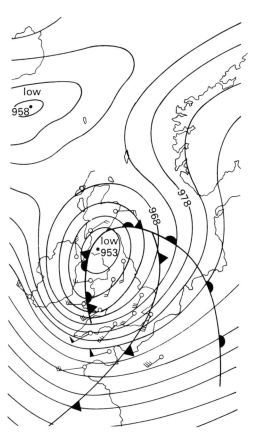

Figure 4.17 *Depression weather map*

78 IN THE AIR

Figure 4.18 *Light aircraft picked up by storm force wind, tossed over club house and deposited on top of a parked caravan, Sheffield, October 1987*

Figure 4.19 *Destroyed Peacehaven caravan site after the Great Gale, 16 October 1987*

Figure 4.20 *Wall collapse in Eastbourne, 16 October 1987*

DECISION MAKING EXERCISE

Forecasting depression weather

The satellite image (Figure 4.21) and the weather map (Figure 4.22) on page 81 show a depression which tracked across Britain on 23 February 1992. This depression moved in from the west at a speed of 40 km per hour.

a) Use the image and weather map to describe the weather over London at the time the picture was taken. Produce a weather bulletin to be broadcast on a local radio station summarising the present situation.

b) Assuming the depression continues on its present track, produce a forecast for London for the next six hours. The forecast should include predictions of windspeeds, wind directions, rainfall, cloud cover, cloud type and temperatures. Use the scale to work out where the depression should be in six hours time, and bear in mind both the time of day and the time of year for which you are making a forecast.

c) Which other organisations do you think would require detailed forecast information for the next six hours? Produce a flow chart, with the Meteorological Office at the start, to say which groups should receive, e.g. rainfall data or windspeed forecasts most urgently. Where does the general public lie in your chart?

d) The Meteorological Office was heavily criticised after the 1987 storm for failing to forecast it accurately. As a weather forecaster, what factors would you argue make an exact forecast difficult?

e) How useful would a storm warning have been for the 1987 storm, given the severity of it?

80 IN THE AIR

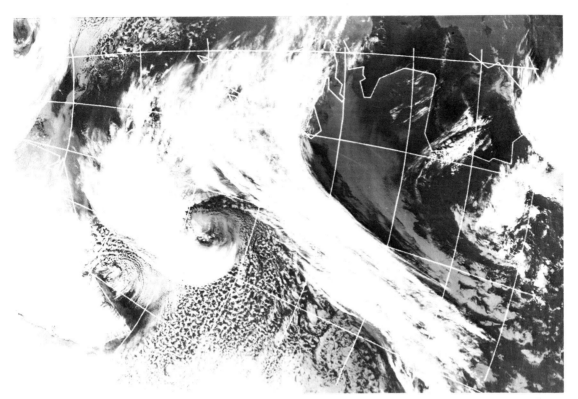

Figure 4.21 *Depression satellite image, 23 February 1992*

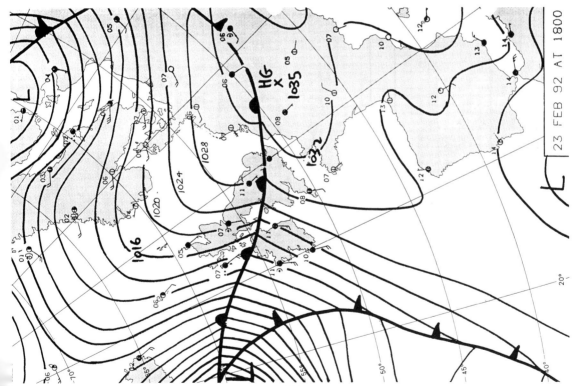

Figure 4.22 *Depression weather map, 23 February 1992*

THAT SINKING FEELING – ANTICYCLONIC WEATHER

Anticyclones are high pressure systems where air converges in the upper troposphere and subsides into the lower troposphere. As the air sinks, it 'piles up' into a dome, away from which gentle breezes flow down a gentle pressure gradient.

Anticyclones are very different pressure systems from depressions, and the weather associated with them is also different. Summer anticyclones bring different weather to winter anticyclones in mid-latitudes but it is possible to summarise their broad weather pattern in a logical sequence.

Q15. Use Figure 4.23 below and put the following statements about anticyclones in a logical sequence.

a) Sinking air warms as it returns to the source of heat at the surface.
b) Dryer air which is stable and sinking means cloud formation is less likely.
c) Less cloud means less precipitation in anticyclones.
d) Slow moving air down a weak pressure gradient results in gentle winds.
e) High temperatures by day give heat wave conditions.
f) Possibility of strong local thermals of rising air.
g) Heat loss at night causes fog and ground frost as the air is chilled to saturation or freezing point.
h) Morning fogs slow to disperse when the sun is at a low angle, causing 'anticyclonic gloom'.
i) Heat loss at night means heavy dew on the grass and morning mists, which clear quickly as the sun climbs to a high angle.

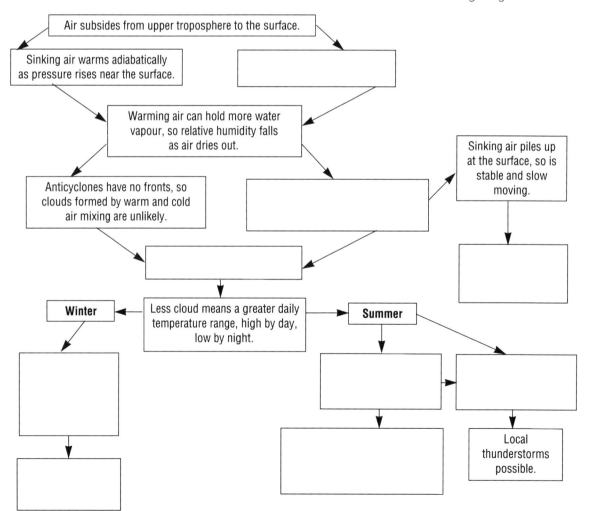

Figure 4.23 Anticyclone weather flow chart

Anticyclones and drought

The fact that anticyclonic weather is not usually associated with a great deal of rainfall can cause the problem of drought when anticyclones dominate the pressure pattern over a region for a long period of time. A drought is not always easy to define because it is not always due to lack of rain; drought may hit farmers because of low levels of river flow or low levels of groundwater.

In Britain, a drought is defined as a period of at least 15 consecutive days with no more than 0.2 mm of rainfall on any one day. One of the most well documented droughts in Britain this century occurred between June and August 1976, when a blocking anticyclone settled tropical continental air over Britain, pushing depressions coming in from the Atlantic to the north over Scandinavia and to the south through the Mediterranean.

The rainfall that these depressions would have brought into Britain was therefore transferred elsewhere in Europe; Scandinavia and the Mediterranean reported rainfall levels to be 150 per cent above average, whilst in Britain rainfall levels were less than 50 per cent of the average. This brought a welcome boost in the trade at tourist resorts around the British coast, but was a major blow to farmers in the south east of the country who watched their crops wither.

Drought in Britain is not always as obvious as the drought of 1976 however. In the period between December 1988 and May 1992, southern Britain had four dry winters and 41 consecutive months of below average rainfall. Anticyclones did not dominate the weather for the whole of this period, but were more common than usual, especially in the winter months. Precipitation was at average levels in the north west of Britain, but below average in the south east. As a result, by the late spring of 1992, boreholes in the Chiltern Hills, a major chalk aquifer to London, were drying up for the first time in recorded history as the water table dropped.

In less developed countries, drought subsequent to a long period of anticyclonic weather is not merely an inconvenience, but a matter of life and death. The Sahel of Africa has been particularly badly hit by drought on many occasions, most notably between 1968 and 1973 when an estimated 100 000 died through crop failure and subsequent famine. The continued domination of the Sahelian climate by anticyclonic weather continues to cause food production difficulties today. In 1983, some 150 million Africans were thought to be affected by drought, but the people are aware that it is not only lack of rain that is the problem.

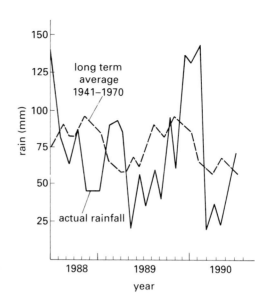

Figure 4.24 *1988–90 drought graph*

In Niger, the Agriculture Minister in 1982 adopted the definition of drought as *'Not as much water as the people need'*, which is a broader definition that accepts lack of rain as only part of the problem. Another cause of drought to crops in Niger is high evapotranspiration rates from the soil, which anticyclonic conditions of clear skies and long sunshine hours encourage. On top of this, the soils are laterites, which means they have a high sand content and a low clay and humus content. As a result, the soil does not hold as much water as it might, and dries out quickly, making it more vulnerable to wind erosion and desertification.

In less developed countries, the link between anticyclones, drought, desertification and famine is not therefore necessarily a causal one. Many believed an increasing occurrence of stronger anticyclonic conditions caused the Sahelian drought. In fact, climatologists studying the pattern of rainfall in Sahelian Africa suggested that drought conditions were frequent historically. The Tuareg of Niger have even given names to past droughts such as *'forget your wife'* and *'the sale of children'* which indicate the hardship they have endured.

> **Q16. a)** *Do anticyclones cause drought?*
> **b)** *Do anticyclones cause famine? If they do not, what other explanations can be used for the continuing famine in the Sahel?*
> **c)** *What are the key differences in the way in which developed and less developed societies adapt to drought?*

Anticyclones as pollution traps

High pressure systems present humans with hazardous weather conditions when they trap pollution beneath them. As air sinks in an anticyclone, it traps the air at the bottom of the troposphere, holding it close to the surface. This trapping of surface air is sometimes aggravated by an inversion layer. An inversion forms when heat is lost from the surface so that the air closest to the ground is colder than the air above it; the upper layer of air then behaves like a lid, preventing the air below from escaping.

Case study – Halkyon days in Athens

Greek meteorologists since the time of Aristotle have recognised the occurrence of high pressure systems each winter which bring sunny, dry weather from mid-December to mid-February. These are called Halkyon days, and in Athens, they bring a daily average of 8.5 hours of sunshine compared to 4.2 hours on non-Halkyon days.

In recent years however, Halkyon days have become associated with severe atmospheric pollution because the vehicle emissions and industrial pollutants in Athens become trapped under the high pressure and incoming radiation intensifies the chemical reactions into a photochemical smog (Figure 4.25).

Any pollutants in the lower troposphere will therefore be held close to the surface, and the absence of strong winds means that they will not be

Figure 4.25 *Ninth day of smog in Athens, 1981*

dispersed. In large cities with a pollution problem, this creates a considerable health hazard, as smogs persist, sometimes for days (page 86).

Q17. a) *Use the data in Figure 4.26 to plot two graphs to compare the concentration of smoke and sulphur in Athens on Halkyon and non-Halkyon days. What do you find?*
b) *Use the graphs to calculate peak smoke and sulphur concentrations for Halkyon and non-Halkyon days. What is the percentage difference in each case?*

Smoke and sulphur concentrations (microgrammes per m³) in Athens on Halkyon and non-Halkyon days 1971–1981					
Smoke class intervals	Frequency Halkyon days	Non-Halkyon days	Sulphur class intervals	Frequency Halkyon days	Non-Halkyon days
1–50	0	9.92	0–100	4.61	29.8
51–100	2.68	29.57	101–200	30.92	42.35
101–150	17.76	27.24	201–300	21.71	18.63
151–200	20.39	15.95	301–400	18.42	7.25
201–250	19.74	9.34	401–500	9.21	1.37
251–300	13.16	4.47	501–600	5.92	0.001
301–350	11.18	1.95	601–700	5.62	0
351–400	5.26	1.17	701–800	1.97	0
401–450	3.95	0.01	801–900	1.32	0
451–500	2.63	0	901+	0.001	0
501–550	1.97	0			
550–600	1.31	0			
601+	0.01	0			

Figure 4.26 *Air pollution in Athens*

Controlling the urban pollution hazard

Air pollution problems like those experienced in Athens are common to many large industrial cities. London had a problem with smog which dated back to the sixteenth century, and the reign of Queen Elizabeth I, as coal began to take over from wood and charcoal as the main urban fuel source.

John Evelyn, a fellow of the Royal Society, wrote in his book *A Character of England* in 1659 that London *'was cloaked in such a cloud of sea coal, as if there be a resemblance to hell on earth, it is in this volcano of a foggy day: this pestilent smoke, which corrodes the very iron, and spoils all the moveables, leaving a soot on all things that it lights: and so fatally seizing on the lungs of the inhabitants that cough and consumption spare no man'*. Evelyn was one of the first to draw attention to the damage that the smoke pollution was doing to buildings and also to human health.

In 1662 John Graunt, a draper by trade, took the issue of pollution impact on health a stage further by gathering information from Bills of Mortality in London. These bills monitored the number and causes of death in London by parish, and although varying in their reliability, began to show that when pollution was heavy the weekly death rate in London increased. Deaths from respiratory problems were particularly common, and asthma, bronchitis, and a disease called tisick, related to asthma, accounted for many deaths.

By 1850, Britain had a Smoke Nuisance Abatement Act passed through Parliament to try to deal with this problem. The main sources of smoke in London's atmosphere at the time were the multitude of small factory chimneys and domestic coal fires and in combination with the fogs of late autumn and winter, it formed dense smog. By the mid-nineteenth century, London became known as the 'Big Smoke'.

Controlling this problem became a major issue. Smog during daylight hours caused an increase in demand for gas lighting; transport in Victorian London was chaotic during fog, as horses and cabs collided, and train accidents were caused by difficulties in signalling. In a December fog of 1873 a group of seven people walked into the Thames, so poor was visibility. Cattle on exhibition at the Great Show in the same year suffocated or had to be put out of their misery.

Into the twentieth century, the chemistry of London smogs was surveyed more thoroughly in order to see how best to prevent the constituents of it from getting into the atmosphere. Alkali inspectors were employed to curb emissions of particular pollutants such as hydrochloric acid, and were often successful in doing so, but the list of pollutants was a long one, and clean air legislation was often slow to be enforced. A variety of smoke abatement laws were passed, but progress in the first half of the century was slow.

The event that caused both a change in the views of the politicians, and legislation to be more rapid was the Great Smog of Thursday 4 to Tuesday 9 December 1952. This was an atmospheric catastrophe. An anticyclone moved over London on the Thursday, and the fog that formed in its lower layers soon became smog. By Friday, the smog had thickened into a 'pea souper', so called because of its yellow green appearance and thick nature. Londoners noticed a choking feeling about the air, their clothes became filthy walking through the smog, and by Saturday, people with respiratory problems were dying, choked by the air they breathed. Deaths continued on Sunday and Monday, and it was only when the anticyclone began to disperse, clearing the fog on Tuesday, that London's agony ended.

An estimated 4000 deaths were attributed to the Great Smog, mainly of people who had chest complaints (Figure 4.27).

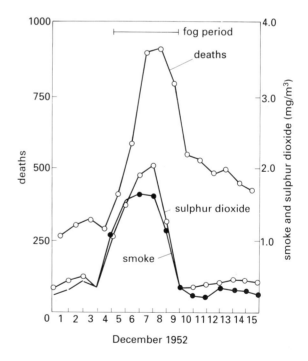

Figure 4.27 *Smoke SO_2 and deaths in the London Smog 1952*

Public outcry about this event forced Parliament to act, and in 1956 the Clean Air Act was passed. This act aimed to control not merely industrial pollution, but also domestic pollution by creating smoke control areas or smokeless zones. A second act in 1968 followed, and at last it seemed that London's air was cleaner; mean smoke concentrations fell from 200 microgrammes per m^3 in 1960 to 30 in 1990, whilst mean sulphur dioxide concentrations fell from 250 microgrammes per m^3 to 70 over the same period.

Out of the frying pan and into the fire

Unfortunately, the decline in industrial and domestic particulate pollution in London coincided with the growth in another type of air pollution, that of vehicle emissions. The growth of motor traffic in London in the last 30 years has been impressive. Between 1972 and 1986, total car miles travelled increased 25 per cent, peak traffic flows 25 per cent, and yet peak period speeds fell by 11 per cent to 12 mph.

Put simply, London has more cars moving more slowly than ever before – a recipe for heavy air pollution.

Some of the products of vehicle emissions include carbon monoxide, nitrogen dioxide, and benzine, and when sunlight produces secondary substances, ozone is created (page 68). As all of these are poisons, their concentration in the air is monitored. WHO recommends guidelines for pollutant concentrations in our air, and these guidelines are still exceeded in British cities today on occasions (Figure 4.28).

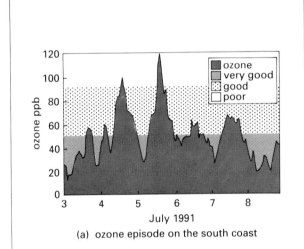
(a) ozone episode on the south coast

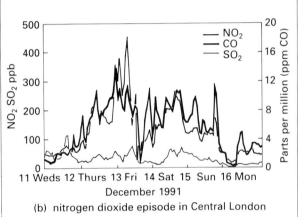
(b) nitrogen dioxide episode in Central London

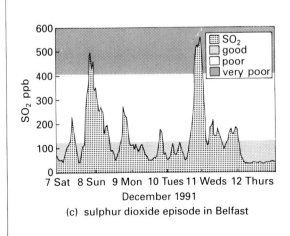
(c) sulphur dioxide episode in Belfast

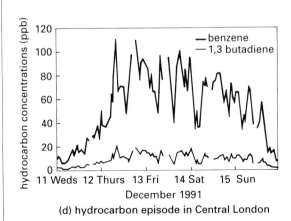
(d) hydrocarbon episode in Central London

Figure 4.28 *Graphs of atmospheric pollutants in British cities*

DECISION MAKING EXERCISE

Urban air pollution

Some of the suggestions for cutting traffic pollution from cars in cities include:

a) higher taxes on petrol and vehicles to discourage car use;

b) schemes to encourage drivers to take passengers to eliminate the 'one person vehicle';

c) heavily subsidising public transport to make the use of it more attractive;

d) stopping citizens from using their cars on any one day in the week (tried successfully in Mexico City).

How would you tackle the second wave of air pollution caused by motor vehicles in London?

THE HUMAN RADIATOR – URBAN CLIMATE

As well as having a more polluted atmosphere than rural areas, urban areas are now known to generate their own climate at a local level, a so-called microclimate. Large towns and cities are recognisably warmer than rural areas because they not only generate heat from their buildings, but also because the tarmac and brick surfaces convert incoming solar radiation during the day into long wave radiation and release it into the local atmosphere. This heat release at night makes cities warmer than rural areas, and creates the so-called **urban heat island**, a peak of higher temperatures in a sea of otherwise uniform temperatures. The reduction of wind speeds and consequent reduced windchill also causes cities to be warmer than the areas around them.

On a May night in 1959, under anticyclonic conditions, Professor Tony Chandler measured the air temperature at regular intervals across London and mapped out its heat island. He found that the difference between the temperatures in the city centre and the city edge was 5.5°C.

Further work showed that the heat island was strongest at night, especially in late summer and early autumn, when long hours of daylight allowed the city to heat up, and then release the heat under a blanket of pollution haze.

In 1992, Derek Lee re-examined London's heat island, and found that it had changed as the city had evolved. Land use in London changed dramatically between the 1960s and 1990s, especially in areas such as the Docklands, and traffic density and fuel consumption also changed. Lee found that by day, London's heat island was reduced in intensity, compared to the 1960s. This could be because London is receiving less heat directly because of the **urban pollution dome** over it. Another finding was that more recently, London's heat island is more

Annual differences expressed as a percentage of rural conditions (all figures are positive unless shown otherwise)			
Climatic factor	Annual difference	Cold season	Warm season
Contaminants	1000	2000	500
Short wave radiation	−27	−34	−20
Air temperature Celsius	1	2	0.5
Relative humidity	−5	−2	−8
Visibility	−25	−34	−17
Fog frequency	60	100	30
Wind speed	−25	−30	−20
Cloudiness	8	5	10
Rainfall	14	13	15
Thunderstorm frequency	16	5	30

Figure 4.29 *Changes in climate caused by urban areas*

intense in winter; differences between central London and Heathrow were more marked at that time of year.

Studies of the microclimates exerted by other cities on their immediate areas shows that other aspects of climate are altered in cities (Figure 4.29).

> **Q18. a)** Use the data in Figure 4.29 to describe the key differences between urban climates and the rural areas around them.
> **b)** Attempt to explain the differences, especially those between the warm season and the cold season.

The changes exerted by a city on the climate are most intense close to the ground. Between the ground and rooftop level, the shelter effect of buildings is great; this is called the **urban canopy layer (UCL)**. Between the rooftops and the rest of the atmosphere lies the **urban boundary layer (UBL)**; a layer where turbulence transfers some of the properties of the UCL into the air above, causing it to be different in composition to the surrounding air. When the properties of the UBL are transferred downwind from the city, an **urban plume** develops, sometimes tens of kilometres in length.

The turbulence of air in cities is caused by two main mechanisms. The first is the friction exerted by buildings, causing eddying effects. Eddies can increase the wind speed in cities locally, especially when wind is funnelled along a street around buildings.

The second mechanism is the uplift of air away from street level caused by the rapid warming of air next to concrete or tarmac surfaces. These thermals of rising air cause dramatic updraughts, especially alongside tall buildings. The updraught of air on one side of a street which has tall buildings (a so-called **urban canyon**), will often form a compensatory downdraught on the opposite side of the street, as air is drawn down to pavement level to replace the rising air.

Small convection cells circulating within the UCL can persist for several hours within cities, especially in a hot summer. If the rising air is very unstable over a large area of the city, it will often continue ascending through the UBL, and form cumulonimbus clouds with thundery rain. Figure 4.30 summarises some urban microclimatic effects.

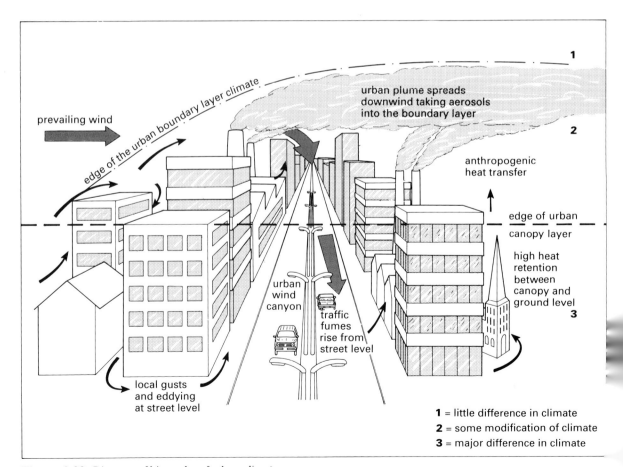

Figure 4.30 Diagram of hierarchy of urban climates

BUBBLES AND BALLOONS – LAPSE RATE THEORY

The behaviour of air rising away from the surface, whether over a town or over a range of hills, can be understood in terms of lapse rate theory. A lapse rate is simply a decline in temperature with altitude. The air at the bottom of the tropopause, or environmental air, usually cools at a variable rate with altitude, depending on the time of day, time of year, and windspeed. The **Environmental Lapse Rate (ELR)** is approximately 6.5°C per kilometre of altitude and can be measured using weather balloon data.

Bubbles of air will ascend through environmental air for two reasons; either they are warmer and less dense, so rise through **thermal uplift**, or they are forced upwards, for example by turbulence, and rise through **mechanical uplift**. Air will continue to ascend through thermal uplift until its temperature cools to the temperature of the environmental air, when it will become stable, with a tendency to sink back to earth. Rising air cools for two reasons. The first is that it is moving away from the source of heat at the earth's surface. The second is that its expansion upon rising into less dense air causes it to cool adiabatically (page 74). Dry air will cool adiabatically at a fixed rate of 10°C per kilometre, the **Dry Adiabatic Lapse Rate (DALR)**, but on reaching saturation at its dew point it will cool more slowly at 5°C per kilometre at the **Saturated Adiabatic Lapse Rate (SALR)**.

When air is cooled to saturation point, the vapour in it will turn to water droplets, so clouds will form above this condensation level. Latent heat is also released as the water changes state, which explains the slower cooling rate of saturated air. If the water droplets then freeze into ice in the upper cloud, latent heat will again be released, and snow may fall.

So when bubbles of air rise through environmental air, we can forecast reasonably well at what height they will become saturated if we know the dew point, and the stability point of the rising air. The dew point indicates when clouds will start to form above the condensation level, whilst the stability point determines the cloud top (Figure 4.31).

> **Q19.** Look at Figure 4.31.
> **a)** At what height is the condensation level?
> **b)** At what height does the rising air become stable?
> **c)** At what temperature does the dry air reach its dew point?
> **d)** Will ice form at the top of the cloud?
> **e)** What has caused the inversion layer on the ELR?

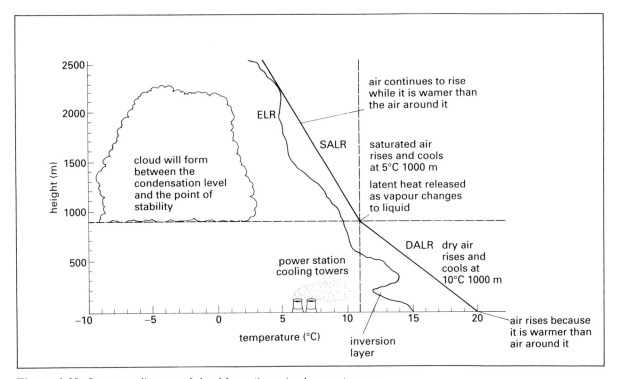

Figure 4.31 *Summary diagram of cloud formation using lapse rate curves*

Lapse rate worked examples

Example 1

The data in Figure 4.32 shows the ELR over a runway at Gatwick Airport in May 1993. A bubble of air rises thermally through this, starting at a ground temperature of 14°C. This bubble of air is known to have a dew point of 9°C.

a) On a sheet of graph paper, plot temperature on the horizontal axis, and altitude on the vertical axis to create a **tephigram** (temperature-height graph).
b) Plot the ELR using the figures provided.
c) Mark on the known dew point of the rising dry air as a vertical line from the temperature axis.
d) Plot the path of the rising bubble of air from its ground temperature. Start with the DALR until the air reaches its dew point, and then use the SALR.

ELR	
Height (m)	**Temperature (°C)**
0	13
100	12
200	10.5
300	9
400	7.5
500	6.5
600	6
900	4
1200	2.25
1500	1
1800	−0.25
2100	−0.75
2400	−2.75
2700	−3.75
3000	−4.5
3300	−5.25
3400	−5.5
3500	−5.75
3600	−6

Figure 4.32 ELR over a Gatwick Airport runway, May 1993

> **Q20. a)** At what height is the cloud base?
> **b)** At what height is the cloud top?
> **c)** Given the thickness and altitude of this cloud, what cloud type would you say it is?
> **d)** How much warmer is the rising air compared to the environmental air at 3 km?
> **e)** At what height might ice form on the wings of aircraft climbing through the cloud?

Example 2

The data in Figure 4.33 shows the ELR over a city in September 1993. A bubble of warm air rises from a road junction through this, starting at 13°C. The bubble of air is known to have a dew point of 9°C.

Follow the same instructions as those for Figure 4.32 to build a tephigram for this situation.

On plotting this data you should find that if thermal uplift is the only mechanism causing the air to rise, then stability will set in at low altitude. Assume mechanical uplift operates to a height of 1500 m.

ELR	
Height (m)	**Temperature (°C)**
0	11
100	10.5
200	10
300	9.5
400	9
500	10
600	9.5
700	9
800	8
900	7.5
1000	7
1100	6
1200	5
1300	4.5
1400	4.25
1500	4

Figure 4.33 ELR over a city, September 1993

> **Q21. a)** At what height is the cloud base?
> **b)** What is the name given to the phenomenon between 400–600 m?
> **c)** What effect does this layer have upon the stability of the air bubble?
> **d)** If thermal uplift was the only mechanism responsible for the bubble of air rising, where would the cloud top be? What name would you give this cloud with its thickness and altitude?
> **e)** If mechanical uplift was forcing air up, where would the cloud top be instead?
> **f)** What type of cloud would you say this would form given its thickness and altitude?
> **g)** Would there be any ice in this cloud?
> **h)** If, instead of starting to rise from a ground temperature of 13°C, the bubble started at 12°C, what would happen then? Would any cloud form at all assuming thermal uplift only?

WIND IN THE WILLOWS

Vegetation also exerts microclimatic effects on its surroundings, and though these may not be as marked as in an urban microclimate, they are still significant. The zone between the ground and the canopy of a forest will experience different climatic conditions to a grassland area outside the forest, because trees will offer more shade from direct insolation, and more shelter from the wind, especially in summer when deciduous trees are in leaf and so offer a greater surface area to wind. Reduced wind speed in forests will cause increased humidity as vegetation transpires water from its leaves and evaporates back intercepted precipitation.

Farmers have recognised for centuries the microclimatic importance of trees as shelter belts in areas prone to soil erosion. The spacing of shelter belts of trees is important, as is their permeability (how much of the wind they allow through them). If the spacing of shelter belts is too great they will not provide adequate protection; a spacing of ten to 15 times the tree height is sufficient to reduce wind erosion.

If the tree is very dense and bushy, it deflects the wind over and around it, but the eddying created downwind can cause soil erosion to increase; shelter belt trees therefore need to allow wind to filter through them, so as to reduce wind speed and minimise eddying. Ideally, they should also have small boles, to reduce the draught of wind at ground level.

Even with a relatively low growing area of vegetation, such as a field of maturing wheat, microclimatic differences can be measured. Small scale differences just a few centimetres above the ground in temperature, humidity, and incoming sunshine can have a noticeable effect on crop growth, and the spread of crop diseases. Mildew and 'rust' in crops occur when rainfall is heavy during growth, and the humidity near the ground is kept high by the shelter that the crop itself creates.

Agricultural research stations worldwide spend considerable time and money in investigating such climatic variations within a crop, because they are now recognised as being of importance. In tropical countries with high evapo-transpiration losses from the soil, a shelter crop is sometimes planted to increase ground level shade and humidity, and so increase the yield of the main crop. This can have adverse effects if the higher humidity encourages pests and diseases, although in the case of coffee bushes in Brazil, planting under shade conditions of other trees reduces the damage by the woolly aphid, a common pest.

The broad microclimatic effects that vegetation, whether a crop, shrub or tree, tends to generate close to the surface are summarised in Figure 4.34.

Q22. a) *Summarise the main changes in microclimate caused by the plant canopy shown in Figure 4.34.*
b) *How would rainfall be affected by such a plant canopy? Draw a further curve, like the ones shown, for rainfall.*

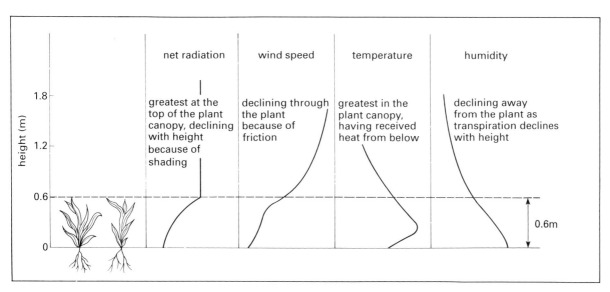

Figure 4.34 *Profiles of microclimate in crops*

Statistical analysis of vegetation microclimate

A sixth-form geography student attempted to measure the microclimatic differences between a woodland in Sussex and an open field of rough pasture.

She monitored readings of maximum and minimum temperature and relative humidity in both environments over a three-week period; the results are shown in Figure 4.35.

Q23. a) *In order to make a comparison of the two microclimates, calculate the following for maximum temperature, minimum temperature, and relative humidity:*
  i) *the maximum reading;*
  ii) *the minimum reading;*
  iii) *the range;*
  iv) *the median reading;*
  v) *the inter-quartile range;*
  vi) *the standard deviation.*

  b) *What do your calculations indicate? Write a summary paragraph to explain what the data shows.*

  c) *The data in the table was collected during the month of November, when many of the oak and beech trees were starting to lose their leaves.*

  d) *What difference would you expect to find if the data had been collected during the height of summer when all the trees were in full leaf? Explain how you think the temperatures and humidity would differ both within the woodland and between the woodland and the field.*

It is possible to look at the connection between the temperature and humidity by using a correlation coefficient such as the Spearman Rank Correlation Coefficient. This will give a good measure of the strength of the connection between variables, though it is important to understand that this statistic does not imply a causal relationship. For example, an attempt to correlate minimum temperature to humidity in the forest would have to allow for the effects of transpiration from the leaves; temperature changes alone might not account for humidity changes.

Q24. a) *Using the data in Figure 4.35, test the following statement by using the Spearmans Rank Correlation Coefficient (see page 187).*

  'There is a relationship between humidity and minimum temperature in either the woodland or the open field'.

  b) *Explain what the results of your correlation test indicate.*

Maximum and minimum temperatures (°C) and humidity % one metre above the ground in deciduous woodland and an open field November 1993						
		Open field			Woodland	
Day	Maximum	Minimum	Humidity	Maximum	Minimum	Humidity
1	10.5	3.5	71	8.5	4	92
2	9	2.5	80	8	2	83
3	11	8	96	10	9	89
4	10	4.5	96	8.5	5	92
5	9.5	4	93	8.5	6.5	93
6	10.5	0.5	83	8	0.5	97
7	11.5	9.5	97	10.5	9	94
8	9	8.5	93	8	7.5	93
9	10	3	88	8	5.5	100
10	9	1.5	81	8	4	96
11	10	7	93	7.5	7	96
12	10	3.5	71	7.5	6	96
13	8.5	1	84	8	3	97
14	9.5	7	81	7.5	6.5	100
15	11.5	8	91	10	8.5	94
16	9.5	5	90	8.5	6.5	93
17	8	3	92	7	6	97
18	7	2	93	7	3	93
19	7.5	2	90	6	2.5	94
20	6	1	87	4	1	96
21	6	2.5	93	5	2	100

Figure 4.35 *Vegetation microclimate data*

PROJECT SUGGESTIONS

A-level microclimate fieldwork exercise

Comparing microclimates in both urban and vegetated areas is an interesting topic to consider if you have to complete a project in geography. The equipment required is not expensive, the fieldwork techniques are straightforward, and access to a variety of sites is rarely a problem. The primary data you collect, such as maximum and minimum temperature or humidity can be compared to the secondary data of the newspaper weather readings. Such data also lends itself to simple statistical handling (mean, median, mode) and mapping. An example of how you could carry out such a project in your school or home area is outlined below.

Aim: to examine and attempt to explain the microclimatic variations within school grounds

The project could begin by explaining how you have observed differences in temperature, wind speed, humidity etc, in your own school grounds and how you wish to examine these further. It is advisable to select a few climatic parameters such as those mentioned, and measure them in detail, rather than to try and measure too much.

Hypotheses

A hypothesis is a reasonable statement which you can test to be true or not. For example, you could hypothesise that temperature and humidity will decline with distance away from school buildings, and collect your data accordingly. Avoid testing multiple hypotheses which could make your project unwieldy.

Equipment

You would need to obtain thermometers to measure temperature, hygrometers for humidity, and an anemometer to measure wind speed. If an anemometer is not available, you can improvise a method of measuring wind speed; a sheet of heavy cardboard hung from a length of dowelling will show how strong the wind is as calm conditions will allow it to hang vertically, a gale will blow it into a horizontal position, and the angle of rest inbetween can be measured against a home made scale.

It is also possible to measure wind speed by using a bubble solution made from washing up liquid. Held in the breeze, a bubble blower will release bubbles of liquid into the air in a five second period in proportion to the wind speed. Be prepared to invent your own field equipment; it need not be expensive or complex to do the job.

Method

Your method of measuring climatic differences in school grounds is very important, because how and where you collect data will determine the results you get. Factors to consider include the following.

- At what height above the ground will you collect data? You should try to standardise this, to eliminate differences that could arise simply because you took, for example, temperatures too close to the ground and so did not measure air temperature properly. A metre rule or taped garden cane will help you do this.
- Will you leave equipment permanently in place or take it around from site to site and take individual readings within a short space of time?
- How many readings will you take? Too few will tell you nothing about variations, a large number might not be possible because of time restrictions. Some statistical tests which you may wish to use, such as the Student's t test, can not be used with less than ten pairs of data, so you need to ensure you have enough for this. Do you want to have detailed readings taken hourly for three or four days, or readings taken daily over a period of several weeks to pick up seasonal microclimatic differences?
- How will you sample your readings? There are three broad sampling strategies, but you could combine them to suit your requirements. The first method is *random sampling*. Here, you could throw a grid over a map of the school, and derive co-ordinates from random number tables to tell you where to take your readings. This may give you readings which are not convenient or physically impossible to reach, so be prepared to be flexible on this. You may also find that your co-ordinates are clustered, and do not cover the whole of your area.

Systematic sampling however allows you to take readings at regular intervals either in an area or along a transect, for example every ten metres. Whilst this ensures that you get a good coverage of your area, it has the drawback of missing sampling points of interest inbetween. A concrete path across a playing field may give interesting temperature variations, but may be missed if it lies inbetween your sample interval.

Stratified sampling overcomes this by selecting 'layers' within the data, and sampling within those layers. For example, you might decide to stratify your readings into; 20 from artificial surfaces; 20 from long grass, and 20 from playing fields. Once you have ensured an even coverage in this way, it would be up to you to decide whether to use random or stratified methods on each surface. Be prepared to justify your choice of sampling.

Results and analysis

When you have collected all your data, it is important to present it in a clear manner, and to analyse patterns within it. You ought to have an idea of how you would plot your data before you start to collect it.

Can you graph it? Temperature variations through time, or over space, or with distance from buildings can all be plotted to see if a pattern exists.

Can you map it? Most data in geography should be mappable. You could plot isolines to show places with the same temperature, humidity, or wind speed on a tracing paper overlay of a map of your school, to identify hot spots, windy areas etc.

Can you use statistics on it? Even simple methods such as calculating mean, median, and modal figures, and calculating the inter-quartile range will give you something from which to draw conclusions. Statistics like Spearmans Rank or the Student's t test might clarify your ideas more, but you need to have collected data in the right way to do this.

Conclusions

Your conclusions should always relate back to your original hypothesis. Do your results prove it? In this case, do temperature and humidity decline with distance from school buildings? Do not worry if the pattern is not perfect, or even if there is no obvious pattern; there are usually good reasons for it which you can discuss at this stage. How might your sampling method have affected your results? Were there unusual weather conditions in the region that might have affected your results? With the wisdom of hindsight, you can usually suggest ways of improving your field technique to allow for such things.

Once you have made your conclusions, would your results be of interest to other groups? Geography is a practical subject and we are all affected by microclimates every day, so how could your findings be applied? Are the places of highest wind speed affecting temperatures and comfort levels near buildings, and what might some well placed windbreaks do to this pattern? Are the shrubs and flowers around the grounds in the warmest place, or getting as much sunlight as they need? Control of global climate patterns may still be beyond us, but control of microclimates is not.

GLOSSARY

Adiabatic a change in air temperature caused by a change in pressure.

Air mass a large body of air of uniform temperature and humidity.

Albedo the reflective nature of the surface.

Anticyclone a stable, slow moving high pressure system in which air subsides from the upper atmosphere in a clockwise rotation in the Northern Hemisphere.

Atmospheric equilibrium the concept that the atmosphere moves constantly in order to achieve a balance of heat and pressure.

Anthropogenic pollution contaminants in the atmosphere generated by human activity, e.g. burning coal in thermal power stations.

Cold front the boundary between cold polar air and warm tropical air at the rear of a mid-latitude in a depression.

Condensation nuclei microscopic particles upon which liquid droplets condense to form precipitation.

Continentality the effect on climate of distance from the worlds oceans giving a high temperature range and short spring and autumn seasons.

DALR or dry adiabatic lapse rate, the rate of cooling of a dry mass of air rising through environmental air (10°C per 1000 m of ascent).

Depression a low pressure system in which tropical air ascends over polar air along the Polar front, creating an anticlockwise circulation of air in the Northern Hemisphere.

ELR or environmental lapse rate, the rate of cooling of environmental air.

Freezing nuclei microscopic particles upon which liquid droplets freeze to form ice crystals.

GCM General Circulation Model, a computer simulation of how the atmosphere circulates.

Greenhouse Effect The trapping of heat close to earth's surface by gases water vapour, e.g. carbon dioxide.

Gyre a major circulation of water around an ocean.

Hadley cell the name given to the thermally driven atmospheric circulation between the Equator and 35° north and south.

Heat budget the balance between incoming and outgoing heat in a part of the atmosphere.

Inversion an increase in temperature with height.

Ionosphere layer of the atmosphere above 80 km in which atomic oxygen absorbs ultra-violet rays.

Isothermal layer a layer of the atmosphere in which temperatures do not change with height.

ITCZ Inter Tropical Convergence Zone, a band of low pressure between the Hadley cells.

Jet stream high velocity winds in the upper troposphere where pressure gradient is very strong.

Long wave terrestrial radiation radiation emitted by the surface of the earth and absorbed by the atmosphere.

Mechanical uplift the forced uplift of air which may be more stable than the air around it.

Mesosphere layer of atmosphere between 50–80 km.

Negative feedback where an input to the system is absorbed or compensated for so there is no change.

Ozone layer a layer of ozone rich air in the upper stratosphere between 20–40 km.

Photochemical smog a layer of pollutants generated by the reaction of hydrocarbons and nitrogen dioxide with sunlight over a city.

Polar front the cloud front along which polar air mixes with tropical air between 40–50°.

Positive feedback where an input causes the system to change and may then cause more changes.

Pressure gradient gradient created by differences in atmospheric pressure along which wind flows.

Rossby waves large meandering waves of air in the middle latitudes which flow around surface high and low pressure systems.

Saturated Adiabatic Lapse Rate coding of a saturated mass of air rising through environmental air (5°C per 1000 m).

Short wave solar radiation radiation emitted by the sun which can not be absorbed directly by the atmosphere.

Specific heat the amount of heat required to raise a mass of a substance through 1°C.

Stratosphere the layer of the atmosphere between 15–50 km in which the ozone layer lies.

Tephigram a diagram which shows variations in temperature and pressure with height.

Thermal uplift the natural uplift of air which is warmer and less stable than the air around it.

Troposphere the lower 15 km of the atmosphere in which most of the water vapour and other atmospheric gases are found.

UBL urban boundary layer, the layer of air above roof level in a city where characteristics of canopy layer below merge with surrounding atmosphere.

UCL urban canopy layer, layer of air below roof level in a city where turbulence, atmospheric composition and temperature are very different from the surrounding atmosphere.

Urban canyon the channel formed by high-rise buildings flanking a road in a city, causing powerful wind funnelling effects.

Urban heat island the warming effect that cities generate in their temperatures compared with surrounding rural areas.

Urban pollution dome the layer of air, usually confined to the UBL, where pollutants and aerosol concentration is high. If this dome spreads downwind, a pollution **plume** develops.

REFERENCES

H Oliver, *Vegetation and Microclimate*, Geography Review, November 1990, p. 2

J Olstead, *Global warming in the dock*, Geographical Magazine, September 1993, p. 12

D Money, *Time and temperature*, Geographical Magazine, September 1993, p. 58

D Money, *Thinking about ozone*, Geographical Magazine, August 1993, p. 52

D Money, *Earth's vital energy flows*, Geographical Magazine, July 1993, p. 33

See also a series of articles on atmospheric motion, composition, moisture, pressure systems, and ozone in Geographical Magazine Analysis, G Bigg, January–July 1992

B Atkinson, *The Atmospheric Heat Engine*, Geography Review, January 1988, p. 2

T Burt, *Measuring the Impact of the Great Gale*, Geography Review, January 1988, p. 6

J Hanwell *Lapse Rates*, Geography Review, January 1989, p. 7

T Burt and P Coones, *Winds of Change*, Geography Review, May 1990, p. 22

H Oliver, *The Ins and Outs of Sunshine*, Geography Review, September 1993, p. 2

R and N Foskett *Weather forecasting – recent advances*, Geofile no: 163, January 1991

R and N Foskett *Urban Climate*, Geofile no: 185, January 1992

H Oliver, *Vegetation and Microclimate*, Geography Review, November 1990, p. 2

D Lee, *Investigating Air Quality* Geography Review, January 1994, p. 33

USEFUL ADDRESSES

Air quality information can be obtained from: Warren Spring Laboratory, Gunnels Wood Road, Stevenage, SG1 2BX Tel: 0438 741122

Satellite image photographs can be obtained from: Peter Baylis, Satellite Station, Dundee University, Dundee, DD1 4HN. Tel: 0382 202575. You will need to give the precise time, date, and type of image required. Each photograph costs £2.00.

Weather learning resources can be obtained from: Mrs Jackie Syvret, Education Officer, Meteorological Office, Bracknell Tel: 0344 854001

SECTION 5

The living world

by Anne Fielding Smith and Roger Smith

KEY IDEAS

- **The nature of ecosystems: plants and animals together with their physical environment.**
- **How ecosystems are structured. Processes occurring within ecosystems.**
- **The nature of soils. Factors controlling soil development.**
- **Variations in ecosystems on a global scale.**
- **The relationship between biomes and soils.**
- **Human activity as a threat to ecosystems.**
- **The management of ecosystems.**

ECOSYSTEMS AND BIOMES

Ecosystems are units in which plants and animals interact with their physical environment. Within each ecosystem energy is transmitted between the living and non-living components of the system. An ecosystem can be of any size from a molehill or a cowpat, to a tropical rain forest the size of Amazonia. Global scale ecosystems are called **biomes**. Two broad factors determine the vegetation characteristic of each biome; temperature and precipitation. Figure 5.1 illustrates the relationship between annual temperature, annual precipitation and global scale vegetation zones.

Q1. *In what conditions are trees absent?*
Q2. *If temperatures remained the same, but average precipitation increased, how would vegetation respond?*

Biome distribution is largely determined by climatic factors, and results in broad zones of

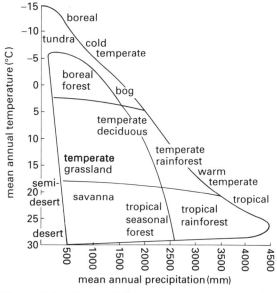

Figure 5.1 *Global relationship between climate and vegetation zones*

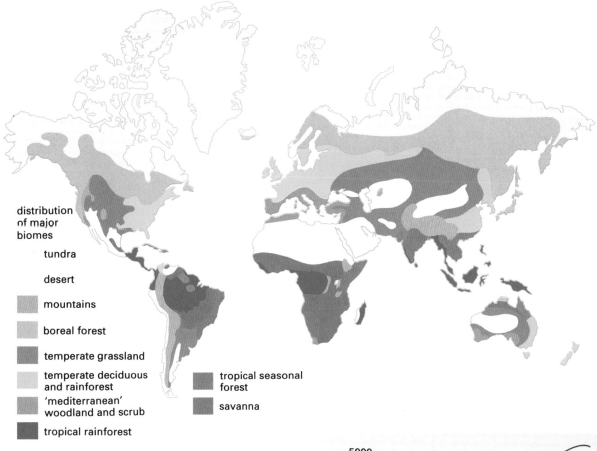

Figure 5.2 *Distribution of major biomes*

distinctive types of vegetation (Figure 5.2). On a local scale, however, many other factors will determine the type of vegetation which grows. These local factors include rock type, soil characteristics, drainage, aspect, altitude and human activity. Human activity is responsible for the present character and distribution of vegetation over much of the British Isles, and in many other parts of the world. The widespread use of fire, which dates back to Palaeolithic times (the Stone Age period in Britain from about 25 000 years ago to 20 000 years ago), the clearing of forests for agriculture which can be traced back to Mesolithic times (the period between the Palaeolithic and the Neolithic, stretching from 20 000 years ago to about 3000 BC), and recent agricultural 'improvement' of land have all contributed to the loss of natural vegetation. Altitude is one example of a natural factor which causes local variations in vegetation. In general, air temperature decreases by 6.5°C for every 1000 m gain in altitude. The effects of this temperature change upon the distribution of vegetation is quite marked, and mirrors that shown in Figure 5.1. Figure 5.3 illustrates the changes that occur in vegetation types with an increase in altitude.

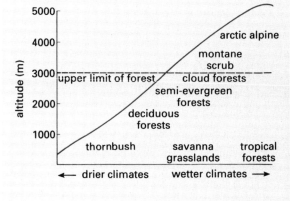

Figure 5.3 *Changes in vegetation with altitude*

> **Q3.** *What factors control the global distribution of biomes?*
> **Q4.** *Describe the effects that changes in altitude have upon vegetation.*

How are ecosystems structured?

Ecosystems may be distinguished on the basis of their structure; that is, by the way in which the organic (living or once living) and inorganic (non-living or abiotic) components of the system are arranged. The organic components of an ecosystem consist of both living and dead matter. Together, living plants, animals and micro-organisms are referred to as **biomass**. Biomass is usually expressed as a dry weight per unit area, often kg/m^2 or tonnes per hectare (t/ha). Each major ecosystem possesses a characteristic biomass. Forests generally have a large biomass, in the range of 20–45 kg/m^2, because they contain large trees. Grasslands, on the other hand, have a lower biomass, in the range of 1–5 kg/m^2. In addition to living material, ecosystems also contain dead and decaying plant and animal material. This is referred to as **dead organic matter** (DOM) and usually consists of surface litter and soil humus. Dead organic matter is also measured as a dry weight per unit area.

In some ecosystems the living components may form distinct layers. Forest and woodland ecosystems, for example, contain several recognisable layers as illustrated in the temperate deciduous woodland shown in Figure 5.4. Here, the canopy layer forms a distinctive, high layer at about 30 m above the ground. Below this a sub-canopy or shrub layer consisting of hazel, holly and brambles is often present. The herb or field layer consists of bluebells, bracken and other herbaceous plants, whilst the ground layer is made of litter, mosses, lichens and fungi.

How does energy flow within ecosystems?

Primary production

The sun is the major energy source for ecosystems. Solar radiation is absorbed by the chlorophyll within green plants during the process of **photosynthesis**. During photosynthesis, carbon dioxide and water combine with solar energy to produce carbon compounds (energy in the form of glucose and carbohydrates) and oxygen. The production of these carbon compounds is sometimes referred to as carbon fixation. Once fixed into plants, the carbon and energy are available to make plant tissue. Since green plants are able to produce their own energy, they are referred to as primary producers, or **autotrophs**. The carbon which has been fixed within the plants (as plant tissue) may then be eaten by animals. Energy is thus passed from the primary producers to primary consumers. Another term for a primary consumer is a herbivore. Herbivores may, in turn be eaten by a secondary consumer, or a carnivore (or an omnivore – an animal that eats both vegetation and other animals). Consumers are referred to as **heterotrophs** since they get their energy by feeding on one another. The total amount of energy which is fixed by green plants is referred to as Gross Primary Productivity (GPP). Some of this energy is used for plant respiration and is lost to the atmosphere as heat. The remaining energy is available to produce new plant material. This is called Net Primary Productivity (NPP). It is usually measured in terms of the weight of new organic matter which is added to a ground area of unit size per unit time, frequently in kg/m^2 per year. Figure 5.5 provides a summary of estimated NPP values for selected biomes.

The rate of production of new plant material depends on the availability of five factors; heat,

Figure 5.4 *The structure of a temperate deciduous woodland, Oakland in Mid-Wales*

Biome	Mean NPP per unit area (kg/m^2/year)	Mean biomass per unit area (kg/m^2)
Tropical rain forest	2.2	45
Tropical seasonal forest	1.6	35
Temperate deciduous forest	1.2	30
Boreal forest	0.8	20
Savanna	0.9	4
Temperate grassland	0.6	1.6
Alpine and tundra	0.14	0.6
Extreme desert	0.003	0.02

Figure 5.5 *Net primary productivity*

water, light, carbon dioxide and plant nutrients. Production is greatest in the tropical regions of the world, where constantly high temperatures and plentiful rainfall allow plant growth to take place for up to 12 months of the year. Similarly, shallow tropical seas and coral reefs reach very high levels of **productivity**. In contrast, productivity is lowest in cold regions where temperatures rarely exceed zero, and in hot deserts where little or no rain falls. The actual rate of production will be controlled by the factor that is in shortest supply. This is referred to as the principal of limiting factors. Carbon dioxide is the least variable of these factors on a global scale, but the other factors vary considerably and thus cause the variations in productivity rates shown in Figure 5.5.

> **Q5.** What is the difference between biomass and NPP?
> **Q6.** Why does productivity vary from one part of the globe to another?

Trophic levels

As energy transfers from plants to animals it passes along a food chain. Each stage in the chain is called a trophic level. Plants represent the first stage in the chain, and are referred to as producers. Some plant material is eaten by herbivores, which represent the second trophic level. Carnivores represent the third trophic level, and in some food chains may also represent fourth or even fifth trophic levels. At each stage of the food chain plants and animals may die and decay. The process of decay will be aided by soil micro-organisms such as earthworms, bacteria and fungi. These are referred to as **decomposers**. Each trophic level represents a feeding level where an exchange of energy takes place. At each trophic level a great deal of energy (around 90 per cent) is lost through respiration, through waste products and through decay, all in the form of heat. This means that less energy is available for each successive trophic level. In most terrestrial (land-based) ecosystems, only ten per cent of available energy is passed to the next trophic level. The rest is lost as heat. Figure 5.6 illustrates how energy flows through a food chain. It is important to recognise that energy flow is a one-way process, and that once energy has been lost (as heat) to the atmosphere, it is lost to the system. In most ecosystems the concept of a food chain is rather simplistic. In reality, many plants and animals exist within complex food webs.

> **Q7.** Draw a simple food chain to show how energy might flow through a forest or grassland ecosystem.
> **Q8.** Explain why energy is lost between successive trophic levels.

How do nutrients cycle within ecosystems?

Unlike energy, which is a one-way flow through an ecosystem, plant nutrients circulate within ecosystems between the living and the non-living components. Nutrients are chemical elements which mostly come from the weathering of rocks and the break down of minerals. Some nutrients enter ecosystems from the atmosphere (from the air and from precipitation) or from the application of fertilisers (either chemical or natural) by humans. At least 18 chemical elements are considered to be essential for plant growth. Nine of these, including carbon (C), nitrogen (N), oxygen (O), hydrogen (H), phosphorus (P) and potassium (K) are required in

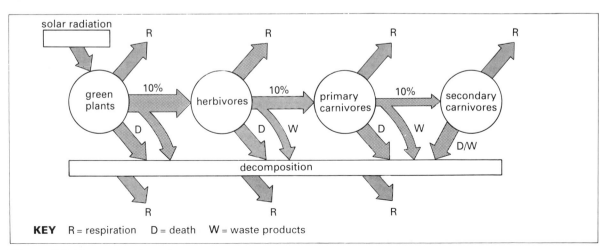

Figure 5.6 Energy flow through an ecosystem

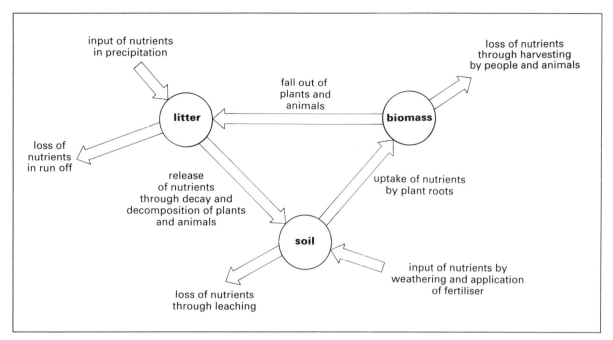

Figure 5.7 *Model of nutrient cycling within ecosystems*

substantial quantities. These are called macro-nutrients. Others, including iron (Fe), sodium (Na) and manganese (Mn), are needed in smaller quantities. These are called micro-nutrients. A third group of elements are required in minute amounts, but are nevertheless essential. These are called trace elements, and include zinc (Zn) and copper (Cu).

Many nutrients are stored in the soil, in the biomass or in the surface litter of ecosystems. Nutrients circulate within the system as illustrated in Figure 5.7. This figure represents a model which may be applied to any land ecosystem. Nutrients are absorbed by plants through a process called cation exchange. The nutrients enable a plant to grow successfully. When a plant dies, or is eaten, nutrients are returned to the soil through the decomposition of dead plant and animal matter, or of waste products. Since ecosystems are not closed systems, some nutrients may be lost to the system by **leaching**, by soil erosion, by gaseous diffusion from soil pores and plant transpiration, and from harvesting by humans. Many natural undisturbed ecosystems are very conservative of nutrients, and few nutrients are lost by soil leaching or other processes. Where, however, ecosystems are harvested intensively, or soil erosion is occurring, then rates of nutrient loss may be greatly increased. In tropical ecosystems in particular a very large proportion of the nutrients are held within the biomass component, and where this is removed, by burning for example, the result is an ecosystem severely depleted in nutrients.

Many nutrients are also involved in longer-term cycling between atmosphere, oceans, sediments and ecosystems. For example, on a global scale, most of the world's carbon is locked within the earth's crust as a reservoir in the form of sediments, coal and oil. Coal is the product of carbon storage by highly productive tropical swamp ecosystems. Less than 0.01 per cent of all carbon currently circulates on a short-term scale from the atmosphere to ecosystems. Although the amount of CO_2 in the atmosphere has increased in recent years by the burning of fossil fuels, the amounts of global carbon involved are minute. They are nevertheless significant for global warming.

Q9. *How does nutrient cycling differ from energy flow?*

Q10. *In what ways does human activity affect nutrient cycling in tropical rain forest ecosystems and in temperature agricultural ecosystems?*

Decomposition

As plants and animals die, they fall to the litter layer (see Figure 5.7). The process of decay involves several stages. In the case of leaves and wood, the primary decomposers (e.g. millipedes, wood lice, beetles and earthworms) attack the litter and break it down into smaller particles. During the second stage these particles, together with the faeces of the primary decomposers, are digested by the secondary decomposers (mites and springtails) to produce even finer particles. The final stage of decomposition of cellulose and lignin, which are the structural materials that contain most of the plant's carbon, can only be performed by fungi and bacteria. Fungi are the chief cause of decay of lignin. Fungal threads called hyphae penetrate between cell walls and excrete digestive enzymes. As these processes take place, the decomposing material is incorporated into the soil by the movement of the soil fauna. The end product is called **humus**, a dark coloured organic material which may remain in the soil for long periods of time. Nutrients released during breakdown become available for uptake by plant roots, although in some situations, they will be lost by leaching. In some ecosystems the process of decomposition is slow. This is the case in low temperatures, such as in the boreal forest ecosystem, or where the input of new material is faster than the rate of decay and peat may accumulate. In warm, humid conditions, such as those found in the humid tropics, decomposition is very rapid. The vegetation of these areas is specially adapted to rapid decomposition, and a shallow, dense root system allows tropical trees to absorb the nutrients. Leaching losses are therefore low and rates of recycling are high.

In temperate ecosystems, earthworms are important in the first stage of decomposition. Earthworms aerate the soil as they burrow and they help mix the dead and decaying organic matter from the litter layer into the upper parts of the soil. As they burrow, some species produce casts on the surface of the earth. It is estimated that earthworms will process between 6.5 and 14 tonnes of soil per hectare per year in temperature ecosystems, and up to three times this amount in tropical ecosystems.

In tropical ecosystems, termites are a very important component of the decomposer chain. Termites are unable to digest cellulose and lignin themselves, but some termites harbour (in their guts) small organisms called flagellates which can digest these substances. A second group of termites feeds on plant material in a different way, by cultivating fungi in their nests. Inside the termites' nests (called termitaria) the fungi colonises the faeces of the termites. The fungi convert food that is indigestible to the termites into a more digestible form. The termites then eat the faeces and the fungal tissues. In some parts of the savanna ecosystem termite mounds (Figure 5.8) occur at a density of up to 600 hills per hectare. Both termites and earthworms are responsible for transferring finer soil particles to the soil surface through the production of worm casts and the breakdown of termite mounds.

Figure 5.8 *Termite mounds in savanna ecosystem, Queensland, Australia*

Q11. *What role do decomposers play in nutrient cycling?*

Q12. *What effect does temperature have upon the rate of decomposition? How does this affect nutrient cycling in tropical ecosystems?*

Succession within ecosystems

During the process of **succession** ecosystems accumulate biomass, dead organic matter and nutrients, and changes occur in the type and number of species present. As succession proceeds changes occur in light availability, space and nutrients, and other local factors and conditions become less favourable for the earlier group of plants, and become more suitable for a different group of species. Towards the end of the process of succession a vegetation community develops which may be relatively stable over a long period of time. The final stage of succession is referred to as a **climax** community. Where this climax vegetation lasts for hundreds, even thousands of years, and seems to be in equilibrium with the environment, it is referred to as a climatic climax. Biomes are examples of climatic climax on a global scale. The whole group of plant communities that successively occupy the same site from initial colonisation to the final stage of succession is called a **sere**. Seres may be subdivided into seral stages which represent the different communities which occupy the site at different times (Figure 5.9).

If succession occurs on ground which has not previously been colonised by vegetation, it is said to be a primary succession. Several types of primary succession are recognised (see Figure 5.10). **Psammoseres** occur on dry sand (usually on sand dunes) and **lithoseres** develop on fresh rock (e.g. after a volcanic eruption, after a landslide or on newly exposed glacial rock or debris). **Hydroseres** are initiated in fresh water environments and **haloseres** under brackish conditions such as salt marshes. If succession occurs on ground which has been previously vegetated, for example abandoned fields, or deforested land, then secondary succession will take place. This will often take place more rapidly than primary succession, as in many cases there will already be a well-developed soil in which plant seeds are present and from which plants can absorb nutrients.

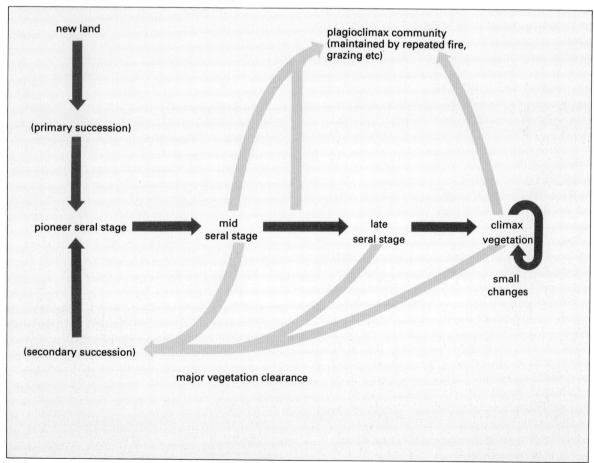

Figure 5.9 *Plant succession under temperate conditions*

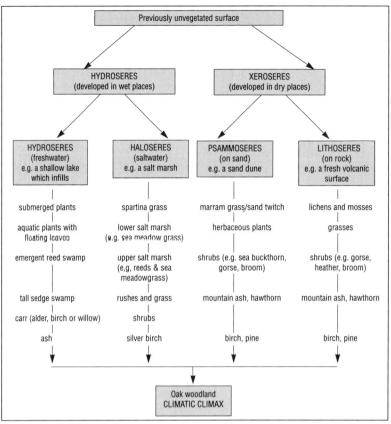

Figure 5.10 *Different types of succession under humid temperate conditions*

The first species to colonise bare ground or rock are known as **pioneer species**. Pioneers need to be able to withstand conditions which are often harsh and low in nutrients. Mosses and lichens will establish on bare rock surfaces, and small plants will colonise bare soil. As the plants die, they contribute humus to the developing soil. Nutrients are gradually added to the accumulating soil, and the soil becomes better able to retain moisture. Shrubby plants and trees may appear later. In time, larger trees may grow and competition for light and nutrients becomes more intense. The final stage of vegetation, which is in equilibrium with the physical environment, is called the climax vegetation.

Many successions share common features: a) the types of plants and animals change in a predictable manner, and start with green plants (producers), followed by herbivores and finally carnivores, but only when primary production has reached a level sufficient to support them; b) biomass (weight of living matter and DOM) increases; c) the diversity of species increases rapidly at first, and then generally decreases again before climax; d) net primary production increases; e) species composition changes from those best suited to cope with harsh conditions and low nutrient status to those species which are best able to compete successfully for nutrients, water and light.

The rate at which succession takes place is determined by local environmental conditions. Succession will normally take place more rapidly under warm, humid conditions than in cold or dry areas (refer back to primary production, page 99). Secondary succession will generally occur faster than primary succession (between 30 and 300 years, compared with over 1000 years or more for some primary successions). This is because there is often a residue of soil and seeds from which secondary succession can begin.

The catastrophic eruption of the island of Krakatoa, near Java, Indonesia, in 1883 provides a good example of rapid primary succession in the humid tropics. During the volcanic eruption of 1883, a layer of ash reaching over 30 m in depth was deposited over the island, killing all traces of life. Less than 50 years later, however, a mature rain forest had established, containing over 250 plant species and more than 720 insect species. In contrast, the emergence of the volcanic island of Surtsey, to the south of Iceland, during November 1963 provided an excellent opportunity for scientists to study primary succession in a cool, moist environment. The island of Surtsey covers an area of 2.5 km^2 and reaches an altitude of 172 m. Half of the island was formed from lava, the rest from tephra (ashes) which gradually hardened into tuff. Within six months of the eruption, plant and animal life had been established. Seagulls, waders and other birds, from the nearby Iceland mainland, were largely responsible for carrying seeds, spores, parasites and insects to Surtsey. In addition, many spores and seeds of mosses, lichens and ferns arrived in the wind together with those of herbaceous plants such as cotton grass and groundsel. Other plants arrived on the beaches of the island having drifted in the sea. The first plants to colonise on Surtsey were bacteria, moulds, algae and herbaceous plants. Moss started to grow on the lava during the third year and lichen became established during the eighth year. Insects were clearly able to survive in the harsh environment by living off algae, bacteria and herbaceous plants, with over 20 insect species recorded in the second year and more than 100 species recorded in the fifth year. Seabirds clearly played an important role in the establishment of both plant and animal life on the island. In addition to transporting seeds and small insects directly, they also supplied the soil with essential nutrients through their excreta and provided shelter, heat and nutrients for other animal life in their nesting areas. The harsh climate, however, means that succession has taken place slowly on the island and that it will be several centuries before a climatic climax of birch is likely to become established.

Q13. *What is the difference between primary and secondary succession?*

Q14. *Describe and explain the processes which occur during primary succession.*

Sub-climaxes and plagioclimaxes

In some circumstances the ecosystem will not proceed to the normal climatic climax. Succession may be influenced by a number of arresting factors. These factors may be topographical (where drainage influences the vegetation) or edaphic (where soil conditions influence the vegetation). These arresting factors will give rise to a sub-climax vegetation in which the plant and animal communities are held in a relatively stable situation by non-climatic controls. Where human activity has halted the normal pattern of succession, for example by grazing or by fire, the sub-climax is referred to as a **plagioclimax**. The grasslands of the South Downs are a good example of a plagioclimax community where centuries of sheep and rabbit grazing helped maintain a species-rich grassland community, in place of the climatic climax forest vegetation.

Five thousand years ago some of the gentle dip slopes of the South Downs were covered with mixed oak-hazel forest, whilst the steeper scarp slopes and escarpment were covered by elm and lime forest. Both forest types represented climax communities, the different tree species reflecting the depth and chalkiness of the soils. Clearance of the forest for agriculture began in the Neolithic period (the period in Britain between 3000 and 2000 BC) and was virtually complete by Roman times. Widespread clearance and cultivation led to severe soil erosion of the previously deep forest soils, and resulted in the formation of the thin rendzina (calcareous) soils that today cover the Downs. In Saxon times and the Middle Ages sheep grazing became an important agricultural use of the South Downs. The sheep produced a turf as a result of their close grazing and trampling. The turf was dominated by fescue grasses and a wide range of short herbs and creeping shrubs such as thyme and rock rose. Taller plants were unable to survive the grazing pressure from sheep, and thus prevented further succession. The soils remained low in nutrients and alkaline, and this, combined with grazing pressure, resulted in an inhospitable environment for tall grassland and woodland species.

During the nineteenth century, as sheep grazing declined due to changing social and economic factors, rabbit numbers increased and maintained the short grassland. Rabbits, however, tend to graze close to their burrows, often leading to bare chalky patches close to the burrows. Further away from the burrows scrub and woodland began to develop. In 1954 myxomatosis was introduced into Britain, which killed the vast majority of rabbits. The vegetation on the Downs responded rapidly. Many species were able to flower (having previously been too heavily grazed) and taller species successfully invaded the area. Quite rapidly the short, species-rich turf (often with up to 40 plant species per m^2) was replaced by a tall grassland community with only five to six plant species per m^2. Woody species such as dogwood and hawthorn appeared to form scrub in which ash and other tree species became established. In addition, changing agricultural practice has meant that much of the remaining open grassland was ploughed up and replaced by rye grass and red clover in order to support beef cattle and sheep. It is estimated that open grassland today represents only about ten per cent of that existing in the 1930s. Figure 5.11 shows a 'typical' Downland scene. In 1987 part of the South Downs was designated an Environmentally Sensitive Area (ESA) and it is possible that natural grasslands will once again flourish, as farmers take advantage of grants and environmental advice.

Q15. *What is meant by an arresting factor?*

Q16. *How has human activity caused the vegetation to change on the South Downs?*

Figure 5.11 *The South Downs*

SOILS AND SOIL PROCESSES

What is soil?

Soil is usually composed of mineral particles with small amounts of organic matter. Water and air occupy the pore spaces between the particles. Soil texture, structure and fertility are influenced by these four components. Figure 5.12 depicts the soil as a system. The mineral component of the soil is derived from weathered rock, whilst the organic component of the soil is provided by the decay of plant and animal matter. As discussed earlier, earthworms and other soil organisms are very effective at mixing together the organic and the non-organic constituents. Water and air enter the soil from the atmosphere by way of soil pores, plant root channels and animal burrows. Nutrients are held in the soil and in clay-humus complexes and some are dissolved in soil water. Nutrients cycle from the soil to the biomass and back again to the soil through decomposition, although some may be lost from the system by leaching.

The main constituent of soil is weathered rock. Rocks are made of minerals of different resistance to weathering and chemical decomposition. As rock weathers, some minerals such as quartz, will remain chemically unaltered and remain in the soil as sand and silt sized particles. Other minerals, such as feldspar and mica, react with water and oxygen to form oxides, hydroxides and clay minerals. Clay minerals are the smallest, and probably the most important mineral constituent of a soil. They consist of minute plate-like particles composed of silicon (Si), aluminium (Al) and variable amounts of oxygen (O) and hydrogen (H). Some clay minerals swell when wet and shrink when dry. They also have the ability to attract and hold plant nutrients, a process known as adsorption, meaning that the nutrients are attached to the surface of the clay minerals by an electrical charge.

Soil texture is determined by the proportions of the three particle sizes: sand (between 2.0–0.06 mm diameter), silt (between 0.06–0.002 mm diameter)

106 THE LIVING WORLD

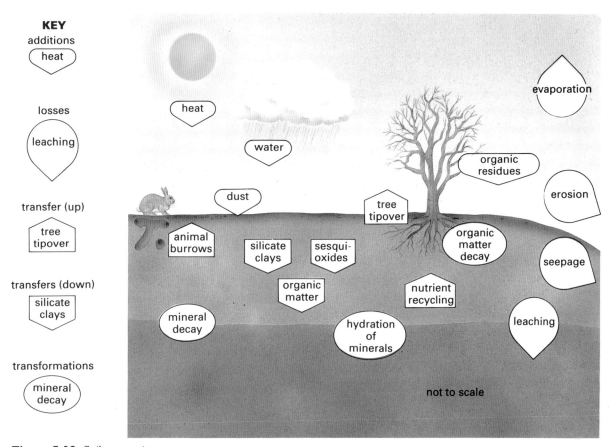

Figure 5.12 *Soil as a system*

and clay (less than 0.002 mm diameter). Figure 5.13 is a triangular graph, sometimes called a ternary diagram, which illustrates the classification of soil texture. A soil which contains a mixture of all three particle sizes is referred to as a **loam**. The presence of even a small amount of clay will change the characteristics of a soil quite considerably, and thus a soil with only 18 per cent clay will contain the word clay within its name. Once texture has been assessed, it is possible to make certain observations about the properties of each soil type. Sandy soils have large pore spaces and are well drained. However, they tend to be poor in nutrients. Silty soils

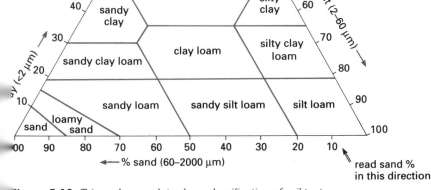

Figure 5.13 *Triangular graph to show classification of soil texture*

often have smaller pore spaces and the particles can pack closely together. They are susceptible to erosion by water and they can be difficult to manage. Clay soils are rich in nutrients, as described above, but can be prone to shrinking and cracking when dry. The small pore spaces, through which water moves only slowly, means that they may become waterlogged in wet weather.

From an agricultural point of view a loam soil is clearly the most desirable. With a small amount of clay (no more than 20 per cent) to retain moisture and to hold nutrients, and equal amounts of silt and sand which promote drainage and aeration, loam soils are easily cultivated and highly productive.

Soil particles are often bound together into large structural units known as peds. This is the concept of **soil structure**. In the upper 25 cm of a soil (the A-horizon) there is often a crumb structure. Here soil particles are held together by organic matter and other cements such as calcium carbonate or iron compounds. Below 25 cm (in B-horizons) structures are usually much larger and typically result from the swelling and shrinking of clay particles. Silty soils generally have prismatic structures in lower horizons, whilst clay subsoils often have blocky structures. Sandy soils have poorly developed structures. Silty soils may be susceptible to the development of platy structures, a dense layer parallel to the soil surface. These result from pressure, for example, from the passage of agricultural machinery. Soil structure is important because a well-developed structure results in an increase in pore space which speeds up the rate of water flow, improves soil drainage and allows the movement of air to lower levels in the soil.

Soil fertility is determined by a number of factors such as structure, drainage, organic matter content and the availability of plant nutrients. As chemical elements are released from rocks, by weathering, or from organic matter by decomposition, they are held within the soil water (in solution), and within a part of the soil referred to as clay-humus complexes. Both humus and clay minerals have a negative electrical charge which attracts positively charged ions (cations) such as calcium (Ca^{++}), magnesium (Mg^{++}), potassium (K^+) and sodium (Na^+). Cations can be passed between clays, humus and the soil water in a process called cation exchange (Figure 5.14). During cation exchange, plant roots come into close contact with clays and humus particles and give out hydrogen ions (H^+) in exchange for nutrient ions. The ability of clays and humus to hold and supply plant nutrients is referred to as the Cation Exchange Capacity (CEC). Humus has a much greater CEC than clay, which is

one reason why humus is a very important component of soils, and a vital component for agricultural soils. Crop growth is dependent upon soil fertility. Problems of nutrient shortages can be solved by the addition of fertiliser, but problems related to poor soil drainage, texture or structure may be more difficult and expensive to deal with.

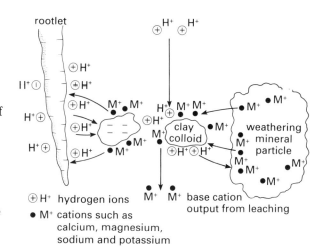

Figure 5.14 *Cation exchange*

Q17. *What are the main constituents of soil?*
Q18. *Explain briefly what you understand by the terms of soil texture, structure and fertility.*

How is soil formed?

Soil forms as a result of the interaction of five factors: climate, parent material (the rock from and on which soil forms), biotic agents (plants, soil organisms, humans), the topographical position of the soil-forming site and time. Figure 5.15 shows the relationship between three of these factors.

Climate is an important influence on soil type on a global and regional scale. Weathering and soil formation occurs more rapidly under warm conditions where water is present. Rainfall is important because abundant rainfall means that water is available to run through soil profiles and leach out soluble materials. Under conditions where rainfall is greater than evapotranspiration, soluble components such as calcium, sodium, magnesium and potassium are leached down through the soil profile and may be removed in drainage water, resulting in acid soils. In the humid tropics, less soluble materials such as iron, aluminium and quartz remain in the upper layers of soils. The oxides and hydroxides of iron contribute to the characteristic red or yellow colour of these soils. Where rainfall is less than potential

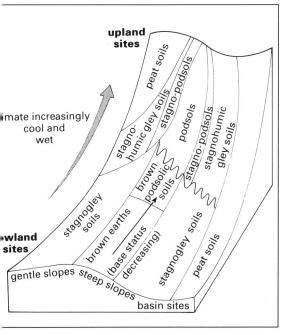

Figure 5.15 *Relationship between soils, climate and relief*

evapotranspiration, for example in semi-arid areas and deserts, soluble components are not removed and may even accumulate to form profiles rich in calcium carbonate, common salt (sodium chloride) gypsum (hydrated calcium sulphate).

Parent material influences soil formation. Sandy parent materials tend to be rich in quartz and are easily leached because water penetrates these profiles easily. Sandy soils are often deep and tend to be acid (and poor in nutrients). Clay is less easily leached, and clayey soil profiles tend to be less acidic. However clay soils often show signs of waterlogging, as the pore spaces between the soil particles are often small and do not allow rapid water movement.

Biotic agents can be important. Grass vegetation puts much organic material into the soil profile through root death and decay. Upper horizons are therefore deep and dark coloured. Forests input more organic matter to the soil surface. The litter and organic layers at the soil surface are often quite deep in forests, except in tropical conditions where rapid decomposition takes place. Where large populations of burrowing animals are present, for example in grassland ecosystems, organic matter tends to be evenly spread through the deep A-horizons. In coniferous forests where earthworms are absent in acid **podsolic soils**, discrete organic surface layers are more evident. Human activity also has an impact on soil development. Ploughing, for example, homogenises the upper layers of the soil, and cultivation often results in soil erosion.

The topographic position of the soil is also important. Soils on flat upland surfaces and in valleys are generally deeper than on steeper slopes. Soils at the base of slopes accumulate soil material by slope erosion, and are also likely to be deep. Soluble components move downslope as water moves through the soil, resulting in more acid soils on upslope positions. Soil catenas occur on slopes. A catena is a sequence of soils down a slope. Catenas are often seen on the same parent material. The soil differences downslope are often related to soil hydrology; upslope positions drain quickly and are oxidised and leached, while downslope soils are wetter and may contain more organic matter. In the tropical savanna ecosystem, for example, leached red acid soils often occur in upslope positions, with darker, wetter soils occurring towards the base of slopes (Figure 5.16).

Time is another important factor in soil formation. In Britain, many soils only started forming at the end of the last Glaciation, about 10–15 000 years ago. These soils are relatively shallow, often less than one metre in depth. In tropical regions, where many soils have developed over a much longer period of time, profiles are often much deeper and more weathered.

Q19. *Identify and describe the main factors which affect soil formation.*

Q20. *Explain the relationship of soil type, climate and relief, as illustrated in Figure 5.15.*

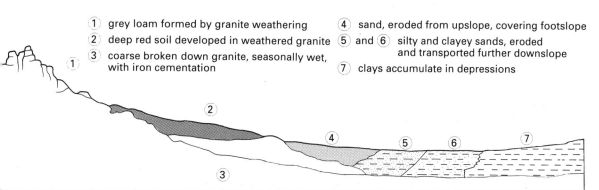

Figure 5.16 *Soil catena in eastern Africa*

FOREST BIOMES

Forests are characterised by high levels of above ground biomass (leaves, stems and animals) compared with the below ground biomass such as roots, animals and micro-organisms (ratio of 1:3 in deciduous forests and 1:4 in tropical rain forests). Deciduous and tropical rain forests also show clear vertical stratification (see Figure 5.4, page 99). This is not generally the case in coniferous forests where shading is more intense.

Net primary production is high in tropical rain forests and deciduous forests (2.2 kg/m^2 and 1.2 kg/m^2 per year respectively) where precipitation is plentiful and temperatures are high enough to allow growth for many months of the year. The colder northern coniferous forests achieve a productivity rate of 0.8 kg/m^2 per year. Coniferous forests are the dominant vegetation type over a wide area of the Northern Hemisphere, whilst deciduous forests characterise eastern and western continental margins (see Figure 5.2, page 98). Tropical forests occur in the continents of South America, Africa and Asia between ten degrees north and south of the Equator.

In tropical rain forests most nutrients are stored in the biomass. Rates of litter fall are high, as are rates of litter breakdown and nutrient recycling. In coniferous forests rates of litter breakdown are slower and more nutrients are stored within decomposing organic matter in the organic layer at the soil surface and within the mineral soil. Deciduous forests are intermediate in character. Figure 5.17 shows the relationship between nutrient stores in different forest types.

Nutrient store	Pine forest	Beech forest	Tropical rain forest
Biomass	1112	4196	11081
Annual litter	40	352	1540
Soil	649	1000	178

Figure 5.17 *Nutrient stores in forest ecosystems in kg ha^{-1}*

Inputs and outputs of nutrients in forest ecosystems have been investigated in the Hubbard Brook ecosystem in New Hampshire, USA (Figure 5.18). The Hubbard Brook ecosystem is a northern hardwood forest ecosystem which contains a mixture of deciduous and coniferous trees including beech, birch, hemlock, spruce and pine. Studies of inputs and outputs of nutrients in the 55 year-old forest,

	Input (kg/ha)	Output (kg/ha)	Net (kg/ha)
Calcium	2.2	13.7	−11.5
Magnesium	0.6	3.1	−2.5
Potassium	0.9	1.9	−1.0
Sodium	1.6	7.2	−5.6
Phosphate	0.11	0.02	0.09
Nitrate	19.0	16.1	2.9

Figure 5.18 *Data showing inputs and outputs of stream nutrients*

which is still growing, show that the system is accumulating nutrients such as nitrogen and phosphorous, and losing bases such as calcium, magnesium, sodium and potassium. Many of these latter nutrients appear to be originating from the weathering of bedrock and underlying glacial till, rather than from the living components of the ecosystem.

Tropical and coniferous forest ecosystems are often characterised by poor, infertile soils. Although tropical rain forests sustain high rates of productivity, this is achieved by conservation of nutrients within the system together with rapid nutrients recycling. In the deeply weathered soils of tropical rain forests, quartz and compounds of iron and aluminium dominate, together with kaolinitic clays which are also rich in silica and aluminium. There are few reserves of weatherable minerals able to supply nutrients such as calcium, potassium or magnesium within the rooting zone. Many soils in tropical rain forests are old. Long continued chemical weathering accompanied by leaching, has removed bases such as calcium, potassium, magnesium and sodium as well as the silica of silicate minerals such as feldspar often to a considerable depth. This process eventually leaves behind a residual soil rich in quartz, iron and aluminium. These soils are referred to as **ferralsols** or latosols and the importance of oxides and hydroxides of iron and aluminium is reflected in their red and yellow colours (Figure 5.19).

In contrast, the weathering of the podsols which characterise coniferous forest ecosystems are dominated by biochemical processes. Slowly decomposing litter layers produce soluble organic acids which break down clay minerals in the upper parts of the profiles and mobilise iron. Bases such as calcium and potassium are rapidly removed from th profile by leaching, and iron and aluminium are transported to lower layers, where they are redeposited. The soluble organic compounds are often also deposited in lower horizons. The result is a podsol profile: an upper horizon of white quartz grains, deficient in clay and without brown surface coatings of iron, overlying a horizon rich in iron and/or humus (Figure 5.20).

Brown soils (Figure 5.21, page 112) are characteristic of deciduous forests in western Europe. Here, soluble organic compounds are broken down by rapid decomposition and are not available to produce podsolisation by mobilising iron. Rates of chemical weathering are slower than in the humid tropics. Many soils are quite young, having developed since the end of the last Glaciation some 10 to 12 000 years ago.

Figure 5.19 *A typical ferralsol in Hawaii*

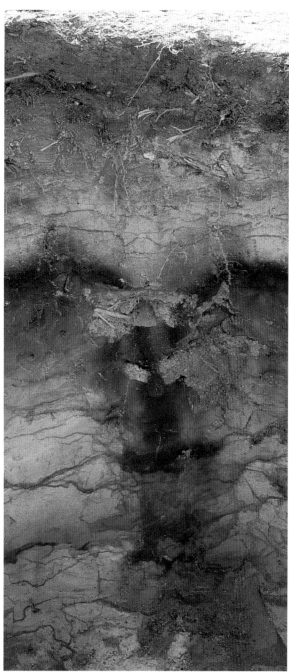

Figure 5.20 *A typical podsol soil profile on Green sand in the UK*

THE LIVING WORLD **111**

Brown soils are dominated by brown hydroxides of iron rather than the redder oxides of tropical soils, and greater reserves of weatherable minerals are present. Soils are shallower than in tropical rain forests, and trees often root more deeply, and bring up nutrients from less weathered layers.

Figure 5.21 A brown soil

Q21. *Use Figure 5.17 on page 110 to show how nutrient cycling occurs in different forest biomes.*

Q22. *Describe and explain the differences between the soils of tropical, deciduous and coniferous forests.*

Forests under threat

Forests are vital to the ecological functioning of the planet and produce almost two thirds of the net primary productivity of all terrestrial ecosystems. Forests and woodlands help regulate the hydrological cycle by intercepting rainfall and regulating river flow. Trees provide organic matter and important nutrients to soil systems and they help to limit soil erosion by protecting soil from rainfall and by the binding effects of roots. Forests also play an important role in modulating global climate. It is estimated, for example, that deforestation may contribute up to 25 per cent of the increased carbon dioxide which is causing the Greenhouse Effect. Forests and woodlands are the habitat of a large number of plant and animal species from which many important drugs and commercial products such as timber, paper, fruits, nuts, coffee, rubber and resin are obtained. Millions of people also live within forests, depending upon them for their livelihood. Forest ecosystems are therefore one of the world's most important resources. However, they are increasingly threatened by activities such as deforestation, industrialisation, dam construction, agriculture and recreation (Figure 5.22). There is an increasing awareness that forest ecosystems must be managed in a sustainable manner and that stewardship of forests should be the responsibility of all who make use of forest resources.

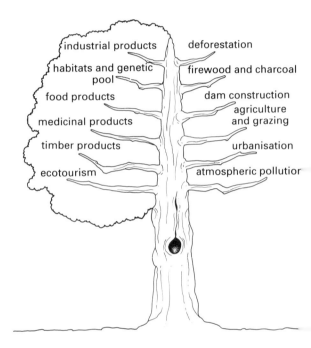

Figure 5.22 *Uses of and threats to forest ecosystems*

Deforestation is one of the greatest threats to forest ecosystems. It is estimated that about 0.6 per cent of the world's tropical forests (11.3 million hectares) is lost annually. In Latin America forest destruction has been especially rapid, with Brazil accounting for an annual loss of 1.5 million hectares, representing 20 per cent of annual global tropical deforestation. In Africa the main losses are due to land clearance for grazing and fodder and for firewood. In West Africa forest removal is also related to increased sales of hardwood. In tropical Asia land clearance for resettlement, agriculture and timber extraction is occurring on a large scale in countries such as Nepal, India, Thailand and Malaysia. In many humid tropical areas deforestation has resulted in accelerated soil erosion. It is estimated that soil degradation in Asia has led to the loss of 300 million hectares over the past 45 years. In Nepal, water running down steep bare slopes has not only caused severe soil erosion, but also increased flood levels in Bangladesh.

Managing forests

Environmentalists stress the need for sustainable development as a means of saving many of the world's resources from over-exploitation. Sustainable development involves using appropriate management techniques to ensure that forest ecosystems are not degraded, but are capable of regeneration and providing a surplus which can be harvested time and again. One technique, agroforestry (Figure 5.23), involves planting crops within the forest rather than **clear felling** the trees. This protects and helps retain the organic levels and nutrient status of the soil. Other initiatives involve helping countries who need to sell timber products to find alternative sources of income. One such initiative is ecotourism, where local people can earn an income from guiding tourists and showing them rare plants and animals, rather than earning money from illegal trapping of animals for skins and furs. In some areas, such as Korup in the Cameroon, biosphere reserves have been created. In these areas local people are given stewardship of National Parks in which sustainable development is encouraged. Another way of helping countries is for rich countries to pay off some of the debts that are owed by developing countries, on the condition that the country relieved of the debt, spends money on setting up and managing nature reserves. These schemes have been carried out in Bolivia and Ecuador, and are sometimes referred to as Debt for Nature schemes. Other initiatives include international co-operation, where members of groups such as the International Tropical Timber Agreement (ITTA) agree to encourage sustainable use and conservation of forests. The success of such organisations is questionable, but other measures such as 'eco-labelling' products made from sustainable sources are being explored by organisations such as the World Wild Fund for Nature (WWF) and Friends of the Earth. In temperate forests, the need for sustainable management was recognised many years ago. In British Columbia the Ministry of Forests is responsible for forest management. Eighty-five per cent of British Columbia is designated as Provincial Forests, with about half of the forests being productive. Replanting of forests is a vital part of the management system and much money is spent on research through its Silviculture Programme.

Q23. *Explain why forest ecosystems are important from both an ecological and a human point of view.*

Q24. *How can forests and their products be used in a more sustainable way?*

Figure 5.23 *Agroforestry in Columbia – coffee beans growing beneath banana trees*

GRASSLAND BIOMES

Grassland ecosystems occur where the dominant vegetation is composed of herbaceous species, in particular members of the grass and sedge families. Natural grasslands occur in places which are too arid for the development of closed forest, but insufficiently dry to result in the discontinuous vegetation communities of deserts. Many grasslands result from human activities such as grazing by domestic animals or fire. In the seasonal tropics and subtropics, trees and shrubs are scattered through grassland to form savanna woodland. Figures 5.1 and 5.2 on pages 97 and 98 show the distribution of grassland on a global scale and illustrate the relationship of grassland biomes to forest and desert biomes.

Precipitation in natural grassland is generally strongly seasonal in character. In temperate grassland, precipitation ranges between 250–750 mm per year, with 600–1500 mm per year in savannas with wooded vegetation. Often in subtropical areas the rainy season is between 120 and 190 days long. Upper soil layers may be intermittently moist, but lower layers are often dry for long periods and, under these conditions, trees cannot compete effectively with grasses. In many grassland areas rainfall occurs during the summer months, which reduces its effectiveness as evapotranspiration rates are very high.

Figure 5.25 *A red savanna soil in Kenya*

Savanna grasslands

Savannas occur in the zone between tropical forest margins and hot deserts. The term encompasses a range of vegetation types ranging from low biomass grassland to high biomass savanna woodland (see Figure 5.5, page 99). Figure 5.24 shows how savannas may be classified according to the percentage of the surface covered by trees. A number of factors determine the type of savanna vegetation that occurs in a given area. These are climate, soil, hydrological and geomorphological factors, fire and grazing. In dry climates grassland predominates, with trees occurring where rainfall exceeds 500 mm per year. Savanna soils are often old and deficient in nutrients, and this influences vegetation type and character. Hydrological factors are important, and treeless grasslands are often associated with wetter, periodically flooded soils near drainage lines. Fire has been considered to be very important in certain savanna areas, and grazing has a significant effect on the maintenance of savannas. In savannas, shrub invasion often occurs where grazing intensity is excessive.

Savanna soils reflect the wide range of climates, parent materials and topographic positions in which they develop. Figure 5.25 shows a red soil in the savanna of East Africa (see also Figure 5.16).

Savanna type	Description of tree coverage
Grassland	Trees less than 1% coverage (i.e. more than 30 tree crown diameters apart).
Savanna	Trees 1–10% coverage (8–2 tree crown diameters apart with occasional thicket clumps).
Dense savanna	Trees or shrubs 10–50% coverage (2–0.3) tree crown diameters apart).
Savanna woodland	Dominant tree canopy 50–90% coverage (trees less than 0.3 crown diameters apart).

Figure 5.24 *Classification of savanna vegetation*

In wetter savannas ferralsols, deep, iron and aluminium rich, strongly weathered soils, may be present. Sometimes hard iron-rich layers called **laterite** or ironstone occur within these soils. Such laterites are generally associated with flat surfaces, and the iron is believed to be concentrated by precipitation from a fluctuating water table in the soil profile. Some laterites in savanna areas seem to be old soils associated with former wetter climates. On the drier margins of savannas **vertisols** are an important soil type. They are rich in swelling and shrinking clay and they crack in the dry season. They are relatively uniform to a depth of one metre because soil material falls down the cracks in the dry season and the upper part of the soil profile progressively turns itself over. This is the origin of the name vertisol – a soil which inverts itself.

Temperate grasslands

In more temperate regions where precipitation ranges from 250–750 mm per year, and temperatures range from well below zero in winter to a summer maxima up to 20°C, temperate grasslands are the dominant form of natural vegetation. Examples of temperate grasslands include the North American Prairies and the Eurasian Steppes (Figure 5.2). Where woody species are absent, temperate grasslands have low ratios of above ground biomass to roots and dead organic matter (Figure 5.26) and there is a high rate of input of dead organic material through the growth, death and annual replacement of roots. Temperate grassland soils therefore tend to contain much more organic matter than woodland soils which results in the deep, dark coloured A-horizons characteristic of these mid-latitude grassland soil profiles. The formation of deep A-horizons is encouraged by high rates of bioturbation, the turning over of the soil by soil organisms such as earthworms, moles and, in some prairie grasslands, a range of burrowing rodents. Until the nineteenth century, many temperate grasslands were grazed by wild and domestic animals. These areas now make a significant contribution to world grain supplies. When first farmed, the abundant nitrogen and other nutrient reserves contributed to their fertility. Artificial fertilisers are now applied, but productivity is encouraged by the deep loamy nature of the soils and the summer rainfall. The **chernozem** is the characteristic soil of temperate grasslands (Figure 5.27). Towards the arid margins of temperate grasslands, as annual precipitation decreases and rates of evapotranspiration increase, vegetation becomes sparse and organic matter levels fall. The soils of these drier areas change to brown or reddish brown rather than black, and are known as **chestnut soils**.

Ecosystem	Fir forests	Steppe (temperate dry)	Steppe (dry)	Savanna (dry)
NPP (kg/m²/year)	0.85	1.12	0.42	0.73
Biomass (kg/m²) of which:	33	2.5	1.0	2.68
above ground (kg/m²)	25.65	0.45	0.15	1.55
above ground (%)	77.7	18.0	15.0	57.8
roots (kg/m²)	7.35	2.05	0.85	1.13
roots (%)	22.3	82.0	85.0	42.2
Ratio of above ground to below ground biomass	1:0.3	1:4.6	1:5.7	1:0.7

Figure 5.26 Ground biomass and dead organic matter

Figure 5.27 A chernozem soil

Q25. Under what climatic conditions do grassland ecosystems develop?

Q26. How and why do savanna grassland soils differ from temperate grassland soils?

Soil erosion

Soil erosion may occur for a number of interrelated reasons, including over-cultivation, overgrazing, poor farming techniques and deforestation. In many grassland areas severe erosion has resulted from the over-cultivation of soils through ignorance, the use of heavy machinery and in some cases political factors. In the semi-arid grasslands of the Great Plains of North America, severe soil erosion occurred during the 1930s, creating the infamous Dust Bowl. A series of drought years coincided with the rapid expansion of wheat cultivation, which had been encouraged by the introduction of tractors during the 1920s. Great dust storms, covering nearly four million km^3, extended from Canada to Mexico. During the 1970s another severe episode of soil erosion in the Great Plains was caused by the ploughing of marginal land as the US Government encouraged farmers to produce as much wheat as possible to export to countries such as the (former) USSR. Similar reasons have been given for severe erosion of many British soils. During the past 20 years British farmers have received grants and subsidies from the European Union's Common Agricultural Policy (CAP) to intensify cereal production. Water erosion of arable soils has increased significantly in Britain as farmers have ploughed up steeper slopes, removed hedgerows and used more powerful machinery. Farmers have been encouraged to sow during the autumn as yields are higher than for spring sown crops. Arable fields are at risk from erosion until about 30 per cent of the ground is covered by the crop. Since October and November are often very wet months in Britain, the fields are still bare and therefore very susceptible to erosion. When erosion occurs, topsoil and subsoil are lost, rills and gullies are cut down hillsides, ditches become blocked, streams and rivers become polluted, and flooding may occur. Soil erosion on the eastern South Downs has been monitored throughout the 1980s and 1990s (Figure 5.28). Eighty per cent of the South Downs is under cultivation, and around 55 per cent of this is currently under winter cereals, requiring autumn ploughing. The severest erosion occurs when intense autumn rain falls on recently ploughed or sown land.

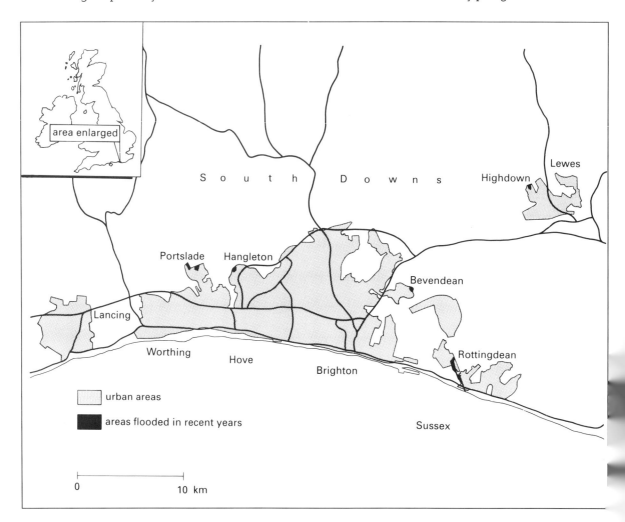

Figure 5.28 *Location of soil erosion and flooding on the South Downs*

Figure 5.29 *Gullying on the South Downs, near Rottingdean, October 1987*

Soil is then carried down wheel tracks, causing deep rills and gullies. Local planning policy has encouraged housing development within the bottoms of dry valleys of the Downs in order to avoid disturbing the downland skyline. However, as cultivation of the slopes has intensified, these valleys have suffered localised flooding by muddy water running off the fields. Figure 5.29 shows the effects of soil erosion on the South Downs in October 1987 when severe rainfall caused a wave of water one metre deep to flow through the village of Rottingdean causing extensive damage to over 40 houses and leaving a bill of over £400 000. It was estimated that over 5000 m³ of soil was eroded from the fields above the village, which had just been sown with winter cereals. A series of temporary earth dams had to be built across the valley floor to prevent further flooding.

Erosion and rainfall in an area of 36 km² of agricultural land on the South Downs, 1982–91			
Year	Max daily rainfall 20 Sept–1 Mar (mm)	Total rainfall Sept–Feb (incl) (mm)	Total soil eroded (m³)
1982–3	42	724	1816
1983–4	22	560	27
1984–5	30	580	182
1985–6	33	453	541
1986–7	38	503	211
1987–8	63	739	13529
1988–9	39	324	2
1989–90	34	621	940
1990–1	51	469	1527
1991–2	25	298	112

Figure 5.30 *Flooding and soil erosion on the South Downs*

Soil conservation and management

Soil erosion may be controlled by a number of means. Mechanical methods such as terracing are employed on steep slopes in many Mediterranean and tropical regions, but can be expensive and time-consuming to build and cultivate. The British Ministry of Agriculture recommends that farmers should cultivate and plant across slopes (contour plough) and should break up soil pans (hard layers of soil) which lie beneath the surface. They suggest that farmers encourage a rough soil surface rather than a fine seedbed which allows water to stand, thus increasing infiltration. It is also suggested that drains are well maintained to prevent flooding and that large fields are divided to channel water away from eroding slopes. Replanting hedgerows or lines of trees could help break wind speed in areas which are vulnerable to wind erosion. In some circumstances artificial embankments or earth dams could be constructed. Crop methods could also be changed. Rates of soil erosion are lowest on areas covered by grass, so farmers should keep a crop cover whenever possible. This could involve a return to crop rotation, and planting a grass ley instead of leaving a field fallow, or leaving crop residues such as stubble for as long as possible. Farmers with vulnerable land could also be discouraged from ploughing and sowing during the autumn months. In cases of severe erosion, it might mean grassland should be established. In some areas, where a variety of crops are grown, it might be suitable to intercrop with a second crop rather than leave large bare gaps between rows of crops. Farmers should also aim to improve the fertility and structure of the soil by careful ploughing, by increasing the organic content and, if necessary, by draining.

In many situations farmers may be damaging soils through ignorance. Education can therefore play an important role, both in the more and less developed parts of the world. Greater investment into research and advice could also lead to a reduction in soil erosion as well as to improvements in food production in the less developed parts of the world. In the more developed areas such as the USA and western Europe, changes to agricultural policies to limit the subsidies paid to farmers and to give subsidies to farmers who farm in a more sustainable manner are now being introduced. Schemes such as the Environmentally Sensitive Areas (ESAs) which was introduced in 1987 are being extended, and recent changes to the CAP may halt some of the erosion which in Britain sometimes reaches levels of more than two tonnes per hectare per year.

DECISION MAKING EXERCISE

Soil erosion on the South Downs

Study the above sections on soil erosion and management, together with Figures 5.28 to 5.30 inclusive. Write a structured report identifying the problems and possible solutions to the soil erosion situation on the South Downs.

1. Describe the events which occurred in parts of the South Downs in 1982, 1987 and 1993.
2. Identify the human and physical factors that contributed to these events.
3. Outline possible solutions to this problem and identify advantages and disadvantages of these solutions.
4. Produce a management plan for this area. The plan will be used by farmers, planners and local residents. You should justify your choice of strategies within your plan.

For further information on this topic see the reference list at the end of this chapter.

DESERT BIOMES

Deserts occur in areas which receive an annual rainfall of less than 250 mm (Figures 5.1 and 5.2). Most of the world's deserts lie either on the western coasts or in the centres of continents between 15–30° north and south of the Equator. Hot deserts such as the Sahara and the Atacama Desert have high day temperatures and a high diurnal range, frequently over 50°C. Temperate deserts such as the Gobi Desert are found in the centre of large land masses or, as in the case of Patagonia, in the rainshadow of large mountain ranges such as the Andes.

Rates of annual net primary production in deserts are low except where underground water is present. Average net primary productivity is 0.009 kg/m^2 per year, but in very dry areas production is less than 0.003 kg/m^2 per year. In arid environments much of the living organic matter is found below the soil surface. Xerophytic plants such as cacti and the creosote bush of the western United States, for example, have wide ranging shallow root systems to collect water from bare areas between individual plants. Phreatophytes, on the other hand, root deeply to underground water, but such plants tend to be restricted to lower areas in desert landscapes where ground water tables are within a few metres of the soil surface.

Soils in desert areas are referred to as aridsols or **xerosols**. These soils typically have poorly developed A-horizons. Calcium carbonate often accumulates in desert soil profiles and may form a hard layer up to one metre thick. Such layers are called **calcretes**. They reflect the fact that rainfall is sufficient to leach calcium carbonate into the soil, but there is insufficient moisture to wash carbonates out of the soil profile. In drier areas gypsum, which is even more soluble than calcium carbonate, may accumulate to form a gypsic layer. Figure 5.31 shows the relationship between the depth of calcrete and gypcrete layers and rainfall in the HaNegev (Negev) Desert in Israel. The depth of the soluble salt horizon reflects the depth to which occasional heavy rainfall penetrates the soil. In desert climates sodium chloride, which is more soluble than calcium carbonate or gypsum, is washed to lower positions in the landscape where it may accumulate. These **solonchak soils** typically occur in depressions, the beds of former lakes and along valley lines.

This potential for high productivity is exploited in irrigation schemes by supplying water, which was the major restriction on production. Salinisation, however, is an environmental hazard particularly where salt-rich groundwater is present at depth. The addition of irrigation water can result in the rapid rise of the groundwater table. In the Murray Valley in south east Australia, groundwater levels rose by nearly ten metres between 1962 and 1974. Once saline groundwater is within one to three metres of the soil surface, soil is drawn up through the soil profile by evaporation and capillary action. The salt is left at the soil surface as the water evaporates and white saline crusts form. Salinity reduces crop yields and in some cases eliminates crops altogether. Salinised and waterlogged soils affect about half of all irrigated areas of Iraq and 65 per cent of those of Pakistan. The only way to remove the salt is to pump out the salty water and replace it with freshwater. This is expensive, time-consuming and can lead to problems elsewhere as the salty water is discharged into rivers. In Pakistan it is hoped that saline water can be diverted into the lower Indus, from where it will be flushed into the sea in the flooding season. It is also planned to divert some of the saline water into evaporation lakes in nearby desert areas to remove the salt.

Irrigation in deserts

Ecosystems in deserts may be very productive where groundwater is present near to the surface.

Q27. Why do calcretes and gypcretes occur in hot deserts?

Q28. What is salinisation, and why does it pose a problem for agriculture in hot arid areas?

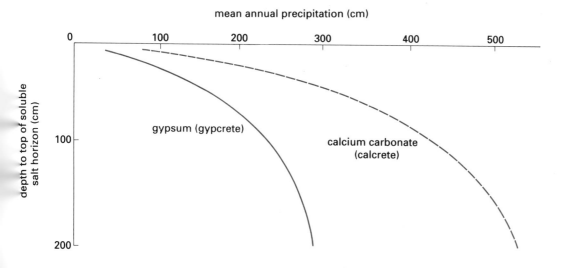

Figure 5.31 Formation of gypsum and calcrete

THE TUNDRA BIOME

Figure 5.32 *Frost polygons in north west Greenland*

Tundra ecosystems are cold with mean summer annual temperatures below 10°C. The growing season is short, frequently between three and four months long. Many tundra soils are underlain by permafrost, a permanently frozen layer about 50–100 cm below the surface. Water therefore drains slowly from tundra soils, and consequently soils are often waterlogged. Soil creep and ground surface heaving are important processes within the tundra and are related to frequent freezing and thawing. These processes lead to the formation of polygonal patterned ground (Figure 5.32). During the warmer summer months the raised edges of the patterned ground generally enclose pools of water which contribute to a mosaic of vegetation communities related to soil wetness and the micro-topography of the ground surface. Cotton grass, sedges, grasses and sphagnum mosses are typical of wet hummocky terrain, with better drained sites having heathland, sedge and grass communities. In some areas, shrub tundra, dwarf birches and willows also occur. Annual freezing and thawing of soils and associated soil heaving prevents clear horizon development. Many soil profiles show evidence of lack of oxygen and grey or mottled colours which are associated with the process of **gleying** (Figure 5.33).

Productivity within the tundra is generally low. Data from Canada, north east Siberia and Sweden shows a range between 250 and 1800 kg per hectare per year. Large amounts of nitrogen are cycled in tundra ecosystems and soils are often peaty,

Figure 5.33 *A gley soil containing iron in its reduced form, apart from in the narrow band just below the humus-rich layer*

especially in depressions. Plants and animals migrate to the tundra to take advantage of the vegetation growth during the short summer season. Large herds of Caribou migrate from the coniferous forests to the North American Arctic during the summer months, and a variety of geese, swans, ducks and waders fly great distances from wintering grounds on the wetlands of Europe and North America.

Current issues within the tundra

One of the biggest threats to the North American tundra is the development of oil and gas fields around Prudhoe Bay by multi-national companies, mostly from the USA. The first pipeline, the Trans-Alaskan pipeline, was completed in 1977 and carries over 17 000 tonnes of crude oil a day along a 1300 km pipeline to Valdez on the Pacific coast.

Further pipelines are proposed to transport gas from Alaska directly to cities in eastern Canada and the USA (Figure 5.34). It is feared that the impact of further development will be far greater than simply the pipelines, as roads, settlements and other services will need to be provided too. Development will damage the fragile ecosystem and bring changes to the traditional way of life for many Innuit and Indian families. One of the main impacts would be to disturb the traditional migration routes of the Caribou, and they have already stopped breeding in the area around Prudhoe Bay. Further development of pipelines across important wilderness areas such as the Arctic National Wildlife Refuge is being challenged by conservationists and local people alike.

Q29. What do you understand by the term gleying?
Q30. How is oil and gas exploitation threatening the tundra ecosystem?

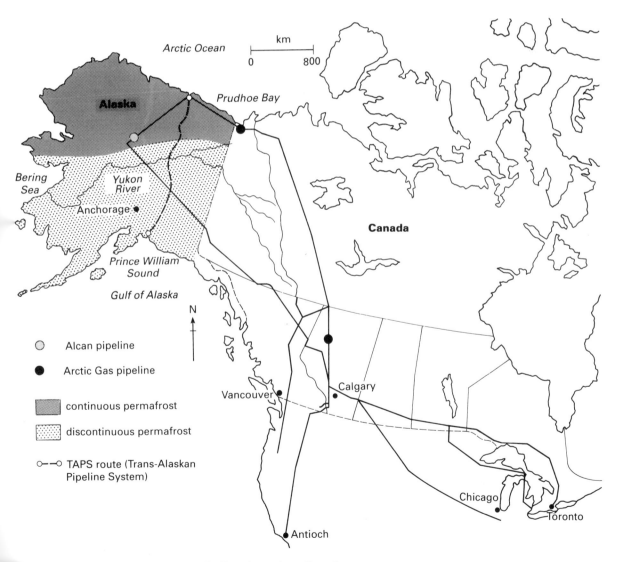

Figure 5.34 Proposed oil and gas pipelines in northern Canada

WETLAND ECOSYSTEMS

Wetlands form a group of ecosystems intermediate between terrestrial ecosystems with aerobic mineral soils and aquatic ecosystems. Wetlands cover approximately six per cent of the earth's land surface and occur in a variety of climatic settings from the tundra regions to the tropics, but are poorly represented in arid areas of the world. Soils of wetlands are characteristically anaerobic (depleted in oxygen), and organic matter may accumulate to form peat where the rate of decomposition, inhibited by wetness and anaerobic conditions, is less than the supply of organic matter through litter production. Some wetlands may be very nutrient deficient, for example blanket and raised bogs. Others may be nutrient-rich if nutrients are brought into the system by water flow. Some wetlands may be extremely unproductive but in the tropics and subtropics cultivated wetlands, often artificially extended, produce high yields of rice and support large human populations. In England the drainage of wetlands in the Fens and the Lancashire coastal plain has resulted in highly productive agricultural soils.

Major wetlands are often found in topographic lows in the landscape and have high groundwater tables. Soils are regularly flooded to the surface.

Often at their seaward margins freshwater wetlands grade into saline swamps or marshes, periodically inundated by tides. The mangrove swamps of the tropics and the salt marshes of Europe are examples of such wetlands. Tundra environments are often wet, and the oceanic climates of western and northern Europe with high levels of precipitation, combined with low rates of evapotranspiration result in the development of peat ecosystems on flat upland surfaces. These extend down to sea level in the west and north of Scotland and Ireland.

Human use and abuse of wetlands

Wetland ecosystems have many functions and economic values. Wetlands can control floods by storing precipitation and releasing runoff gradually. Wetland vegetation stabilises river banks and helps prevent erosion. Wetland areas are also effective in removing nutrients, and are used in countries such as Uganda and Florida, USA to help absorb sewage and purify water supplies. In tropical wetlands such as the mangrove swamps of India and Bangladesh,

Figure 5.35 *Lake Baikal, eastern Siberia*

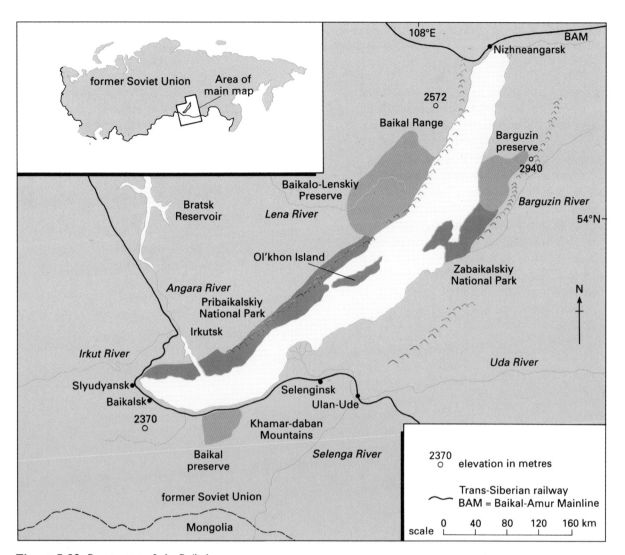

Figure 5.36 *Pressures on Lake Baikal*

firewood is grown commercially and fisheries support many villages.

Wetland ecosystems everywhere are under threat from agriculture, urbanisation, pollution and dam construction. In Ireland 80 000 hectares of bog have been drained since 1946 and in the Philippines 24 000 hectares of mangroves were lost between 1967 and 1975 for wood chip production. In eastern Siberia, the oldest and deepest lake on earth, Lake Baikal (Figure 5.35) is threatened by pollution from industry and agriculture. Lake Baikal is 1637 m deep and 25 million years old. It covers an area of 34 000 km² and is very rich in both biomass and number of species. It is estimated that there are 1550 animal species and 1085 plant species and the lake supports an important fishing industry. Logging around the lake began during the beginning of the twentieth century, causing soil erosion and landslides. An estimated three million tonnes of soil now enters the lake every year. Unwanted and lost logs decaying in the water have depleted oxygen levels and nitrates and partially treated sewage and industrial effluent have been deposited in the lake from the Selenga River which passes through an industrial region (Figure 5.36). The chemistry of the water has been altered and has affected the fauna and flora of the lake. Since 1987 thousands of endemic seals have died because they are at the top of the lake's food chain, and have consumed many toxic pollutants in the fish they eat. Despite a public outcry and numerous attempts to monitor and improve water quality, pollution continues although the lake has now been adopted as a United Nations Educational, Scientific and Cultural Organisation (UNESCO) World Heritage Site and several National Parks have been created in the vicinity of the lake.

Q31. *Under what conditions do wetland ecosystems occur?*

Q32. *Identify some of the pressures facing wetland ecosystems.*

ECOSYSTEMS IN THE FIELD

Field investigations of small-scale ecosystems such as sand dunes and woodlands can be very successful. It is possible to investigate both structure and change in ecosystems by using a range of simple techniques such as a quadrat analysis, transects, soil analysis and human impact. A good way to start planning your investigation is to identify one or more hypotheses, for example 'human impact in Plumpton Wood decreases with distance from the main gate' or 'ground vegetation is most abundant in areas where the trees are smallest'. It is also a good idea to create a data logging sheet to take into the field with you (Figure 5.37).

Quadrat analysis

A quadrat is a square of known size. The size of the quadrat depends upon the nature of the vegetation under investigation. For most A and AS level studies a quadrat with 50 cm or one metre sides is usually sufficient and very easily constructed. The quadrat can be further divided into areas of 10 cm^2 with pieces of thick string or wire. The quadrat may either be placed randomly (using a random numbers table), or systematically (every five or ten metres along lines), or in a stratified manner if several types of vegetation need to be measured. The more quadrats that are sampled, the more representative the results will be of the total ecosystem.

Transects

A belt transect is carried out by placing two tapes at a known distance apart (one metre, for example) and then recording the vegetation within the transect, either continuously or by using a quadrat at suitable intervals. For added detail, slope angle could be measured, using a clinometer. This method is particularly suitable when studying sand dunes, as quadrats can be located to sample each of the main ridges and slacks and a scaled cross-profile can be drawn afterwards. A line transect is carried out by placing a tape across the feature to be investigated (a path or a slope, for example), and a record made of all the vegetation which touches the tape. Recording may be done continuously or at regular intervals (e.g. every five or ten metres). Again, further information may be obtained by measuring slope angle with a clinometer.

Title of investigation: the number of ground species varies with the age of the coppiced trees. Date: 27 May 1995
Location: Wilderness Wood, Hadlow Down, East Sussex.
Transect number: 1 of 4

Category	Quadrat or station number														
	1	2	3	4	5	6	7	8	9	10	11	12	13	14	15
Height of tree															
Girth of tree															
Type of tree															
Age of tree															
% bare ground															
Light meter reading															
Soil temperature															
Number of ground species															

Figure 5.37 *Data logging sheet for use in the field*

Investigating soils

Acidity, colour and texture are three properties of soils which are relatively easy to measure or estimate in the field. Acidity is measured by adding distilled water to a small amount of soil and testing the solution with pH paper. Most British soils have a pH in the range of 4 to 7. Soil colour may be described using a soil colour chart called a Munsell Chart. In general, soils are darker towards the top, reflecting the presence of organic matter. Colour changes reflect the processes in operation within the soil process. Soil texture refers to the proportion of sand, silt and clay sized particles within the soil. It is possible to take samples back to the laboratory/classroom for analysis with sieving apparatus. In the field it is possible to feel the general texture by rubbing soil between your thumb and forefingers and rolling the soil between your palms. Sandy soils are gritty and even when wet they do not stick together easily. Clay soils are sticky when wet and will roll into smooth, thin ribbons. Loamy soils feel slightly gritty and when moist will roll into weak ribbons.

Investigating human impact

Ecosystems are easily disturbed by human activity. Measuring the extent of this impact can be done by recording the number of people to pass a particular spot over a fixed period of time, by using a questionnaire, by comparing compaction and infiltration rates of soil along footpaths and adjacent, relatively undisturbed vegetation and by noting the presence of litter and management strategies such as coppicing, replanting, fences and information boards. Bi-polar scales can be devised to record some of these impacts and impact scores calculated.

Analysing data

It is important to use appropriate methods to illustrate your data so that you can make observations and draw conclusions. Transects may be drawn as scaled cross-sections on graph paper then annotated with important information. Species counts may be illustrated by using histograms. Scattergraphs with lines of best fit or regression lines could be used to plot changes in soils, human impact or vegetation characteristics. Impact scores could be plotted as proportional symbols, or analysed by constructing an isopleth map. Statistical tests such as Spearman's Rank Correlation Co-efficient or the Chi-Squared test could be used to test for significance and association.

Using the Chi-Squared test

The Chi-Squared test is a test of significance which provides a valuable means of comparing sets of data and tells us whether the differences between two or more sets of data are statistically significant. The data must be collected in such a way that it can be grouped into classes and a null hypothesis stating that there is no relationship between the two variables should be set up. A data table should be set up and the Chi-Squared formula applied. Having completed the calculation, the result should be compared with a significance table supplied in most standard techniques books. The null hypothesis can then be rejected or accepted. The main limitation of the test is that it becomes invalid if any of the expected values fall below five.

The example shows the results of a survey of the number of blackbirds observed in woodlands of different types in Sussex. The type of woodland is determined by the underlying rock and soil type. In order to see whether there is a statistical association between the presence of blackbirds and different types of woodland, a Chi-Squared test was carried out.

The null hypothesis states that the distribution of blackbirds is not influenced by the type of woodland. The data table (Figure 5.38, page 126) shows the calculations that were carried out, and assumes that if the null hypothesis is correct, we would expect equal numbers of blackbirds in each of the four woodland types, i.e. 70. The formula for Chi-Squared is:

$$\chi^2 = \frac{\sum (O - E)^2}{E}$$

where $\sum$ = the sum of, O = observed values and E = expected values. The final value of Chi-Squared, 16.6, then needs to be compared with a Chi-Squared significance table (Figure 5.39, page 126). First, the degrees of freedom (n − 1) should be calculated. In this case there are four categories of woodland, so n = 4 giving 3 degrees of freedom. If the table of critical values in Figure 5.39 is then studied it can be seen that the critical value for 3 degrees of freedom (at the five per cent significance level) is 7.81. As our value is larger than this we may reject the null hypothesis and accept that the distribution of blackbirds is influenced by the type of woodland. The final task is to analyse the result, and to consider why blackbirds occur more frequently in coastal and clay woodlands rather than in woodlands of sand or chalk areas. When combined with a study of the structure of these different woodland ecosystems, some interesting ideas might emerge.

Rock type	Observed	Expected	(O–E)	(O–E)²	(O–E)²/E
Chalk	59	70	–11	121	1.73
Sand	48	70	–22	484	6.91
Clay	89	70	19	361	5.16
Gravel	84	70	14	196	2.80
Totals	280	280	—	—	16.6

Therefore $\chi^2 = \chi \frac{(O-E)^2}{E} = 16.6$

Figure 5.38 *Data table for Chi-Squared test*

χ^2 Test Degrees of freedom	(5%) 0.05	(1%) 0.01	(0.1%) 0.001
1	3.84	6.63	10.83
2	5.99	9.21	13.81
3	7.81	11.34	16.27
4	9.49	13.28	18.47
5	11.07	15.09	20.52
6	12.59	16.81	22.46
7	14.07	18.48	24.32
8	15.51	20.09	26.12
9	16.92	21.67	27.88
10	18.31	23.21	29.59
11	19.68	24.73	31.26
12	21.03	26.22	32.91
13	22.30	27.69	34.53
14	23.68	29.14	36.12
15	25.00	30.58	37.70
16	26.30	32.00	39.25
17	27.59	33.41	40.79
18	28.87	34.81	42.31
19	30.14	36.19	43.82
20	31.41	37.57	45.31
21	32.67	38.93	46.80
22	33.92	40.29	48.27
23	35.17	41.64	49.73
24	36.42	42.98	51.18
25	37.65	44.31	52.62
26	38.89	45.64	54.05
27	40.11	46.96	55.48
28	41.34	48.28	56.89
29	42.56	49.59	58.30
30	43.77	50.89	59.70
40	55.76	63.69	73.40
50	67.50	76.15	86.66
60	79.08	88.38	99.61
70	90.53	100.4	112.3
80	101.9	112.3	124.8
90	113.1	124.1	137.2
100	124.3	135.8	149.4

Figure 5.39 *Critical values of Chi-squared*

PROJECT SUGGESTIONS

1. Make a detailed study of a small-scale ecosystem such as a local woodland or local nature reserve. Identify the structure of the ecosystem using annotated field sketches and photographs. Use transects along which to place quadrats and estimate approximate proportions of different species. Identify and quantify ways in which human activity is affecting your chosen ecosystem and prepare a management plan using all the information you have collected.
2. Examine the relationship between soil characteristics and vegetation (either semi-natural or agricultural) in a particular area. Use an appropriate sampling technique to collect your data and use a Chi-Squared test to compare your results.
3. Using the suggested reading list, carry out a library based investigation into one of the following: desertification, soil erosion, disappearing wetlands or threats to forest ecosystems. You could either look at one of the topics in general, or you could focus on one particular case study in more depth.

GLOSSARY

Autotroph an organism is autotrophic if it is able to manufacture its own food from inorganic material.

Biomass the weight of living material expressed as a dry weight per unit area e.g. kg/m^2 or t/ha.

Biome a global scale ecosystem with a broadly uniform climate and vegetation.

Brown soil characteristic soil of temperate deciduous woodland, typically brown.

Calcrete an accumulation of calcium carbonate which produces a hard layer up to one metre thick, associated with arid soils.

Chernozem characteristic soil of temperate grassland ecosystems such as the Prairies, typically black reflecting high humus content.

Chestnut soil characteristic soil of dry temperate grassland ecosystems such as the dry margins of the Prairies.

Clear felling the practice of cutting down all the trees on a site, leaving the ground unprotected against erosion.

Climax the final stage of ecological succession in which a relatively stable fauna and flora are in equilibrium with the physical environment.

Dead organic matter (DOM) dead and decaying plant and animal matter consisting of surface litter and soil humus.

Decomposers organisms such as millipedes, earthworms, bacteria and fungi which break down organic matter and convert it to humus.

Ferralsols tropical soils with a high content of iron and aluminium formed by vigorous weathering under hot, wet conditions.

Gleying the reduction and mobilisation of iron from its red-yellow ferric form to its blue-grey ferrous state.

Halosere a vegetation succession initiated in brackish conditions.

Heterotroph an organism which obtains food from other organisms.

Humus dark brown to black, highly decomposed, amorphous organic material within a soil profile.

Hydrosere a vegetation succession initiated in shallow freshwater conditions.

Laterite a soil horizon which is hard or will harden on exposure and is composed mainly of oxides of iron and/or aluminium, with varying amounts of quartz and kaolinitic clay.

Leaching the removal of solutes downwards, sideways and sometimes out of a soil profile by moving water.

Lithosere a vegetation succession initiated on bare rock.

Loam a soil containing a mixture of sand, silt and clay which provides optimum conditions of soil aeration, drainage and fertility.

Photosynthesis the manufacture of organic compounds by green plants using light energy from the sun to fix carbon.

Pioneer species species adapted to colonise ecosystems early in the sequence of plant succession.

Plagioclimax a plant community checked in its progress towards climax by human activity, for example the grasslands of the South Downs.

Podsol a strongly leached, acidic soil typically occurring under heathland and coniferous forest vegetation.

Productivity the rate of biomass production within an ecosystem; Gross Primary Production (GPP) is the total amount of energy fixed by green plants. Net Primary Production (NPP) is the energy remaining once those green plants have consumed some of this energy in respiration.

Psammosere a vegetation succession initiated on bare sand.

Sere a sequence of vegetation types characteristic of plant succession in particular environmental conditions.

Soil fertility a measure of the ability of the soil to sustain plant growth (particularly of crop plants) related to the chemical and physical character of the soil.

Soil structure the aggregation of the primary soil particles of sand, silt and clay into larger secondary units.

Soil texture a measure of the proportions of sand, silt and clay grains within a soil.

Solonchak soil soils with a high salt content, associated with topographic depressions in arid environments.

Succession a developmental sequence of ecosystem changes associated with biomass increase. Primary succession occurs on new land surfaces. Secondary successions are the result of disturbance, and take place on previously vegetated sites.

Vertisols a dark coloured soil, rich in swelling and shrinking clay and characteristic of base rich sites in semi-arid climates.

Xerosols a desert soil showing little horizon development but sometimes containing substantial accumulations of soluble salts such as calcium carbonate or gypsum.

REFERENCES

J Boardman (1992), *Agriculture and Erosion in Britain*, Geography Review, Volume 6, No 1

A M Mannion and S R Bowlby (1992), *Environmental Issues in the 1990*, Wiley

G O'Hare (1988), *Soils, Vegetation and Ecosystems*, Oliver and Boyd

D Robinson and R Williams (1988), *Making Waves in Downland Britain*, Geographical Magazine, Volume 60, No 10

D Robinson and J Blackman (1990), *Water Erosion of Arable Land on the South Downs*, Geography Review, Volume 4, No 1

P R St John and D A Richardson (1990), *Methods of Statistical Analysis of Fieldwork Data*, The Geographical Assocation

J Boardman (1984), *Erosion on the South Downs*, Soil and Water Volume 12 pp 19–21

J Boardman (1989), *The drain on fertility*, The Guardian, June 6th 1989 p 38

N Foskett and R Foskett (1991), *Alaska – a threatened wilderness*, Geofile 161, January 1991

A Goudie (1989), *The Nature of the Environment* (Second Edition), Blackwell

B J Lenon and P G Cleves (1983), *Techniques and Fieldwork in Geography*, Collins Educational

A M Mannion (1991), *Global Environmental Change*, Longman Scientific and Technical

R Prosser (1992), *Natural Systems and human responses*, Nelson

P R Pryde (1991), *Environmental Management in the Soviet Union*, Cambridge

D A Richardson and P R St John (1989), *Methods of presenting fieldwork data*, The Geographical Association

J O'Rieley and S E Page (1990), *Ecology of plant communities*, Longman Scientific and Technical

D A Robinson and J D Blackman (1990), *Soil erosion and flooding*, Land Use Policy January 1990 pp 41–52

L E Rodin and N I Basilevic (1968), *World Distribution of plant biomass*, UNESCO Terrestrial Ecosystems, edited by Eckardt

P Sarre (ed) (1991), *Environment, Population and Development*, Hodder and Stoughton

J Silvertown and P Sarre (eds) (1990), *Environment and Society*, Hodder and Stoughton

A Stephens (1986), *Soil Erosion – Case Studies on the South Downs*, Geographical Educational Materials for Schools, University of Sussex

D Waugh (1990), *Geography: an integrated approach*, Nelson

SECTION 6

Economic activity

by Gary Phillips

KEY IDEAS

- **Economic activity can be classified into categories such as primary, secondary, tertiary and quaternary.**
- **The location of manufacturing industry is influenced by many interconnected factors, e.g. raw materials, government policies etc.**
- **A series of theories and models have been proposed to attempt to explain the changing location of industry.**
- **Increasingly, the global scale of analysis has become important for understanding patterns of manufacturing industry.**
- **The United Kingdom has experienced de-industrialisation.**
- **Service industries show a considerable variety of type and have experienced many of the processes that have influenced manufacturing industry.**

Economic activities are very varied and can be classified in a number of ways. One of the most commonly used classifications is the distinction between primary, secondary and tertiary activities. Primary economic activity involves the production of basic raw materials through, for example, farming, forestry, mining and fishing. Secondary activities involve the production of manufactured goods from raw materials or components. This sector includes activities such as steel making, brewing and car assembly. Tertiary activities involve the provision of a service of one sort or another, e.g. retailing. In recent years a fourth type of activity, quaternary, has been identified which involves the exchange of technical information and capital. This section is concerned with secondary and tertiary economic activities only. Primary economic activities are covered in other sections.

The relative importance of these broad categories of economic activity varies significantly from place to place at all geographical scales. There are large differences in the employment structure of individual countries based on the primary/secondary/tertiary classification. The situation in a selection of countries is shown in Figure 6.1. This shows that the different categories of economic activity also make varied contributions to the Gross National Product (GNP) or **Gross Domestic Product** (GDP) of a country. A study of Figure 6.1 suggests that the relative importance of the different categories of economic activity (as measured by their importance in employment terms) varies according to the level of economic development in that country. A generalised sequence of changes has been proposed to describe the changes that many countries experience. This is shown in Figure 6.2.

Country	% workforce employed in			% of GDP contributed			GNP per capita US Dollars
	Primary industry	Secondary industry	Tertiary industry	Primary industry	Secondary industry	Tertiary industry	
Nepal	94	3	3	58	14	28	170
Tanzania	82	6	12	66	7	27	120
India	72	11	17	32	29	39	350
Brazil	40	24	36	9	43	48	2550
Italy	13	46	41	4	34	62	15150
Japan	12	40	48	3	41	56	23730
UK	2	22	57	2	37	61	14570
USA	2	32	66	2	29	69	21100

Gross National Product (GNP) is a measure of a country's total production of goods and services including the net income from overseas.
Gross Domestic Product (GDP by) is a measure of domestic production and excludes net overseas income.

Figure 6.1 *The contributions made by different types of economic activity to employment and wealth creation in selected countries 1993*

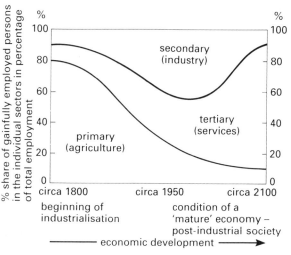

Figure 6.2 *Fourastie's model of employment changes*

Q1. Draw a series of scattergraphs to investigate some of the relationships contained within Figure 6.1. Analyse the relationship between GNP per capita and the contribution that secondary industry makes to employment and GDP in the countries in Figure 6.1.

Q2. What general relationships are suggested by the scatter of points on the scattergraphs? Are there any countries which do not fit into the general pattern? If so, can you explain why?

MANUFACTURING INDUSTRY

Manufacturing is the production of goods by a wide variety of industrial processes. It is a very important form of economic activity because it produces wealth and creates employment. In the UK, manufacturing contributes about 40 per cent of GDP and employs about 22 per cent of the workforce. In other countries the contribution to both employment and GDP is considerably smaller.

Manufacturing industries include a wide variety of different activities. In the UK the Standard Industrial Classification (SIC) is used to classify all forms of employment. Manufacturing industry consists of three of the ten divisions. At this stage the distinction between a factory and a firm must be made. Firms are economic and legal units and may consist of a large number of factories or plants as well as office units and research laboratories. Plants or factories have individual locations and because of this geographers have in the past spent more time studying individual factories rather than firms. However, this has changed in recent years as geographers have tried to understand industrial locations more fully. Many firms are becoming increasingly multinational or transnational in character. They range from small family works employing a handful of workers to large corporations employing many thousands of workers in several different countries. Many manufacturing industries are becoming increasingly **footloose** due to changes in manufacturing and transport technology.

A systems approach to manufacturing

A system may be defined as a combination of elements or components which are held together by links. These links work in such a way that any alteration to one part of the system produces a series of changes in other parts of the system. All systems have inputs, throughputs and outputs. In a typical factory the raw materials or components used would be classified as inputs and the finished goods as outputs. The actual manufacturing process itself changes the inputs into throughputs and then to outputs. The inputs and outputs have to be transported to and from the factory. Thus various costs are incurred in the assembly, manufacture and distribution of the finished goods. The sale of the goods to a customer produces revenue and when this is greater than the amount spent on the various costs, a profit is made. At a different level of analysis, the concept of a system could be applied to a large corporation such as the Ford Motor Company. Such a company is made up of many different activities, each of which often possesses an independent location. These various different locations are then linked together. A firm may consist of a group of manufacturing functions including such activities as research and development, component manufacture, end-product manufacture and sales and marketing. When we talk about a manufacturing location, we need to be clear about which specific part of the manufacturing group we are talking about since different locational factors may influence different parts of the firm's manufacturing system.

Factors affecting the location of industry

The location of individual factories and of firms both large and small is often the result of the interaction of many factors. The relative importance of these factors varies over time and from one type of industry to another. Today's industrial landscape is the result of a complex interrelationship between past and present. In attempting to explain present-day location patterns we may have to consider the original location factors at the time of the establishment of the firm and also the factors which operated to permit the firm to continue in its location.

Raw materials

Few manufacturing industries use raw materials directly. Those that do are called processing industries. Such industries include oil refining and steel making. Where there is a large loss of bulk on processing, access to raw materials at source becomes an important locational consideration. A good example of this is the processing of sugar beet in the UK. This is a crop that is grown mainly in eastern England and it is here that the main sugar beet refineries are located. This is because there is a 90 per cent loss of its weight on manufacture. In the past the main sugar refineries were located in the ports of Liverpool and London, because prior to the use of sugar beet, imported sugar cane was used as the raw material. This illustrates the importance of a port location. This is a break-of-bulk location where processing takes place at the point of import.

Energy resources

Access to energy resources was a vital factor in industrial location in the past because power was the least mobile of all resources. The iron and steel industry changed its location twice in the search for energy supplies. It used charcoal initially and later coal and coke. Large industrial concentrations grew up on the coalfields of Europe in the nineteenth century. However, technological improvements in the manufacture of steel reduced the amount of coal needed relative to iron ore. This reduced the 'pull' of the coalfields. This was further reduced when electricity replaced coal in the twentieth century. Energy then became a **ubiquitous** resource. This is one reason why many industries are more footloose than they once were. However, some industries are still energy-intensive and are thus influenced by the availability of energy resources. Such industries include metal smelting, e.g. the production of aluminium which is attracted to locations with abundant hydroelectric power such as Norway.

Transport

Transport's importance to industrial location lies in the costs of moving raw materials and components to the place of production, and the finished or semi-finished product to market. During this century the relative importance of transport costs has declined from about 20 per cent of total costs on average, to about five per cent. This decrease has occurred because of technological improvements such as bulk cargo handling and containerisation. Nevertheless the influence of transport and accessibility is still apparent in a variety of ways. Motorways, such as the M4, in the UK, have become a focus for industrial developments, as have ring roads and by-passes in urban areas. Transport junctions or nodes often attract industry because of their break-of-bulk attraction. Most ports are also important industrial centres. Airports have proved less attractive than seaports in the past, but with the increasing use of air for freight transport an airport location is becoming more attractive.

Factory site requirements

This factor operates at the local scale. Large-scale operations such as an integrated iron and steel works require a large area of relatively flat land and are consequently attracted to floodplains or coastal locations where such sites are naturally available. Factories are taking up more space because the development of conveyor belt technology favours single storey buildings. In addition, the increased use of cars for the journey to work necessitates large spaces for car parks. Thus sites are increasingly being modified through land reclamation schemes.

Land costs

The cost of land for building factories can and does vary at a variety of scales. Traditionally in Britain the most expensive land has been found in the south east and in large cities. Firms which are particularly demanding in the space they require are forced to look elsewhere for locations. This has encouraged the dispersal of industry away from inner cities and from the south east of England in recent years. (See section on the urban-rural shift of industry, page 146).

Market

Industries located at or near their principal markets are said to be market-orientated. This may occur for a variety of reasons. If the manufacturing process adds bulk or weight to the product, the costs of transferring the finished product increase and therefore need to be minimised by a market location. Breweries and soft drink makers tend to locate close to their main markets. In Nigeria, for example, virtually every major city has its own brewery. If a product is perishable a market location is preferable as it may deteriorate or break in transit. As a result, the baking industry in many urban areas, is locally based to serve the immediate district. Today most factory locations have a market orientation although it may not be obvious at first sight. Large companies often consider their markets to be international which partly explains the move towards multiplant production.

Capital

The supply of capital is the lifeblood of all industry. It is used to acquire buildings and machinery and to buy raw materials and pay other costs, e.g. labour. There are two sorts of capital.

Financial capital is money available for investment. A feature of these funds is their mobility. As large multinational corporations dominate the world economy, they move capital around the world, investing in regions where profits are high and disinvesting where profits are low. Geographical differences in capital supply are greatest at the global scale. Shortages regularly occur in poorer countries and are a major obstacle to their development. Even in countries such as the UK some forms of capital are not equally available everywhere. Venture capital groups, dealing with investment in high risk businesses, are concentrated in London in the case of the UK. The supply of capital involves an investment risk and so there is a tendency for such capital to be loaned to firms which are close to the loaning institution, where their progress can be easily monitored. Governments try to counter this process by using regional development policies to fund industry in peripheral locations.

Fixed capital is the investment tied up in buildings and machinery and is immobile. It is the major reason for the survival of past industrial patterns. The greatest influence on the future location of industry is likely to be its present location. Industrial inertia occurs when an industry remains concentrated in an area long after the original advantages for the location have disappeared. Recently the growing tendency of industry to lease rather than purchase factory space has removed one of the major restrictions on relocation. Without extensive capital tied up in the site, industry can afford to be much more flexible in its location.

Labour

Labour varies spatially in quality and quantity. It also varies in its ability and in its reputation for militancy. For some industries labour forms a high proportion of total costs, and may exert a strong locational influence e.g. defence equipment manufacture requires very high skill levels which push up labour costs. The cost of labour can vary at a number of scales. At the global scale costs differ enormously between countries. The very low wage rates in many developing countries, such as Taiwan and Singapore, have attracted multinational consumer electronic equipment companies to set up production there. In 1993 average hourly wages in Germany were US $25 whereas in Portugal they were only US $5. Within the UK there are geographical differences but these are relatively small due to unions negotiating national pay rates. Variations in productivity can more than offset differences in wage levels between regions. Given two areas equally advantageous from all other points of view, a manager is more likely to locate in the area where strikes, unrest and absenteeism are less common. This helps explain why many industrialists switched production away from the so-called Golden Triangle in north west Italy to the area known as the Third Italy in recent years. The relative lack of spatial mobility of labour is reflected in the fact that government regional policy in the UK has traditionally concentrated on taking work to the workers rather than vice versa.

> **Q3.** *Attempt the following essay question using material from this section and other sources you have access to.*
> *Examine the effect on industrial location of the availability, cost, skills and reputation of labour.*

Scale of production

Plant size can have an important effect on the costs of manufacturing a product. Large plants usually have lower unit costs than small ones. These savings produced from within the plant itself are termed internal economies of scale. They occur because of the use of specialised machinery and labour and bulk purchase of raw materials. Some industries can benefit from these savings to a greater degree than others, e.g. oil refining and cement manufacture. Plant size can have an important influence on location because of the size of site which is required and the size of the workforce which needs to be found within commuting distance of the plant.

Other savings can be made as a result of external economies of scale or **agglomeration economies**. These are achieved by forces operating outside the individual firm and favour the concentration or agglomeration of industry and the consequent emergence of industrial regions. (See industrial regions, page 143.)

Government policies

The type and extent of government involvement in industrial location depends on the political system in any country. In command economies state control is rigid and absolute. Governments exert an influence in many different ways and at different scales. By providing **infrastructure**, both local and national governments try to create an environment conducive to industrial growth and development. In developing countries manufacturing industry is often concentrated in those regions where the infrastructure is at its most advanced, usually around the main cities. In Britain the amount and type of intervention depends on which political party is in power. When Labour is in power there has usually been more direct involvement in industry through regional policy and the use of state subsidies for particular industries. Successive Conservative governments in the UK since 1979 have followed different policies (see the changing industrial geography of the UK, page 140). In the last 20–30 years policies at the international scale have become very important. In theory members of the European Union co-operate in matters of regional policy and also in some industries, such as aerospace, which require large amounts of capital for research and development.

Theories and models of industrial location

Because of the unevenness of the distribution of manufacturing industry across the earth's surface, it is not surprising that geographers have long been interested in attempting to formulate theories to explain these locations. It is not the purpose of this section to investigate all the models in depth but to show how theory has evolved up to the present day. As things stand at the moment, there are three main approaches in industrial geography: the external environment approach; the internal environment approach; and the structural approach.

Classical location theories such as those of Weber and Losch provided the mainstay of the external environment approach. They used economic concepts such as costs, revenue and profit to explain the location of industry. This involved making a series of assumptions about the behaviour of industrialists, the cost of transport and the existence of an **isotropic surface**. Weber's model was developed in 1909. He was a German economist and thought that the selection of a site to minimise costs, especially those of transport was of paramount importance. He developed a technique for mapping differences in transport cost (through the construction of isodapanes; lines of equal transport cost). However, Weber recognised that the availability of a pool of cheap labour and agglomeration economies could attract industries away from a location with the least transport costs (the **optimum location**). Weber's model has been criticised for placing too much emphasis on transport costs and for ignoring the revenue aspect of profitability. Despite its shortcomings it was the first real attempt to explain industrial location and stimulated others to come up with alternative explanations.

Losch produced a complementary model which came to the conclusion that firms seek a location which offers the largest possible market area and thus the highest possible profit. Each firm will avoid overlap with competing firms and the result will be a series of hexagonal trading areas similar to those associated with Christaller's Central Place Theory. This has been criticised for only looking at one aspect of profitability, i.e. revenue, and for virtually ignoring costs. In 1971 David Smith attempted to look at both costs and revenue through the concept of space-cost curves, and the concept of margins of profitability instead of the idea of a single optimum point as envisaged by Weber and Losch. This introduced the idea of sub-optimal locations which are selected for a variety of reasons such as the availability of sites or government subsidies. Smith's neo-classical theory is much more useful. However, in the 1960s and 1970s dissatisfaction arose because these so-called economic theories could not provide explanations for much of what was observed in the real world.

The internal environment approach dates from this period and is still being developed. It suggests that it is not the geographical distribution of factors of production that creates the pattern of industry but the way each firm interprets the environment it finds. This is in keeping with the development of behavioural geography. In 1967 Alan Pred put forward the idea of a behavioural matrix which allows industrial decision makers to be classified on the basis of their ability to use information and on the amount and quality of the information at their disposal. This can be combined with David Smith's spatial margin idea. However some have criticised this approach as being of limited value as it distracts attention from the structural features of the economy to which firms react. Four main internal influences thought to be important are ownership, organisation, linkage and age of buildings. Different kinds of ownership will clearly have an important influence on location decisions. Small, single factory companies controlled by an individual will usually be located close to the home of the owner. In a multiplant company the distribution of the other factories it already owns will strongly influence the decision about where to locate the next factory. The spatial significance of the internal organisation of a company is best appreciated by recognising that manufacturing involves several different kinds of activity, e.g. administration, research and development, assembly etc. These different operations often take place on different sites in a multiplant firm. This fits in with the idea of the product cycle which is discussed later in the section. A consideration of different types of **linkage** and of the age of industrial premises is relevant to the understanding of industrial location.

Finally, the structuralist approach views the location of industry as being directly related to the structure of society. The pattern of industries found in a capitalist, western society will be different to that found in a centrally planned, command economy. The distribution of power within a society is all important. This is the most recent of the three approaches to demand attention.

These different approaches, though very different, do complement each other to a certain extent.

Global manufacturing change: patterns and processes

'We live in a world of increasing complexity, interconnectedness and volatility; a world in which the lives and livelihoods of every one of us are bound up with processes operating at a global scale' (P Dicken, *Global Shift – Industrial Change in a Turbulent World*, 1992).

Explanations of patterns of industrial location are increasingly being made with reference to the global scale. The days of being able to explain the opening or closure of a factory with reference to local conditions alone have long gone.

Economic activity (and manufacturing industry in particular) is becoming more internationalised and increasingly globalised. Internationalisation refers to the geographical spread of economic activities across national boundaries. It is not a new phenomenon. Globalisation is a more recent and complex form of internationalisation which implies a degree of functional integration between internationally dispersed economic activities. However, these processes work at different rates in different places. National boundaries no longer act as watertight containers of the production process. In the past many industries were protected from competition by geographical distance. Today a growing number of economic activities have meaning only in a global context.

A new global division of labour has developed – a change in the geographical pattern of specialisation at the world scale. The straightforward production of manufactured goods in the more developed countries (or core countries) and the production of raw materials in the peripheral or non-industrialised countries is being transformed into a much more complex structure involving the fragmentation of many production processes and their geographical relocation on a global scale. Factors such as telecommunications, corporate organisation and the technology of production are responsible for these

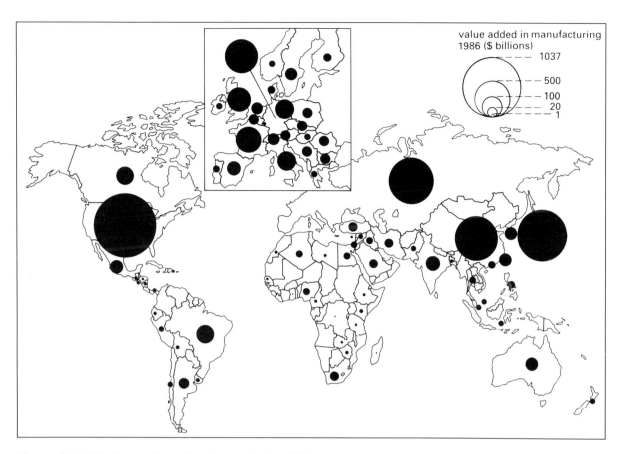

Figure 6.3 *Global map of manufacturing production 1986*

changes. These are also having an effect on the patterns of service activities, especially international finance (see service industries, page 151).

The traditional international economy of traders is giving way to a world economy of international producers. A study of Figure 6.3 shows a very uneven picture. Though manufacturing activity is widely distributed, the majority of production is concentrated into a small number of countries. The USA, western Europe and Japan dominate the world scene. However, against this backdrop of apparent continuity there are several important changes occurring at present. The USA has shown a substantial decline in its share of world manufacturing production from 40 per cent in 1963 to 24 per cent in 1987 and although it retains leadership of all countries, its position is increasingly being challenged by Japan. Also, a small group of developing countries has begun to make a real impact on the world manufacturing scene. These include countries such as South Korea, Taiwan, Brazil and Spain. Manufacturing production is no longer almost exclusively a core-region activity as it had tended to be for the last 200 years. The centre of gravity of the world manufacturing system has begun to shift towards the Pacific. The global economy is now multi-polar – three clear regional blocs are discernible: North America; The European Union; and South East Asia.

Q4. *Explain the difference between the increased 'globalisation' and 'internationalisation' of manufacturing activity.*

Q5. *Try to find out why most African countries have found it very difficult to develop manufacturing industries. (Read the section on Kenya, page 149, to help you here.)*

Many important forces have shaped the trends previously described. Three of the most important – transport and communication, production technology and the growth of multinational or transnational companies – will be briefly considered. New types of transport technology are crucial to the changing geography of manufacturing production. Through the use of containerisation and long-haul jumbo aircraft transport has become faster, cheaper and more flexible. It has become an 'enabling' rather than a 'determining' factor. Such improvements have allowed the physical separation of distinct production stages, and has encouraged multinational firms to exploit appropriate locational advantages for each function. Faster communications now mean that time is worth more money. The effect of this has been to

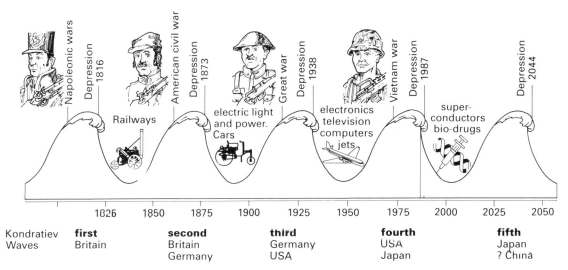

Figure 6.4 *Kondratiev waves*

encourage the clustering of suppliers in central but less congested areas within a few hours' easy travel of major assembly plants. Satellite communication has also stimulated the expansion of global banking (see service industries, page 151).

Technology affects manufacturing in many ways but two concepts have been considered of importance in recent years: Kondratiev waves and the product cycle. In 1925 a Russian economist, Nikolai Kondratiev, put forward the idea that the world economy had experienced a series of long, predictable waves. Each wave lasts just over 50 years and there have been four waves since the Industrial Revolution (see Figure 6.4). New technologies 'energise' each wave, yield groups of new industries and may shift production advantage to new places. The Kondratiev cycle theory attempts to contribute to the understanding of entire economies. The product cycle theory, by contrast, is concerned with the changes associated with individual industries. The model (see Figure 6.5) was originally proposed by R Vernon in 1966. It assumes that all products pass through a series of stages of development, from the initial development of the product through to eventual obsolescence. At each stage of development the relative importance of the various production factors changes, e.g. the amount of capital required and the type of labour. As a result, different types of geographical location are relevant to different stages of the product cycle. In the initial stages a location in a core region or country is often required because of access to scientific know-how at universities and research laboratories. Mass production often takes place in branch-plants in peripheral regions or countries so as to take advantage of low labour costs. Product cycle theory

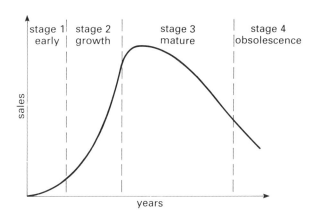

Figure 6.5 *Product-cycle and production inputs*

is an example of a more 'dynamic' sort of industrial location theory. (See industrial location theories, page 133).

The final factor which helps explain some of the global changes in manufacturing activity is the increasing role of multinational or transnational companies (TNCs). The tendency to concentrate business into larger and larger firms is very strong in market economies, despite the recent boom in small enterprises in some sectors of the economy. Firms become TNCs when they produce in two or more countries. Manufacturing TNCs emerged with the third Kondratiev wave (see Figure 6.4) and giant TNCs have grown to become an important part of the increasing globalisation of economic activity. The largest TNCs in 1970 and 1987 are shown in Figure 6.6. The USA still dominates as the national home for the biggest companies but its position is being threatened by other countries, especially Japan. American TNCs such as Ford brought new manufacturing capacity to Europe in the 1960s. In the 1990s Japanese rivals such as Nissan followed suit. The UK has gained foreign direct investment from some foreign-owned TNCs who perceived it to be a low wage offshore platform from which to supply continental Europe (see the changing geography of UK manufacturing industry, page 140). Since 1970 cheaper energy and labour became the prime motive for TNCs investing in Asia and Latin America. The geography of TNC activity and foreign direct investment is very dynamic. The most spectacular growth in TNC activity is in the USA, especially with investment from Japan. Similarly, the prospect of a foothold in the European Union after 1992 has attracted much Japanese investment to the UK, e.g.

Nissan in Sunderland and Toyota in Derby. The mix of multinational enterprise with the high capital mobility that it brings in the flows of capital, management, new technology and different production systems, has been a key feature of global economic change.

Q6. *What are the advantages to companies of becoming transnational or multinational in character?*

Q7. *How has the composition of the 20 largest industrial corporations changed between 1970 and 1987? Calculate percentages for each nationality in each year.*

Q8. *Name the corporations which have changed their rank order (upwards or downwards) by the greatest amount.*

Q9. *Using information provided in Figure 6.8, page 139, list the three main factors that multinational companies take into account when deciding on location for investment.*

The car industry at the global scale.

Car manufacture provides a very good illustration of the many influences at work in the global economy in recent years. The significance of this industry lies not only in its sheer scale (approximately four million people directly employed) but also in its immense spin-off or **multiplier effects** through its linkages

rank	1970	1987
1	General Motors (USA)	General Motors (USA)
2	Exxon (USA)	Shell (UK–Netherlands)
3	Ford Motor Co (USA)	Exxon (USA)
4	Shell (UK–Netherlands)	Ford Motor Co (USA)
5	General Electric (USA)	IBM (USA)
6	IBM (USA)	Mobil (USA)
7	Mobil Oil (USA)	British Petroleum (UK)
8	Chrysler (USA)	Toyota Motor (Japan)
9	Unilever (UK–Netherlands)	IRI (Italy)
10	ITT (USA)	General Electric (USA)
11	Texaco (USA)	Daimler-Benz (West Germany)
12	Western Electric (USA)	Texaco (USA)
13	Gulf Oil (USA)	AT&T (USA)
14	US Steel (USA)	Volkswagen (West Germany)
15	Volkswagen (West Germany)	Du Pont (USA)
16	Westinghouse Electrical (USA)	Hitachi (Japan)
17	Chevron (USA)	Fiat (Italy)
18	Philips (Netherlands)	Siemens (West Germany)
19	British Petroleum (UK)	Matsushita Electrical (Japan)
20	Nippon Steel (Japan)	Unilever (UK–Netherlands)

These rank positions were calculated across the 500 US industrials and the 500 non-US industrials. Therefore some minor inaccuracies might be present since affiliates of these firms which did not make either of the top 500 list would not have been counted.

Figure 6.6 *20 Largest industrial corporations 1970–87*

Figure 6.7 *Nissan car plant in Sunderland*

THE MULTINATIONALS ERA

According to a recent report from the United Nations, direct investment by transnational corporations is becoming a very important component in the world economy. In the past couple of decades it has increased by on average, 13 per cent per year. This trend is the result of a combination of several factors, for example the development of telecommunications and a growing acceptance of transnational companies by governments.

Foreign direct investment is helping to integrate national economies. This is illustrated by the operation of car companies such as Ford who are developing their world car. Components for these will be made at single locations and shipped around the world.

The growth of transnational or multinational companies has started to raise concerns about the future validity of national laws and regulations and the continued transfer of assets from the developed world to the developing world.

An increasing amount of foreign direct investment is going into services and high-technology manufacturing industries. In the developing world most is concentrated in South East Asia and Latin America.

Most of Africa has been unable to benefit from this sort of investment. Transnational companies are looking for many things: a skilled local workforce, good infrastructure and a welcoming attitude from the government. Simple cash incentives and cheap labour are not enough on their own. South Korea illustrates this situation. In 1992, the country experienced a net outflow of foreign direct investment because it made life difficult for Japanese companies in terms of taxation, and real estate law. Attitudes have recently changed and the government is now trying to be more helpful to foreign investors. Like it or not, too much depends on them.

September 1993

Figure 6.8 The multinationals era

with other industries. The world car industry is predominantly an industry of transnational corporations. Car manufacturers are some of the most influential TNCs in the world (see Figure 6.6). The industry is market-orientated to a high degree. Between 1948 and 1989 world car production increased from 4.6 million to 34 million. Since 1960 major changes have occurred in the global distribution of the industry. The most dramatic development has been the spectacular growth in Japan, and the relative decline in the USA and UK. Car production has also increased recently in some **Newly Industrialising Countries** (NICs) such as South Korea.

For a variety of reasons geographical differences between individual national markets have begun to be reduced, leading to the existence of regional or even global markets. Within such markets, however, there are signs of greater segmentation and fragmentation, i.e. for customised versions of a general model. This can be satisfied as a result of the dramatic changes taking place in the way cars are made. Until the 1970s, the car industry was dominated by the 'Fordist' mass-production method of manufacture. It used the economies of scale principle to mass produce. However, it was a very rigid method. It encouraged a dispersal of the industry in the sense that component suppliers were often located long distances from the point of final assembly so that they could take advantage of low cost production. Improvements in transport and communications allowed this to happen. Since the 1970s, however, major changes in the way cars are developed and manufactured have been occurring. These new 'lean production' methods were initiated by Japanese manufacturers especially Toyota. They combine the best features of 'craft' and 'mass' production methods. They also encourage geographical proximity between component supplier and car assembly plant. The car industry is becoming more market-orientated as a result of these new 'just-in-time' production methods as opposed to having stock piles of 'just-in-case' goods.

Since 1980 Japanese car manufacturers have adopted a policy of relocating increasing amounts of their production capacity into the USA and UK in an attempt to capture a greater market share in those parts of the world. Many of these new transplant factories occupy **greenfield sites** and take advantage of financial incentives provided by the host government through its Regional Policy. British and American producers have tried to imitate the Japanese production methods but this has produced huge job losses as a by-product of these restructuring programmes.

Another solution has been through increased participation in joint venture schemes with Japanese firms, e.g. Honda and Rover have a collaborative agreement in the UK. Eastern Europe and China seem to offer potential for future car manufacture and selling. Car ownership is at a low level in most East European countries and in 1993 it had hourly labour costs of only £3, compared to £9 in Britain.

Q10. Why is the car assembly industry market-orientated to such a high degree?

Q11. Why have all the main car assembly companies adopted lean production methods?

Q12. *How have lean production methods affected the location of component suppliers in relation to the assembly plants?*

The changing geography of manufacturing industry in the UK – continuity and change

The last 30 years have witnessed some profound changes in the form and type of manufacturing activity in more developed countries and in the UK in particular. The UK has experienced extensive **de-industrialisation**, which is defined as an absolute loss of jobs in manufacturing industry. This is sometimes associated with a drop in output and therefore with industries' contribution to GDP. Between 1966 and 1988 there was a 40 per cent fall in employment in manufacturing from about 8.4 million to about five million (see Figure 6.9). For the first time since the Industrial Revolution the country's balance of payments in manufactured goods went into the red (from 1983). This process is not uniform across all types of manufacturing industry nor across all regions. It is both regionally and industry selective. The process has been most severe in industries such as steel making, shipbuilding and textiles. Other manufacturing industries such as furniture manufacture and high-technology industries have fared less badly (see Figure 6.10). Regions which were over-dependent on the first group of industries have been particularly badly hit (see Figure 6.9).

At the sub-regional level, further redistribution took place. Major cities and conurbations lost manufacturing jobs, especially from inner cities, and smaller towns and some rural areas gained as a result (see industry in urban areas, page 145).

Changes also occurred in the geography of individual industries. Also since 1970, manufacturing production has been increasingly concentrated in large enterprises. This has geographical significance because many large companies have a chain of production, with one of their factories making one component, which is passed on to another factory of the same company for sub-assembly and so on. Within large companies functions are often geographically separated, with control functions becoming increasingly concentrated into south east England and the larger cities of other regions. Smaller settlements have become increasingly concerned simply with production where jobs are more unstable and badly paid than those associated with control functions.

Q13. *Explain the link between de-industrialisation, re-industrialisation and Kondratriev cycles.*

Q14. *Using Figure 6.10 plot the data for the following categories of employment as a series of line graphs (use the same vertical and horizontal scales for each line graph): textiles; metal manufacture; agriculture; and insurance/finance.*
How do the different industries compare in terms of their rates of change at different time periods?

As mentioned previously, new industries have emerged to replace old ones. This process is termed **re-industrialisation**. This has happened partly due to massive inward investment by foreign firms, especially Japanese. Many of these firms are in the electronics and motor vehicle industries and most of this investment has taken place in peripheral regions such as South Wales and Scotland. Many jobs were created in new small businesses in the 1980s, in high-technology and related sectors. The area around

Region	1966	% of GB total*	1988	% of GB total*	% change 1966–1988
South East	2363	28	1321	26	−44.1
East Anglia	173	2	218	4	+26.01
South West	429	5	364	7	−15.15
East Midlands	631	7	493	10	−21.87
West Midlands	1197	14	697	14	−41.77
Yorkshire and Humberside	860	10	443	9	−48.49
North West	1251	15	600	12	−52.04
North	461	5	261	5	−43.38
Wales	317	4	213	4	−32.81
Scotland	726	9	385	8	−46.97
Great Britain	8408		4995		−40.59

* Figures may not add up to 100 because of rounding up. Figures in '000s.

Figure 6.9 *Changes in manufacturing employment 1966–88*

Industry (Numbers in employment, '000s)	1951	1961	1971	1981	1991
Agriculture, forestry and fishing	1126	855	432	360	272
Mining and quarrying	841	722	396	332	—
Food, drink and tobacco	727	704	770	632	563
Chemicals and allied trades	435	499	438	395	307
Metal manufacture	616	626	557	326	300
Engineering and electrical goods	1601	2031	2028	1730	1289
Shipbuilding and marine engineering	277	237	193	144	—
Vehicles	735	838	816	636	—
Other metal goods	458	525	576	428	—
Textiles	986	790	622	363	186
Leather, leather goods and fur	78	60	47	31	—
Clothing and footwear	676	546	455	313	260
Bricks, pottery and glass	314	321	307	216	—
Timber, furniture	326	304	269	227	226
Paper, printing and publishing	515	605	596	493	480
Other manufacturing industries	264	295	339	265	—
Total: manufacturing	9975	9958	8841	6891	4822
Construction	1388	1600	1262	1132	962
Gas, electricity and water	357	377	377	340	—
Transport and communications	1704	1673	1568	1440	1349
Distributive trades	2689	3189	2610	2635	—
Insurance, banking, finance	435	572	963	1233	2693
Professional and scientific services	1524	2120	2916	3695	—
Others, miscellaneous	3485	3519	3379	3993	—
Total: service sector	11582	13050	13075	14468	15754

Figure 6.10 *Employment trends in major UK industries 1951–91*

Cambridge in East Anglia is a good example of such growth. Much of this has taken place on purpose-built **science parks** in the city itself or in surrounding small villages.

The third process to have affected the UK in the last 30 years is that of tertiarisation. This involves the growth of service sector employment and is linked directly to de-industrialisation. The process includes many different types of service activities including tourism and producer services.

Before an individual industry and region are studied to illustrate some of these processes, it is important to remember that they have not totally transformed the basic geographical distribution of industrial activity in the country. Urban areas still dominate the pattern of manufacturing activity in the early 1990s as they did in the mid-1960s.

The changing geography of the UK car industry

The UK has a long history of car making. The industry grew rapidly in the inter-war years so that by 1948 the country was the second biggest producer of cars after the USA. The industry can trace its origin back to the production of bicycles and horse-drawn

Figure 6.11 *Cambridge Science Park*

carriages and traditionally has been strongly market-orientated. Production expanded rapidly after the 1939–45 war and reached a peak in the early 1970s.

From the very early days large overseas car producers have been involved in the UK industry. Ford opened its first factory in Manchester in 1911 but switched production to Dagenham in 1931. Most of the growth up until 1960 took place in a wide belt stretching from London through the West Midlands to the north west of England. However in the 1960s a dispersal of the industry took place to peripheral regions such as Scotland, where Chrysler opened an assembly plant at Linwood near Glasgow. Regional policy was behind this trend (see Figure 6.12).

The 1970s was a problematic era for the industry. Both British Leyland (as it was then) and foreign-owned car producers experienced poor labour relations and a lack of investment in new technology. Car production in Britain fell due to these reasons and because of a series of economic recessions which hit the country during this period. More cars were being imported from overseas and all major car producers carried out rationalisation

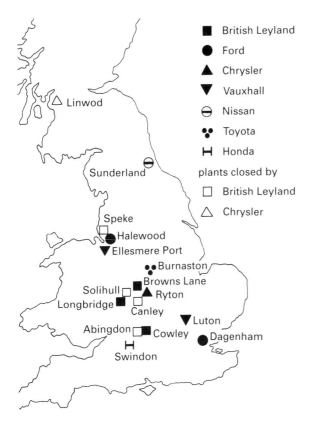

Figure 6.12 *Location of Britain's car assembly plants*

programmes which were designed to make their companies more efficient. This included the closure of several peripheral sites opened in the previous decade. At this time the major car companies began to invest in new technology and new production systems in an attempt to compete with the Japanese car producers.

From the mid-1980s onwards the Japanese producers themselves began to invest in car production in Britain in order to secure a greater market share in anticipation of controls on car imports into Britain and other member countries as a consequence of the 1992 Single Market Treaty. Nissan opened a factory in the north east of England in 1986, followed by Toyota and Honda in 1992. In the late 1980s car production increased in the UK as a result of these Japanese investments, though Britain is still not in the same league as a car producer as it was 40 years ago.

The future of the industry is debatable. The problems of inefficiency and bad labour relations which were apparent in the 1970s have largely disappeared. Labour costs are lower than in Germany but not as low as they are in eastern Europe. There is evidence to suggest that many car producers are switching investment to countries such as Poland and Hungary. This might well be at the expense of the UK.

Q15. *Why did the pattern of car manufacture in the UK become more dispersed in the 1960s?*

A case study of regional economic change: the East Midlands

The East Midlands has a very broad base of traditional industries such as coal mining, textiles, shoe making, railway engineering and chemicals. As a result it has not been affected to the same extent by de-industrialisation as has its neighbour, the West Midlands. Between 1966 and 1988 the region lost 22 per cent of its manufacturing employment compared with 42 per cent in the West Midlands. In the case of the West Midlands this is because of the over-dependence on the car industry and associated industries. Up until 1974 the two regions shared similar unemployment levels but between 1974 and 1983 unemployment was much higher in the West Midlands as a consequence of the rationalisation programme implemented by major car producers (see Figure 6.13). Since 1983 the levels of unemployment have converged but unemployment today is still proportionally lower in the East Midlands.

The East Midlands has also benefited from re-industrialisation to a greater extent. Many new, small firms have been set up in both the manufacturing and service sectors. Some have been established by foreign firms. The Toyota Motor Corporation opened a new car assembly plant at Burnaston, near Derby in 1992. It will employ 3000 people when fully operational and will help reduce the effect of redundancies in the engineering industry created by Rolls-Royce in the city of Derby. The pattern of job

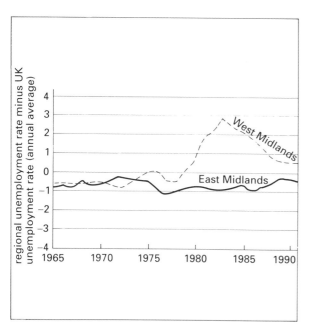

Figure 6.13 *Unemployment rates in East and West Midlands 1965–92*

losses is quite complicated (see Figure 6.14). Between 1981 and 1987 the East Midlands lost 102 100 jobs in manufacturing industry. Three district types were least affected: rural districts; districts to the south and east of Nottingham; and districts which had special help from the government. The Corby area is an example of this latter kind of district. Its steelworks shut down in 1980 making 8500 people redundant and male unemployment rose to over 30 per cent. Assisted Area status was given with access to EU funds to help solve the problem. Large amounts of public money were spent on improving the local infrastructure and on advance factory units, built in anticipation of likely demand. The town also acquired an **Enterprise Zone** in 1981. This offers simplified planning regulations and other benefits and has been very successful. By 1991 Corby had attracted some 700 new companies which together employed 11 000 people. The employment structure of the town is now much more diversified than it was before. The policies have been so successful that the town lost its Development Area status in June 1993 when the government reviewed its Regional Policy.

Although big cities and industrial towns fared the worst they are still the main industrial employers. Over the same period jobs in services increased in suburban and rural districts. It all added up to an urban-rural shift in employment.

Industrial regions

A pronounced feature of industrialisation is the tendency for manufacturing industries to cluster together to form distinct industrial regions. The development of industry in the UK over the last two centuries illustrates this process very well. By 1851 a number of clearly identified industrial concentrations had developed. Most of these were based on or close to coalfields in an attempt to minimise transport costs for the main source of energy at the time. Apart from the presence of a natural resource such as coal, there are several other reasons why industries cluster together. These include linkage, association, agglomeration economies and inertia.

Firms may be linked together in different ways. Vertical linkage occurs when a raw material goes through several successive processes. Horizontal linkage occurs when an industry relies on several other industries to provide its component parts. The assembly of cars is a very good example of horizontal linkage. Weber realised the significance of linkage in influencing industrial location. He realised that some industries are drawn away from the lowest cost location because of the influence of potential links with other firms. Linkage also exists between service industries and manufacturing industries. Any assembly industry usually displays a high degree of linkage with other industries.

Sometimes firms are attracted to the same location to share a common factor of location, e.g. many firms may congregate on an **industrial estate** to share access to a cheap site and good accessibility. This is the idea of association and firms displaying this are not necessarily linked together in any way.

Often firms cluster together to benefit from external economies of scale or agglomeration

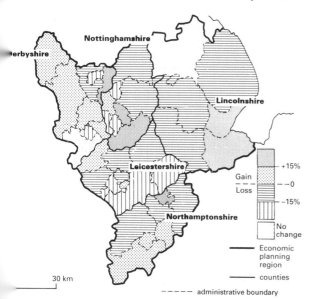

Figure 6.14 *East Midlands: changes in industrial jobs 1981–7*

ECONOMIC ACTIVITY 143

Figure 6.15 *Modern industrial estate in Hampshire, England*

economies. These savings can be divided into localisation economies and urbanisation economies. Localisation economies occur when firms which display a degree of linkage locate close together so as to reduce transport costs, to facilitate just-in-time delivery and to allow for greater personal contact between firms. Urbanisation economies are the savings that are made from locating in an urban area. All manufacturing industries require access to transport facilities and other forms of infrastructure.

The final reason for the development of clusters of industries is the influence of inertia. Industries find it difficult to change their location because of the large investment in fixed capital, such as factory buildings and machinery and the acquired advantages that have built up over many years. If a large firm or type of industry is successful it may attract more development through the multiplier effect. This process is termed **cumulative causation** and was first proposed by Myrdal, a Swedish economist, in the 1950s.

Some people argue that industrial regions pass through a series of cyclical stages or phases (see Figure 6.16). The West Midlands region illustrates this sequence of changes.

The evolution of an industrial region: the West Midlands

The West Midlands has been a major centre of manufacturing industry since Tudor times. The production of iron, metal goods and pottery during

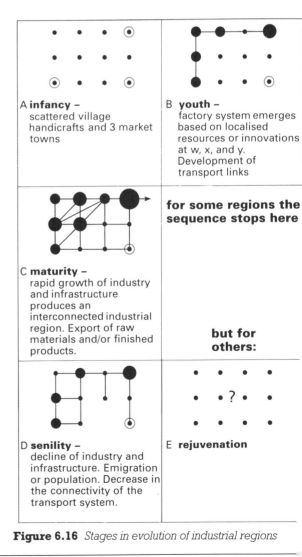

Figure 6.16 *Stages in evolution of industrial regions*

the Industrial Revolution was based on local supplies of coal and ironstone. In the nineteenth century metal workers turned their skills to the manufacture of bicycles, cars, and machine tools. This has continued to the present day and the region has developed a reputation as a centre for metal working and engineering. The different industries are linked together but the starting point of this system is the metal making industry. This is concentrated in the area around Dudley and Sandwell in the Black Country. It grew up here using raw materials and fuel produced on the South Staffordshire coalfield. Metal making has declined in recent years and the last large steelworks closed at the end of 1982. The remaining steel furnaces are small and make steel by melting down scrap metal. Most iron and steel now has to be imported from other areas of Britain, such as South Wales. The metal making industry led to the growth of other linked industries in the area. There is an obvious relationship between the pattern of industries shown in Figure 6.17 and the sequence of industries shown in Figure 6.18. Linkage is expressed in a geographical pattern.

The region was very badly affected by de-industrialisation in the 1970s and 1980s. Between 1966 and 1988 the region lost 41 per cent of its jobs in manufacturing industry (see Figure 6.9).

Stage	Process	
1	Metal making	Smelting steel and non-ferrous metals: refining and alloying. The products are metal for use in the next stage.
2	Metal shaping and processing	Making castings, forgings and stampings. These are mostly sent on to stage 3, but there are some finished products like chains, brassware and saucepans.
3	Metal using	Metal is now made into recognisable products, e.g. heating radiators, or into parts for use in stage 4, e.g. car components, gas rings, cooker panels and doors.
4	Assembly	Finished products are put together, e.g. cars, cookers, washing machines, lawn mowers, locks.

Figure 6.18 *Industrial linkage system in metal industries of West Midlands*

Unemployment increased significantly and by 1984 large parts of the region qualified for Assisted Area status for the first time in its history. Job losses in the car assembly industry were magnified through the multiplier effect working in reverse; for every job that was lost directly in the car assembly industry several others were lost in component suppliers, i.e. metal using and metal shaping industries.

Industry in urban areas

Until recent decades the vast majority of manufacturing industry in economically developed countries was to be found in large urban areas. Industrialisation and urbanisation went hand in hand. However, in recent years, there is increasing evidence that this historical relationship is being modified. Large cities and conurbations are increasingly unattractive to modern industries. This is not confined to the UK but is a feature of all other economically developed countries.

key

- stage 1 (metal making) and stage 2 (metal shaping and processing) are concentrated in this area
- stage 3 (metal using) and stage 4 (assembly industries) are dispersed over a wider area but are mostly found in the West Midlands conurbation
- ■ industrial satellite towns linked with the West Midlands metal and engineering industries
- ● other industrial towns where the main industries are not linked with the metal and engineering industries

Figure 6.17 *Location of industry in West Midlands*

ECONOMIC ACTIVITY **145**

The urban-rural shift in the UK

Since the 1960s there has been a marked shift of manufacturing activity and employment from the conurbations and big cities to small towns and rural areas (see Figure 6.23, page 149). Generally, the bigger the settlement the greater the decline in manufacturing employment. Between 1960 and 1981 London lost 688 000 jobs in manufacturing (−51 per cent) whereas rural areas gained 128 000 (+26.3 per cent). This urban-rural shift can also be detected in the employment trends of individual industries, e.g. high-technology industry (see Figure 6.19). There is also evidence of the process operating in the East Midlands in the 1980s (see Figure 6.14).

	Employment	Employment change 1981–87	
	1981	No	%
Conurbations (8)	462 336	−70 861	−15.3
More urbanised counties (14)	310 583	−13 180	−4.2
Less urbanised counties (21)	364 300	+11 113	+3.1
Rural counties (20)	89 542	+6 057	+6.8

Figure 6.19 *Urban–rural shift of high-technology industry in UK 1981–7*

There is no simple explanation for the urban-rural shift. It is the result of a variety of factors; people increasingly prefer to live in small towns and rural areas (significant for highly qualified scientists and managers in high-technology industries); availability and cost of land in large settlements and availability of cheaper, less unionised labour in rural areas. Most of the evidence suggests that lack of expansion room in working areas has become very important as new production methods require more and more space per worker employed. This Constrained Location Theory was first proposed by two land economists in 1982. Public bodies such as the Development Board for Rural Wales have conducted advertising campaigns to encourage firms to locate in peripheral rural areas of the country and in 1992 planning rules were relaxed which previously prohibited the use of agricultural land for urban and industrial development. This is happening at a time when agriculture, as the dominant rural land user, is increasingly being questioned because excessive production is being curtailed. The urban-rural shift of manufacturing industry has a parallel in the **counterurbanisation** process which has affected the distribution of population in the UK and other developed countries in recent decades. Urban-rural shift of manufacturing industry has not occurred on anything like the same scale in countries in the developing world.

Intra-urban manufacturing change

The decentralisation process, previously referred to as the urban-rural shift, has been a feature of the geography of manufacturing industry within large settlements since 1945. Outer suburban areas have increased their share of industry at the expense of inner city areas. Bales' model of intra-urban industrial location in developed world cities is being rapidly changed (see Figure 6.20). The twentieth century has

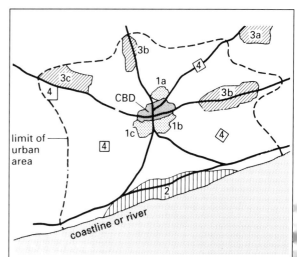

Note: it is unlikely that smaller towns would contain such a variety of industries or industrial areas. Also, waterfront industries would be missing from some inland towns.

1 centrally located — e.g printing
2 waterfront — e.g. flour milling
3 suburban — e.g. chemical manufacture
4 randomly located — e.g. local newspaper production

Figure 6.20 *Intra-urban industrial locations in city in developed world 1981*

seen a steady erosion of the advantages that inner city areas had for the location of manufacturing industry – access to raw materials via railway termini, access to labour before the days of efficient public transport, and agglomeration economies through links between different groups of industries. Today inner city locations are high cost locations for industry and their demise has been attributed to many factors. These include the loss of skilled workers through increasing suburbanisation of populations, declining accessibility especially by road transport, the reduction of inter firm linkages and the effects of urban renewal policies. The progressive outward shift is well illustrated in Figure 6.21 which shows the development of industrial estates in the City of Leicester. Since the early 1980s successive British Governments have tried to stem the loss of jobs from inner city areas by implementing a variety of policies. These include the use of **Urban Development Corporations**, Enterprise Zones and **National Garden Festivals**.

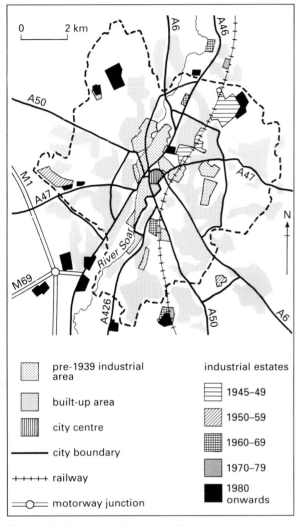

Figure 6.21 *Industrial estates in Leicester*

Industrialisation in the developing world

'Most Third World countries are committed to changing their rural-based agricultural economies to urban-based industrial ones. There may be differences in the level of industrialisation they wish to achieve, the speed at which they wish to industrialise, or in their industrialisation strategies, but nearly all of them are strongly committed to their goal of industrialisation.'

(Rajesh Chandra, *Industrialisation and Development in the Third World*, 1992)

Industrialisation has historically been regarded as the only means of developing (see Figure 6.2, page 130). It is the process by which a country increases the proportion of GDP contributed by manufacturing industry (or alternatively its share of employment). Industrial growth itself is not sufficient because other sectors of the economy may increase their contribution to output at the same time. It is necessary for the manufacturing sector to increase its relative importance in the economy more rapidly than other sectors. The process must not be seen in purely economic terms because it includes social, political and cultural changes too and is usually linked with urbanisation.

The fact that most developing countries were colonies of the present industrialised countries is very important in explaining the degree and type of their contemporary industrialisation. The main function of a colony was to supply the mother-country with raw materials and to buy its manufactured goods. As a result, industrialisation only began in earnest after political independence took place. However, the provision of physical infrastructure such as railways and political institutions did aid industrialisation when and where it occurred.

Q16. *Explain the difference between industrialisation and industrial growth.*

While there has been significant industrial growth in developing countries, their share of total world production and trade has not improved very much since the 1960s. The developing countries are a diverse group and it is not surprising that there are marked contrasts in levels and rates of industrialisation from country to country (see Figure 6.1, page 130). African countries generally have very low levels and rates of industrialisation. Latin America, because of its early independence and proximity to the USA has done reasonably well, e.g. Brazil and Mexico. The most rapid growth has occurred in South East Asia and more specifically from the four 'Asian Tigers'; Hong Kong, Singapore, Taiwan and South Korea. Reasons for these differences in economic performance are many and varied. They include the fact that some countries became independent earlier than others and have a bigger domestic market than others. Most foreign direct investment has been concentrated on a small number of countries. Mexico, Brazil and Indonesia account for about one-third of all foreign investment. Lastly, investment in literacy programmes and 'appropriate' government policies have been important.

Recognition of the success of a small group of countries came in the early 1970s when a number of

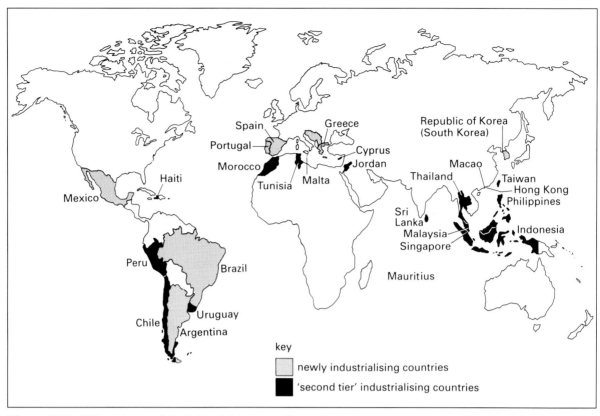

Figure 6.22 *NICs and second tier industrialising countries*

international organisations, including the World Bank, began to identify **Newly Industrialising Countries** (NICs). A second wave of such countries was identified in the early 1980s (Figure 6.22). Countries which have low growth rates and where most production is consumed within that country do not come into either category, e.g. India and China. In some of these countries most of the manufacturing takes place in the **informal sector** of the economy. These are family businesses operated in the shanty towns and villages. The statistics for this sector do not appear in official figures for output and employment yet they make a major contribution to the domestic needs of the country.

The state and multinational corporations both play vital roles in most developing countries. TNCs are a source of capital, technology, expertise and markets. However, they also present problems since their objectives may not complement national development goals. TNCs conduct extensive locational analyses before deciding in which country to invest. Factors such as size of domestic market, security of capital and stable government policies seem to be very important in such decisions. The state, as in the developed world, plays a crucial role in the industrialisation of developing countries, directly and indirectly.

Most developing countries began their industrialisation with import substitution so as to be less dependent on foreign imports. This has involved developing **consumer goods** (which are relatively labour intensive relative to the amount of capital invested) and **heavy capital industry**. Only a few countries have been successful with this latter type because of insufficient capital, energy resources and a limited local market but South Korea provides an example. For many countries over-borrowing has brought massive balance of payment problems and debt. As a consequence they have had to reappraise the role of industrialisation in their economy. Economic viability has had to be put before political expediency.

The location of manufacturing industry in developing countries

The dominant feature is the concentration of most manufacturing industry into a few centres. The largest concentrations of population have attracted most industry, and through the operation of agglomeration economies and cumulative causation this situation becomes self-perpetuating. Apart from these economic factors, other processes have been important. The adoption of import substituting industrialisation meant that industries tended to locate where the main market was to be found and the best infrastructure is usually found in such locations. In more established industrial regions such as South East Brazil, the process of dispersion is beginning to take place for similar reasons as in the developed countries (see industry in urban areas, page 145). Most dispersion tends to only be to the edge of cities where greenfield sites and industrial estates are to be found. Some longer distance decentralisation, to growth poles in peripheral regions, has also taken place where countries have adopted a regional development strategy. The aim is to reduce economic and social disparities which have increased as a by-product of industrialisation (see Figure 6.23).

Q17. With reference to Figure 6.23 explain in your own words the differences between developed and developing countries with respect to the location of industry.

Q18. With reference to the UK show how the motivating factors 1, 2 and 3 in Figure 6.23 have been responsible for the recent changes in the location of manufacturing and service industries.

Industrial development in Kenya and South Korea

As can be seen from Figure 6.24 on page 150, these two countries are very different in many respects. This is mainly due to the different industrialisation experiences that each has had over the last 40 years.

There was very little development of manufacturing industry in Kenya until independence from Britain in 1963. Then it was decided to pursue a policy of import substitution, and industries such as textiles, food processing and the assembly (not manufacture) of motor vehicles developed. This policy was not very successful in later years because of the rapid decline in domestic demand and as a result several foreign companies withdrew their investment during the 1980s. Industrial policy changed partly as the result of an agreement with the International Monetary Fund (IMF) in 1988. The policy is now to encourage those industries which manufacture for export and to promote the very small industries of the informal sector. The setting up of

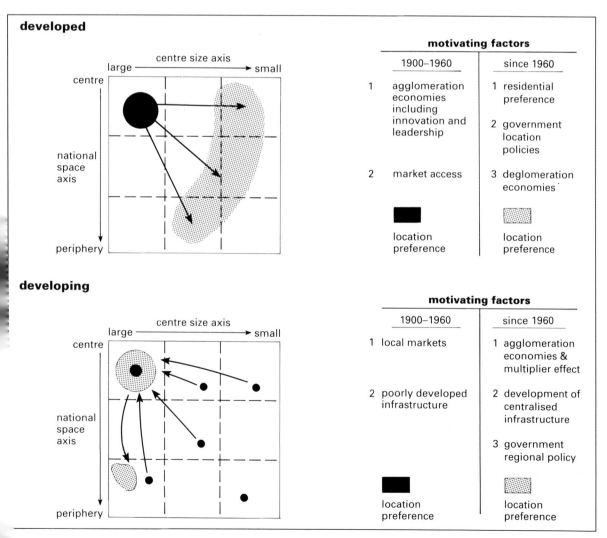

Figure 6.23 Changing industrial location factors in ELDCs and EMDCs

three **Export Processing Zones** confirms the policy of encouraging export-orientated industries.

	Kenya	South Korea
Population in 1991 (in millions)	24.9	43.2
Birth rate (births per thousand)	47	15
Death rate (death per thousand)	10	6
Urban population %	24	72
Adult illiteracy %	31	4
% of GDP contributed by manufacturing industry	20	44
GNP per capita in 1990 (US dollars)	370	5400

Figure 6.24 *Comparison of Kenya and South Korea*

The problems that Kenya faces in trying to industrialise are well illustrated by its sugar processing industry. In the 1970s and 1980s five government refineries and several private refineries were opened in the north west of the country. The region is very well suited to growing sugar cane and the theory was that the refineries could produce a surplus for export. Unfortunately inappropriate subsidies and mismanagement in the government-run refineries led to sugar being left to rot in the fields and the closure of most of the refineries. Today Kenya has to import increasingly large amounts of a product that it should be capable of producing itself. The most likely hope for the country is to encourage the growth of small-scale, sustainable industries which fit in with the environment and use labour from Kenya's ever-growing population.

South Korea is not a large country in size or population yet it has achieved remarkable economic growth since the end of the Korean War in 1953. Like most other developing countries it began its drive for industrialisation with import substitution. This involved the use of protective trade barriers against foreign competition. The early opportunities to produce low-level goods were quickly exhausted. Unlike many other developing countries it decided at an early stage to pursue an export-orientated policy. This was done by overhauling the system of incentives leading to a reduction of protection for domestic industries. The currency was also devalued which helped make Korean exports more competitive. The success of this policy can be seen by the fact that manufacturing contribution to GDP increased from 19.5 per cent in 1973 to 30.4 per cent in 1984. Industries such as textiles and clothing manufacture had a **comparative advantage** because of the country's cheap labour. Other factors which helped the industrialisation process included massive economic aid from the USA and a strong government which ruthlessly crushed opposition and prevented trade unions from taking strike action. A feature of the rapid economic change that the country has experienced is the emergence of a few giant family owned business groups, known to Koreans as *chaebol*, e.g. Samsung and Hyundai. These have grown from almost nothing to become world scale organisations within the space of 20 years.

In the 1970s the emphasis changed to more capital-intensive industries such as electronics, shipbuilding and steel. Since the early 1980s the emphasis has changed again, this time to high-technology industry. This reflects the transition from being a cheap producer of goods to a mature industrial economy.

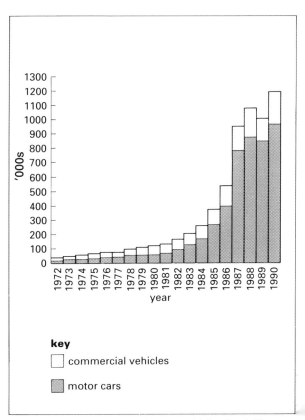

Figure 6.25 *Production of motor vehicles South Korea 1971–90*

The changing economic fortunes of the country are reflected in the growth of its motor vehicle industry (see Figure 6.25). It has grown from 1965 to become one of the top ten producers by 1992. The Hyundai Company at Ulsan has designed its own range of cars which have sold well throughout the world. In a dramatic reversal of the traditional pattern of the world vehicle industry, Hyundai opened an assembly and components factory in Toronto, Canada. By 1988 over 500 000 South Korean cars were being sold in the USA. However, by 1990 sales were falling rapidly because of an increase in labour costs, variable quality of the product and exchange rate fluctuations. Many analysts question whether South Korea can support nine manufacturers of cars and commercial vehicles. However, the government continues to support expansion plans and aims to make South Korea one of the top five producers of cars by the year 2000. Yet it might soon find its Asian market weakening as China builds its own cars. Sales at home could also stagnate because the nation's roads are becoming clogged with traffic.

In the early 1990s South Korea's progress slowed down as it has been affected by world recession, inflation and rapidly rising wage costs, which have damaged the country's international competitiveness. The military regime started to lose its grip on wage rates and as a result wages went up by 130 per cent between 1988 and 1993. The Korean miracle is fading slightly but the best of the Korean industrial giants are now robust enough to tackle world markets head on. The case of South Korea provides considerable inspiration to other developing countries such as Kenya.

> **Q19.** *Why has South Korea been able to develop its manufacturing sector of the economy more easily than Kenya?*

Future prospects

The future industrial prospects of developing countries depend both on domestic policies and on international factors. The perceived threat to developed country industrial powers by a handful of countries such as South Korea, Brazil, Singapore and Taiwan has provoked calls for greater protectionism. The question of international debt is also very important. Foreign debt repayments can have a crippling effect because in order to repay debts countries cannot import essential raw materials and energy sources. This often leads to a reduction in industrial production.

Recent technological developments could also harm the future of developing countries. Increased use of robots and computer-controlled manufacturing means that cheap labour, one of the major advantages of manufacturing in less developed countries, could quickly evaporate and the economic centre of gravity could once again move towards the developed world.

If developing countries are to make continued progress in industrialisation there will need to be an international effort to assist them: there is only one world system and what happens in one part of it ultimately affects other parts. A reduction in the rate of population growth will mean that instead of capital being used to meet the social requirements of the population, it could be used to invest in industry. Kenya, with a population growth rate of 3.8 per cent clearly illustrates this dilemma.

SERVICE INDUSTRIES

One of the most significant developments in the global economy has been the rapid growth of the service industries. They account for the largest share of GDP in all but the poorest countries and are increasingly the major source of employment in all the developed market economies and in many developing countries as well (see Figure 6.1). The service sector has also been attracting an increasing share of world foreign direct investment, and without doubt is being increasingly internationalised.

The service sector is extremely diverse. As a result of this it has become common to separate out tertiary services from quaternary services. The latter group can be regarded as high order services involving research and finance which have the transmission and reception of information as core activities.

An alternative approach suggests that services can be categorised into consumer services and producer services. Consumer services provide output which goes directly to consumers, e.g. retailing, whereas producer services provide output which is consumed or used exclusively by other industries, e.g. business consultants, wholesaling or market research. Inevitably, however, there are services which do not fit easily into either category. Some services are just as likely to provide for the requirements of other industries as for individual consumers, e.g. banks and legal services. In theory consumer services should look for locations accessible to the sources of final demand whereas producer services should be more interested in access to other firms.

The growth of the service sector

Since the 1960s employment in service industries has increased in all the developed market economies. There has been an increase in both absolute and relative terms. Some people regard this as evidence

ECONOMIC ACTIVITY **151**

of shift towards a **post-industrial society** (see Figure 6.2). Among the less developed countries there are differences between the very low income countries, e.g. Nepal, and the more affluent countries, e.g. Brazil. The increase in the service sector in the NICs is related to the growth of the manufacturing sector, which produces demand for services such as transport and administration. Increases in urbanisation also indirectly affect the demand for services of all sorts.

Service activities have always been important to the UK economy. At the turn of the century over 40 per cent of the workforce was employed in services and by 1900 this had increased to 67 per cent. Two million new service jobs were created in the UK between 1980 and 1989 at a time when large numbers of jobs in manufacturing industry were lost. This was the era of de-industrialisation, re-industrialisation and tertiarisation in the UK. Yet in spite of the growing dominance of service employment we live neither in a post-industrial age nor a purely service economy. With a nine per cent increase in total service employment between 1981–87, the rate of increase varied considerably from one sub-sector to another with some service industries utilising part-time employment and the use of female labour. Alongside these changes have been processes operating which have changed the spatial nature of services in the UK. These spatial changes will be discussed later in this section.

Principles of service industry location

Service industries have been created in response to demand. For each service activity there is a market which is expressed by means of supply, demand and price. Markets for most services vary in size and intensity over space and time. A location must be chosen which allows the client or customer to reach the source of the service. Access involves the consumer in time and money costs which will influence the decision about where to go to obtain a service.

Service industries can be considered in terms of a system in the same way as manufacturing industries. The main inputs of service industries are capital and labour. Information could also be regarded as an input and an output but because it is intangible, it is difficult to measure its contribution to the location behaviour of service firms.

Central place theory also helps to explain the location of consumer services. Central places are settlements that provide services for people living in them and in the surrounding area. Christaller used the concepts of service thresholds and service range to help explain why some settlements contained more services than others. Any service needs a minimum number of customers in order to generate sufficient revenue to stay in business. This is the threshold population of a particular service. Linked with this idea is that of the range of a service. This is the maximum distance customers are prepared to travel to obtain a particular service. For a given service to exist in a particular locality its threshold population must exist within an area which has as its radius the range of that service. The result of the interaction between threshold and range is a hierarchical system of service centres (i.e. settlements) where there is a small number of large settlements with the full range of services available and a larger number of smaller settlements which have fewer services on offer.

Other factors have also affected the location of consumer service activities. Until the 1960s the location of most consumer services was based on maximising consumer access. However rationalisation programmes based on the concept of economies of scale have steadily affected the location of such facilities. In the area of hospital services, for example, a large number of small cottage hospitals have been closed down to be replaced by a small number of large regional hospitals. Such changes invariably produce a reduction in accessibility for some places and groups of people.

Population distribution is a less reliable indicator of the location of producer services as these rely on links (e.g. information exchanges) with other industries. The value and importance of the links to other producer services will determine the significance which they attach to ease of contact and thus choice of location. 'Information-rich' producer services will tend to group or cluster together. Other factors that influence service location include type of labour required, the influence of planning controls and the scale of operation.

The changing geography of consumer services – the case of retailing in the UK

Employment in retailing has increased only slightly in recent years despite substantial increases in output, e.g. sales volume output increased by 38 per cent between 1980 and 1988. Like other branches of the service sector, retailing has been affected by structural adjustments which have effects on the

	1950	1961	1971	1980	1986
Number of single shop businesses ('000)	375	370	330	215	217
Number of businesses with:					
50 shops and over	362	430	330	300	220
10–49 shops	1407	1470	940	960	636
Number of establishments ('000) in businesses with:					
50 shops and over	53	66	60	55	48
10–49 shops	28	30	19	19	12
Percentage of sales in businesses with:					
50 shops and over	24	31	36	45	52
10–49 shops	12	9	8	8	8

Figure 6.26 *Retailers in UK 1950–86*

spatial pattern of retailing in the UK. As can be seen from Figure 6.26 there has been a decrease in the number of businesses and an increase in the number of large businesses. This process has not been evenly felt across all types of retailing or across all the country. Non-specialist retailing, e.g. general food retailing, has been affected most and specialisation has provided the small firm with some protection. The decreases have been greatest in inner suburban and fringe areas of central shopping districts and in small towns and rural areas. In some cases retail provision has fallen to levels at which there is now concern over access to basic services by those consumers who are unable, unwilling or cannot afford to travel some distance to purchase basic products.

Store types have become more customer focused. This is one of the reasons for a polarisation of shop size in the UK. Small shops have been used to target very specific consumer groups (e.g. convenience stores for time-pressured consumers) whilst large stores have been used to provide a wide range of items in a broad product area to all types of consumer. There has been a rapid growth in various types of **superstore** especially for the retailing of food, DIY and furniture. The development of large retailing units such as superstores and **hypermarkets** has contributed to the modification of typical intra-urban shopping hierarchies such as that of the City of Leicester (see Figure 6.27).

The new forms of retailing have generated demand for new types of retail property. The small

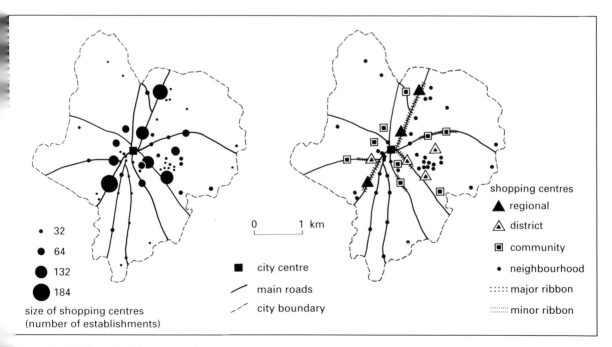

Figure 6.27 *Shopping hierarchy in Leicester*

number of major retailers are increasingly attracted to prime sites in city centres thus pushing rent levels up in these locations. This has also contributed to the redevelopment of many city centres. Pressure on prime sites has also encouraged the development of new out-of-town **regional shopping centres**. An additional factor has been the continued suburbanisation of population in all developed world cities. By 1993 four regional shopping centres existed in the UK: the Metro Centre in Gateshead; Thurrock Lakeside off the M25; Merry Hill near Birmingham; and Meadowhall near Sheffield.

In the early 1980s groups of superstores or retail warehouses began to congregate on suburban **retail parks** which share infrastructure such as car parking. By the early 1990s almost 250 such schemes had been developed. Such developments threaten already struggling poorer quality shopping areas on the fringes of high streets. In 1993 the UK Government introduced new policies on the future of such developments. This reflects the fear of ministers that many high streets are becoming run down and people are being forced to travel further and further to reach new facilities. New out-of-town schemes will only be allowed as long as they do not undermine existing facilities. Regional shopping centres of more than 152 500 m^2 will only be allowed in areas where there is likely to be significant growth in population or retail spending.

> **Q20.** Look at Figure 6.27 on page 153. Why is such a hierarchy an efficient way of providing retail services? Why do some shopping centres in Leicester have a ribbon or linear form?
>
> **Q21.** Look in an atlas at maps of population distribution and the position of motorways in the UK. Using this information make a prediction as to where the next two regional shopping centres in Great Britain might be built.

The third locational trend has been the growth of retailing in a variety of unconventional locations to serve specific consumer needs such as at transport termini and at petrol filling stations. There has also been evidence of UK retailers such as Marks & Spencer being involved in establishing branches in mainland Europe. Much more international activity is likely in the future.

The changing geography of office-based services

Until 30 years ago there were few signs that offices would be located anywhere other than in city centres. There were a number of reasons for this. A central location enabled the working population to be brought together easily using public transport. Face-to-face contact was also facilitated by a central location close to other offices and central locations were regarded as being rather prestigious. This is the operation of agglomeration economies in the service sector. Increasingly, however, the need for a central location has been challenged by a variety of telecommunication innovations, starting with the telephone, which allows contact to be made without the need for face-to-face contact. For most office workers the majority of their contacts outside their firm are programmed, in that they deal with routine matters and therefore do not require face-to-face contact. However, face-to-face contact is still highly valued in some commercial activities. From the 1950s onwards an increasing number of companies have chosen to relocate at least some of their routine office functions away from the congested and expensive central areas of cities such as London. Agglomeration diseconomies have been in the ascendancy in recent decades.

Decentralisation of office activities has occurred at both the regional and local (urban) scales. In the case of the City of London many insurance companies moved out in the 1980s, e.g. Commercial Union moved to Basildon in Essex in 1985. Many firms moved to other sites in London if they could get permission from the authorities. Croydon was a popular destination. The biggest firms tended to locate outside London but mostly in the south east or fringes. Towns such as Bracknell, Peterborough and Cambridge attracted firms because of their lower rents and more pleasant living conditions. The City of Cambridge has itself experienced problems as a result of excessive growth of offices in the 1980s and illustrates the sorts of decentralisation that has been experienced by all large towns and cities (see Figure 6.28). At the urban level a variety of new office concentrations have emerged. These include expansion into inner suburbs via conversion of residential properties, the development of planned office centres in inner suburban locations, e.g. La Defense in Paris, and **Business parks** alongside motorway junctions or ring-roads. Nonetheless there are still many offices which value the contact environment of city centres. They will remain in such a location along with others who remain because of inertia and tradition.

> **Q22.** Look at figure 6.28. How do you explain the fact that Waterbeach is expanding in size yet is losing retailing services?
>
> **Q23.** Which groups of people will be most affected by these changes and why?

The internationalisation of service activities – the case of financial services

The last 20 years has witnessed a revolution in the way financial services are organised. The global financial system has become very complex and volatile. Technological developments in communication and information have been very important in the internationalisation of financial services. The 'raw material' and 'product' of these industries, i.e. information, needs to be transmitted as fast and far as possible if there is to be a significant added value. The development of satellite communications systems has been especially important to the development of international financial markets. It is now possible for financial service firms to engage in global 24-hour trading. The 24-hour trading is currently limited because of national government restrictions and regulations, though these are being removed. In 1986 the so-called Big Bang removed barriers in the UK which had existed between banks and securities houses and allowed the entry of foreign firms into the Stock Exchange.

Technological developments in communications systems release financial services companies from the spatial constraints on the location of their activities. Financial services firms are particularly footloose. However at both global and national scales these activities are often more highly concentrated than virtually any other kind of economic activity apart from those based on highly localised raw materials. New York and London dominate world financial activity (see Figure 6.29). There are 25 major cities which control almost all the world's financial transactions. These cities are control points of the global economic system. They tend also to be the location of the headquarters of the world's transnational corporations.

The potential for other cities to develop as significant centres of finance is likely to be very limited – so called front-office functions must be close to the customer. However, the essence of all office-based services is the transformation of massive volumes of information. Much of that activity is routine data processing performed by clerical workers. Such back-office activity can be separated from the front-office functions and can be performed in different locations. The introduction of microcomputers and networked computer terminals has encouraged the dispersal of such back-office functions in much the same way that branch plants are separated from headquarter functions in manufacturing industry. A geographical division of labour exists in the service sector as well as the manufacturing sector.

Villagers to fight offices 'invasion'

RESIDENTS of Waterbeach fear the heart of their village is being swamped by commercial interests as offices replace shops.

"Already Tony's Meat Fayre has permission for office development. Now there is an application to turn the old Co-op in Greenside into either offices or a creche.

"Many people in the village have said how worried they are and we fear that once a change of use to offices occurs, our shops will be gone for ever."

Residents think another site could also become offices. This had been set aside by Cambridge and District Co-op for a new supermarket before Tesco opened a store at Milton, making such a venture unviable.

Mrs Dorothy Cattell, chairman of the village's over-60s club, which has about 73 members, said: "The situation is becoming impossible.

"We have asked the parish council to act by passing on our concern over the way shops are disappearing and offices are taking over.

"Our village store and newsagent's shop carry a reasonable range of goods, but our old people cannot get all they need – nor can they afford the prices that an independent trader has to charge."

When the Waterbeach Co-op closed last year the parish council asked Tesco for a free shoppers' bus to the Milton store. But this was refused because there is a public service route.

Mrs Cattell said: "This doesn't help our old people. They have to walk from the bus to Tesco's, then back to the bus again with heavy shopping and some of them just can't do it."

Unlike many villages which lose their shops, Waterbeach is expanding. New homes are being built and the village is also a permanent home for the 39 Engineer Regiment.

Figure 6.28 *The Cambridge Weekly News*, 1 February 1990

Q24. *Explain the difference between front-office and back-office functions.*

Figure 6.29 *London's Telecom Tower*

DECISION MAKING EXERCISE

Four-wheel drive in the USA

Mercedes-Benz, the German car manufacturers, are looking to open a new car assembly plant somewhere in the USA to manufacture a landrover-type vehicle. It will employ 1500 workers and will produce 60 000 vehicles each year, two-thirds of which will be exported to other countries.

1 Using all the information provided and Figures 6.30 and 31, produce a short list of five states, in rank order, where the company should consider locating their new plant. Produce a written justification of your choices, stating which factors you considered to be of greatest importance in arriving at your decision.
2 Apart from the information provided, what other sorts of information would the company need to consider before making a decision?
3 Why do you think that Mercedes decided to locate its new plant somewhere in the USA in preference to other countries?

Find out what actually happened (the Financial Times Supplement on Locating in North America, 28 October 1993 will provide you with the answer!). How does this compare with your short list?

State	Total population (1983) (m)	% population with at least four years of college education	Hourly wage rates in manufacturing industry (US dollars)	Well-being Index
Alabama	3.95	12.2	10.18	−183
Alaska	0.47	21.1	10.75	—
Arizona	2.96	17.4	12.08	−32
Arkansas	2.32	10.8	10.18	−179
California	25.17	19.6	12.08	65
Colorado	3.13	23.0	11.29	86
Connecticut	3.14	20.7	11.98	156
Delaware	0.60	17.5	11.98	51
District of Columbia	0.62	27.5	11.98	—
Florida	10.68	14.9	9.85	−80
Georgia	5.73	14.6	9.85	−160
Hawaii	1.02	20.3	11.61	—
Idaho	0.98	15.8	12.82	70
Illinois	11.49	16.2	13.10	40
Indiana	5.48	12.5	13.10	−3
Iowa	2.90	13.9	11.44	117
Kansas	2.42	17.0	11.44	70
Kentucky	3.71	11.1	10.18	−95
Louisiana	4.43	13.9	10.18	−141
Maine	1.10	14.4	11.98	−2
Massachusetts	5.70	20.0	11.98	124
Maryland	4.30	20.4	11.98	6
Michigan	9.07	14.3	13.10	48
Minnesota	4.14	17.4	11.44	108
Mississippi	2.58	12.3	10.18	−264
Missouri	4.97	13.9	11.44	−21
Montana	0.81	17.5	11.29	27
Nebraska	1.59	15.5	11.44	40
Nevada	0.89	14.4	12.08	−15
New Hampshire	0.96	18.2	11.98	78
New Jersey	7.50	18.3	11.98	91
New Mexico	1.39	17.6	10.94	−47
New York	17.60	17.9	11.98	90
North Carolina	6.08	13.2	9.85	−153
North Dakota	0.68	14.8	11.44	45
Ohio	10.75	13.7	13.10	36
Oklahoma	3.29	15.1	10.94	−34
Oregon	2.66	17.9	12.82	88
Pennsylvania	11.89	13.6	11.98	43
Rhode Island	0.95	15.4	11.98	59
South Carolina	3.26	13.4	9.85	−197
South Dakota	0.70	14.0	11.44	7
Tennessee	4.68	12.6	10.18	−125
Texas	15.72	16.9	10.94	−88
Utah	1.61	19.9	11.29	121
Vermont	0.52	19.0	11.98	34
Virginia	5.55	19.1	9.85	−74
Washington	4.30	19.0	12.82	90
West Virginia	1.96	10.4	9.85	−66
Wisconsin	4.75	14.8	13.10	104
Wyoming	0.51	17.2	11.29	67
U.S. Average			11.45	

Note: This index was calculated by Professor D M Smith and used information on income levels, employment, housing, health, education and crime. A score of 0 represents the national average.

Figure 6.30 *United States characteristics*

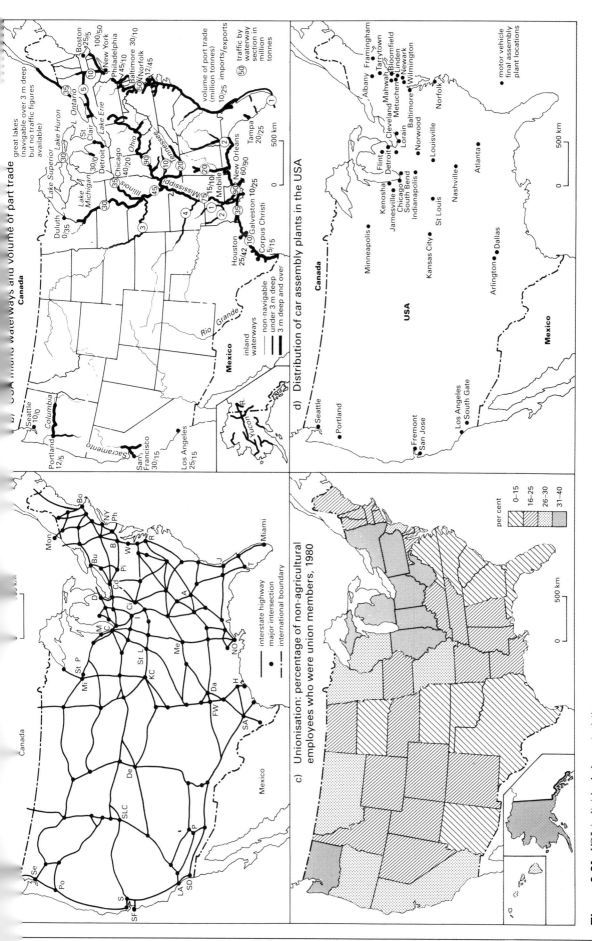

Figure 6.31 *USA individual characteristics*

PROJECT SUGGESTIONS

Manufacturing industry

1. Study the distribution of different industries within a large town or city. Use commercial directories such as Yellow Pages together with postcode information and a large street map. Make a comparison of old established industries and more modern industries to see if they display similar or different locational patterns. Construct a matrix to show the number of each type of industry in each postal district or ward. Proportional symbol or dot maps could then be constructed to compare this information. Use the Chi-Squared test to compare the two distributions. (You must check your own data against the requirements of the test before you use it.) Look for explanations of the patterns that you discover and suggest reasons for your results.

2. Use old directories, such as Kelley's Directories, to examine the changing distribution of an industry in a town or city. This could be related to the growth of the settlement by referring to old maps and by using the results of past censuses.

3. Study different sorts of industrial estates to see if they attract different types of industries. For example, an old inner city industrial estate could be compared with a modern business park located on a greenfield site. Carry out a questionnaire survey of a sample of firms on each site to find out characteristics of the firms, such as type of activity, number of workers etc. Measure the impact of the estate on the local area by undertaking traffic counts, bi-polar surveys and by looking at the catchment area for workers.

4. Some manufacturing industries give rise to negative externalities (negative side-effects) which can have an affect on people who live close by, e.g. the production of noise. Measure the extent of these negative externalities using simple objective methods. Do they display a distinct distance-decay pattern? How is the extent affected by meteorological conditions, e.g. wind speed and direction? Do the companies do anything to minimise the effects, e.g. provide free double glazing for nearby residents? Compare the actual externalities with the perceived externalities as revealed by a questionnaire of residents. Care would need to be taken with the wording of questions and with the sampling framework used.

Office activities

5. Use directories to study the distribution of office-based activities within an area.

6. Devise a questionnaire to measure the strength of inter-group and intra-group linkage of various types of office-based activities. Compare the results of this to the actual geographical pattern of the various types of office activities. Activities which have a high degree of intra-group linkage should be located close to other firms of the same type. This could be measured using techniques such as the nearest neighbour index or index of dispersion.

Retailing

7. Measure the impact of a newly opened superstore. This is not as easy as it sounds because in order to judge the effects one would need to know what things were like prior to the opening of the new development. Look at the impact from a variety of viewpoints such as the impact on existing shops and shopping centres, the effects on traffic flows etc. The impact on shopping behaviour could be investigated by conducting a questionnaire survey of shoppers.

8. Identify a shopping hierarchy in a town or part of a large city. Visit each centre and count the number of each type of shop present. Use this information to calculate the centrality index of each centre (an index which takes into account the frequency and type of shops present in a centre). Use these to classify centres into levels of the hierarchy. Further characteristics of the centres could be investigated, e.g. size and shape of their catchment areas, by using a questionnaire survey.

GLOSSARY

Agglomeration economies these are savings which arise when industries cluster together to facilitate linkage. They are also known as external economies of scale. Agglomeration diseconomies can also arise, e.g. land prices may become very high due to excessive demand.

Business parks developments of low density office accommodation usually set in landscaped surroundings.

Comparative advantage a principle used in economics, that a region or country will produce those goods for which it has the best combination of the factors of production. This leads to specialisation and the development of industrial regions.

Consumer goods goods in a finished form that can be used by the domestic purchaser immediately, e.g. electrical appliances.

Counterurbanisation the movement of population in developed countries away from large conurbations and cities to small towns and remote rural areas.

Cumulative causation the self-generating process by which a core region develops. It was first proposed by Gunnar Myrdal, a Swedish economist.

De-industrialisation the decline of the relative importance of manufacturing in the economies of developed countries such as the UK.

Enterprise Zone small areas of industrial decline and inner city deprivation in the UK where financial incentives are available to encourage job creation. Exemption from many of the normal rules of planning legislation is also given.

Export Processing Zones this is an extension of the industrial estate. They are often enclosed zones within which firms are given a wide range of benefits and incentives. They are used by developing countries to try to attract foreign investment.

Footloose industries industries which have no particular market or source of raw materials and are free to locate almost anywhere.

Greenfield site an industrial or commercial site usually located on the edge of a built-up area which was not previously used for an urban land use.

Gross Domestic Product the value of a country's total production of goods and services usually in a given year. It is usually expressed on a per capita basis so that countries of different sizes can be compared.

Heavy capital industries usually capital-intensive industries with a large ratio of capital investment to workers, e.g. steel making.

Hypermarket large units used for retailing of food. They have a sales area of at least 4647 m^2 and are usually free-standing (as opposed to being part of an existing centre).

Informal sector small-scale manufacturing and retailing activities found in developing countries. They are not officially recognised by governments in the sense that they are unlicensed and unregulated.

Infrastructure the basic services which industry and commerce needs to be able to function effectively, e.g. electricity, water supply etc.

Industrial estate purpose-built areas created to house a variety of industries. They can be publicly or privately financed and built.

Isotropic surface a surface on which every point is identical in all respects, i.e. a totally uniform plain.

Linkage the ties between allied industries or services.

Multiplier effect an initial investment which sets off a chain reaction of further growth (positive multiplier) or decline (negative multiplier).

National Garden Festival a scheme first used in Europe for revitalising inner city areas. It involves large scale landscaping and reclamation in advance of a festival. After this has finished most of the land is used for housing and commercial use. These have been held in the UK since the early 1980s.

Newly Industrialising Country (NIC) a country which has experienced rapid and successful industrialisation in the last 30 years.

Optimum location the location which maximises profits through cost minimisation (Weber) or revenue maximisation (Losch).

Post-industrial society a situation which exists when most people are employed in service industries.

Regional shopping centre covered shopping precincts or malls containing as many as 200 shops but on greenfield sites. A handful have been created in the UK since the 1980s.

Re-industrialisation the development of new industries and businesses in developed countries. They are often based on high-technology products.

Retail parks groups of retail warehouses with additional parking facilities built on the edge of towns, often on greenfield sites.

Science parks a collection of high-technology industries which have links with a major university. The first science park in the UK was opened in Cambridge in 1973.

Superstore food retailing units which range in size from 2323–4647 m^2. They are usually integrated into existing shopping centres, e.g. ASDA.

Ubiquitous materials materials which are available everywhere and therefore have no influence on location.

Urban Development Corporations government appointed bodies (Quangos) in the UK responsible for the revitalisation of inner city areas, e.g. London Docklands.

REFERENCES

The following books contain more detail on some of the topics covered in this section.

P Dicken (1986), *Global Shift: Industrial Change in a Turbulent World* (Second Edition), Harper and Row

R Chandra (1992), *Industrialization and Development in The Third World*, Routledge

M Raw (1993), *Manufacturing Industry: The Impact of Change*, Collins Educational

G Humphrys (1988), *Changing Places*, Geography, Vol 73, Part 4

M Bateman (1988), *Shops and Offices: Locational Changes in Britain and the EEC*, John Murray

SECTION 7

Food for thought

by John Lifford

KEY IDEAS

- **To what extent does food production depend on physical factors in the environment?**
- **How do economic and social processes affect agriculture and agricultural change?**
- **How far does the level of economic development influence food consumption patterns?**
- **In what ways do the farming systems of the poorer and richer nations differ, and how are they linked?**
- **How can farming systems be managed in a sustainable way?**
- **What links are there between farming and the countryside?**

A WORLD VIEW

Producing enough food to eat

Generally speaking the less economically developed a country is, the greater the proportion of people whose main task it is to produce food for that country. This is because of the lower economic efficiency of farm systems in Economically Less Developed Countries (ELDCs). The reasons for this are complex and involve the following.

- LOW INPUTS of fertilisers, pesticides, herbicides. This leads to ...
- HIGH CROP LOSS to pests (locusts, monkeys, etc.) and disease (e.g. witchweed);
- LAND FRAGMENTATION due to inheritance customs, leading to small plots of land;
- LOW LEVEL of TECHNOLOGY due to lack of capital;
- LABOUR SHORTAGES due to the migration of the young to towns, leading to ...
- LACK of FARM INNOVATION, the older farmers often relying on the city income of relatives to support financially marginal farming;
- TENANT FARMING can lead to the expropriation of crops by landlords;
- DECLINING FALLOWS and thus DECLINING SOIL FERTILITY, due to population pressure on land or lack of recycling of organic matter (e.g. dung made into fuel cakes);
- POOR MARKETING FACILITIES due to remoteness and also political pressure to keep food prices low for the towns;
- LABOUR INTENSIVE in many areas.

Food for all?

The previous section implies that food production is poorly distributed in the world especially as low productivity groups often coincide with a rapidly growing number of people. Figure 7.1 shows this inequality, especially in terms of the balance between food and population. Note that it does not show absolute production levels but relative ones; that is, production compared with a previous date.

> **Q1.** Describe the pattern of change shown in Figure 7.1.
> Explain the pattern of change using atlas maps of population change and farming types to help you find reasons.
>
> **Q2.** Use the axes labels in Figure 7.2 to describe the farming groups for each area. Suggest reasons for the placing of each area. You should find that atlas maps of farming types, climate, and population density will help.

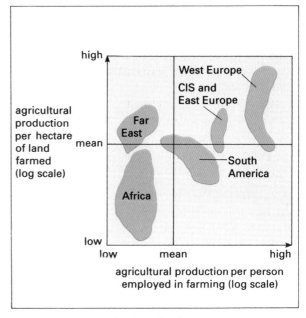

Figure 7.2 World food production groups

Global food consumption patterns

Only about 12 per cent of food produced in the world is traded because the vast majority is consumed within the country of origin. Therefore patterns of consumption should closely reflect patterns of production. The global mean is 2670 average calories per person per day and the means for ELDCs and Economically More Developed Countries (EMDCs) are 2400 and 3400 respectively. Thus there is considerably inequality. Although plant foods dominate agriculture worldwide, animals are a far more important source of protein in the EMDCs (78 per cent of total protein intake) compared with ELDCs (42 per cent). The reason seems to be that it is more expensive to produce animal foods than plant foods because:

Figure 7.1 Food production change (per person 1976–83)

- plant foods are eaten directly by people so there is no great loss of energy between the **trophic levels** of herbivore and carnivore that occurs in a meat eating system (a reason for being vegetarian!);
- plant foods take up six to seven times less land to produce the equivalent number of calories that animal foods do;
- plant foods produce more protein per hectare and more calories and protein per hour of labour.

Not surprisingly therefore, as people's income increases so their diet becomes more diverse and they eat more costly foods such as vegetables, fruit, and livestock.

Furthermore as consumers' disposable income increases, the proportion spent on food declines, although the absolute amount increases.

> **Q3.** Write an explanation of this last sentence, being careful to distinguish between a relative change (percentage change) and an absolute one.

Population density affects the distribution of consumption of types of livestock: pigs are ideally suited to high density areas such as South East Asia because they can scavenge on waste food, and areas such as western Europe because they can be reared in intensive systems and are highly efficient converters of cereal feed to meat protein. Indeed, some 60 per cent of cereals in EMDCs are used in animal feed. In low population density areas cattle (and to some extent sheep) tend to dominate livestock type.

The production and consumption balance

A snapshot indicator of the balance between production and consumption is the measure of self-sufficiency. Figure 7.3 shows the volume of production of cereals as a percentage of consumption. A value of 100 would indicate exact self-sufficiency, values over 100 indicate surplus, and those under 100 show deficits.

> **Q4.** Present the information in Figure 7.3 graphically to show levels of surplus and deficit clearly. Use a mirror graph in which the x axis (=100) is near the middle of the y axis. Write some factors that might explain the levels.

Another approach to the production and consumption balance is through the Malthusian debate on the link between resources (production) and population growth (consumption). It started in 1798 when Thomas Malthus proposed the model shown in Figure 7.4a. Many people see this as

Volume of production of cereals as a % of consumption	
Sub-Sahara Africa	79
North Africa + Near East	74
Far East	96
Asian planned economies	96
Latin America	92
IS	88
North America	200
West Europe	110
Australasia	280

Figure 7.3 Cereal self-sufficiency 1988

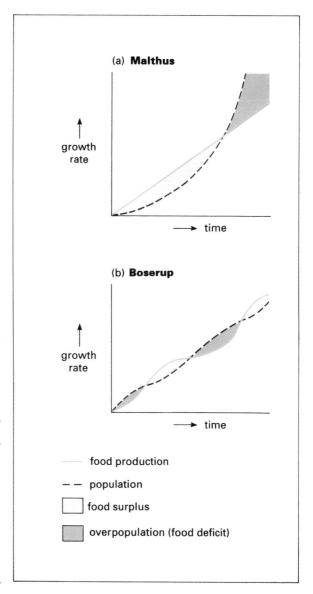

Figure 7.4a and b Population models 1 and 2

applicable to ELDCs with their rapidly expanding populations. The alternative model (called Anti-Malthusian) came from Ester Boserup in 1965 (Figure 7.4b). She argued that high population growth stimulates agricultural changes which, in turn, lead to increased production. The Green Revolution in South East Asia, a package of scientific advances in mechanical and biotechnology applied to farming, might be used to support this theory.

The main conclusion from the debate is that whether a country experiences Malthus or Boserup evolution, depends on one main determining factor; poverty. Where poverty exists there is little money for a farmer or a government to invest in improving food output. Reducing population growth would undoubtedly help ELDCs to balance food supply and food demand better, but by itself population control will not solve malnutrition and famine.

Q5. *Consider Figure 7.5 and decide which facts reflect a Malthus or Boserup view.*

Between 1961–09 world cereal production doubled. World population grew by two-thirds in this time.	ELDCs' cereal production rose from 189 to 249 kg/person, 1985–90.	In ELDCs the average price of maize fell by $19 from 1985–88.	Crop land per person declined from 0.30 to 0.21 hectares in the 1980s.
In Asia and South America the majority of countries showed a per person increase in food production between 1978–87.	Food production per person in Africa declined 1978–89 in 35 out of 41 countries, and in 14 out of 16 in Central America.	If crop yield improvements of 1960–88 can be continued, African cereal yields will match the West in 130 years time.	

Figure 7.5 *Selected world food trends*

FOOD ISSUES IN ECONOMICALLY LESS DEVELOPED COUNTRIES

The famine issue

Famine is a severe shortage of food leading directly to a large number of deaths. It is now almost exclusively a sub-Saharan problem. The causes of famine are complex. Some have argued that physical factors are crucial – especially drought, flooding, pests, diseases, and soil erosion. Lack of rain is certainly a problem, but its unreliability is the key. In the Sahel of Africa, for example, there is considerable rainfall variation annually, seasonally, daily, and spatially. This is exacerbated by the fact that the lower the rainfall total the greater the rainfall variability. This makes farming at best a high-risk enterprise, and at worst a disaster. Why, then, do farmers attempt agriculture here? The answer is partly to do with the need to find increased grazing and crop land but may also be influenced by the cyclic nature of rainfall patterns: farmers do relatively well in the parts of the cycle with above average rain and this encourages them to overfarm in much drier weather.

Q6. *Use Figure 7.6 to make a spatial correlation between the two variables of annual rain and rain variability. One way to do this is to draw a grid on tracing paper to put over one of the maps. Number the rows and columns and, using a random number table, generate a series of points (n = 30 at least) to sample the total and variation in rain. Put this data into a correlation or Chi-Squared test.*

Some droughts occur in areas such as Australia and south west USA where there is no famine. It is, therefore, the combination of drought and poverty which is most likely to plunge a country into famine. Similarly erosion is, on closer inspection, only a cause of famine through misuse (albeit unintended) of the soil resource. Sudden downpours of convectional rain dislodge soil particles into the surface voids thus lowering infiltration capacities and encouraging rill action and gulleying via overland

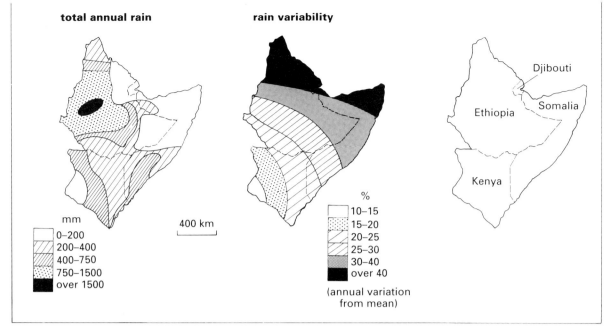

Figure 7.6 Environmental indicators in east Africa

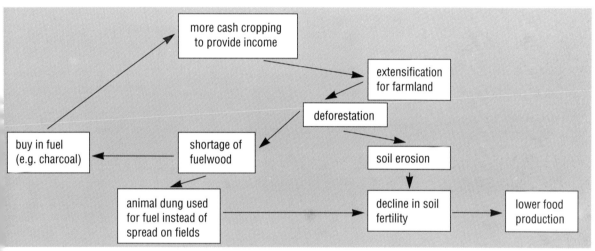

Figure 7.7 Environmental interactions in Ethiopia

low. But the important triggers here are deforestation, overcultivation, and overgrazing which leave the surface unprotected. In the central highlands of Ethiopia the forest cover has been reduced from 40 per cent to around five per cent in less than 100 years, and on arable land six millimetres of humus is lost each year. Figure 7.7 illustrates the impacts of this.

Case study: Ethiopia

Geographers increasingly study the economic and social structures of countries with food shortages and their trading power relationships with EMDCs. Ethiopia serves as a good case study to highlight the role of politics in the development of famine.

Firstly, tenant farmers have traditionally been in a feudal relationship with their landowners. This can mean that up to three-quarters of their crop has to be given to the farm owner. Secondly, up to 1974 the development of a capitalist economy meant that much of the more fertile land was used for mechanised farms which yielded greater profit for landlords than renting land to peasants. The latter were forced to graze further up the slopes, triggering soil erosion. For example, the Afars of eastern Tigray were excluded from their grazing land near the banks of the river Awash so that British, Dutch and Israeli companies could grow cotton and sugar. In grazing the higher slopes, much of the forest cover was destroyed. Thirdly, with the introduction of a socialist government in 1974, land reform has given each peasant household the right to ten hectares of land. However, the government's decision to reduce the share of the market price that it passes on to producers from 63 per cent to 27 per cent was a bitter blow for 700 000 farmers growing coffee. In addition, a disproportionate amount of

FOOD FOR THOUGHT **165**

government resources goes to the large state farms: these cover only four per cent of the total cultivated area but attract 80 per cent of the agricultural credit, 82 per cent of imported fertiliser and the bulk of foreign aid projects devoted to farming. The government's priorities seem to be to feed the armed forces and the urban population centres (where political power is held).

Fourthly, about 46 per cent of Ethiopia's budget has, until recently, been spent on military hardware to fight civil wars in Tigray and Eritrea (70 per cent of famine victims are from these regions). Not only does this divert funds away from farming, but crops are destroyed because of the cover they provide, infrastructure is disrupted, fields are mined, farmers flee their land, and food convoys are held up and looted. In a final twist to this tale, after the civil war stopped in 1991, 840 000 demobilised soldiers and over a million refugees from neighbouring Somalia and Sudan needed feeding.

Q7. *Draw an annotated flow diagram to show the interrelationships between the factors that contribute to famine in Ethiopia.*

Managing famine

Famine can be regarded as a quasi-natural hazard in that it has both environmental and human causes. Management can be divided, therefore, into four broad types. Firstly there is famine relief. This is relatively short-term and aims at trying to cope with the immediate human suffering by the offer of food aid. Clearly, whilst this is necessary for humanitarian reasons and also possible because of the world's surplus grain, there are a number of problems involved: supplies may not reach the most needy due to corruption at unloading points; civil war can prevent distribution (as in the Somali crisis of 1993); food may flood the home market and depress prices for indigenous farmers; and political and economic dependence on the donors can be created.

Secondly there is impact reduction. Once a food crisis is perceived, governments can reduce the impacts. Recently three successful ways of doing this have been shown. In Gujurat in 1985–6 the Indian Government moved fodder from surplus areas to potential famine ones and prevented an increase in livestock deaths. Consequently families were able to keep a source of milk and meat as well as a store of wealth. Another approach is to buy in food into government stocks. In 1984 Bangladesh imported over 400 000 tonnes of rice from Thailand during the first few days of the Monsoon floods and this helped

avert a major famine when further floods occurred later on. Finally various Food for Work programmes have been set up in which famine-prone people receive food payments for various community work schemes such as building grain stores, road improvements, and soil conservation measures.

A third strategy is prediction. A feature of each of these successes has been some measure of prediction of a potential food shortage. Accurate meteorological data, harvest predictions, and a government organisation that can react quickly are essential. Fourthly there is prevention. This is the most effective management because it is long-term. In Botswana, for example, prevention has taken two parallel directions: increasing the arable sector of the economy by trying drought-resistant crops, soil moisture conservation, and fertilisers; and secondly diversifying the rural economy with small-scale industry, forestry, and wildlife farming. While the early stage of research handicaps the former, the country is making progress and the main reason for this is political stability. A government that is not fighting a civil war is likely to be economically stable and thus able to concentrate on programmes that raise the incomes of the poor so they are less vulnerable to drought or flood.

INCENSE OIL PRODUCTION IN SOMALILAND, 1993

Civil war in the 1990s in this area of Northern Somali, has meant that coastal traders do not have enough long-term capital to invest in frankincense gum collection. Before the war this money was loaned to gum collectors so they could buy food to live for the six to nine months of the gum collection period (trees need scratching for 4.5 months before gum flows, and it is therefore a high value crop).

The gum collectors' response has been to sell livestock to the traders to purchase food, but this is killing their agricultural base, so they become famine-prone. In 1993 livestock prices were low because of the large numbers of farmers selling to traders.

Figure 7.8 *Incense oil production, Somaliland*

Q8. *Use the information in Figure 7.8 to explain how and why one year's food supply from an aid agency (TEAR fund) helped manage the famine in Somaliland. What type of management was it?*

Food and technology in ELDCs

The high-tech approach

Many ELDCs have been persuaded to try and improve their food production systems by applying modern technology derived from the EMDCs. Where successful the hope was to provide a (peaceful) Green Revolution to stimulate agriculture. Developments in biotechnology are a key part of this, both in **high yielding varieties** (HYVs) of hybrid seeds which are bred to be disease-resistant and quick growing, and in **agrochemicals** such as fertilisers and pesticides for weed and pest control. Improvements in mechanical technology have been designed to support these changes, particularly tractors and ploughs, irrigation equipment such as pumps, wells and tanks, and harvesting and threshing machines. The economic impacts of such a package have often been impressive. Wheat production increased seven per cent each year from 1950 to 1980 in India. Rice has also shown increases, although other crops like maize have not performed as well. Most of the successes have been in South East Asia. However, many farmers have been forced into increased dependence on creditors from whom they borrowed money to buy the necessary inputs. This situation can escalate as shown in Figure 7.9.

They become landless and either work as hired labour for other farmers (often becoming poorer in the process because of the seasonal nature of the work), or join the growing numbers trying to find work in urban areas. For the poorer farmers the problem with the Green Revolution is that it has to be taken as a package and is therefore costly: HYV seeds cannot tolerate even short drought so irrigation must be arranged; seeds are hybrid so they need to be bought each year; and fertilisers and pesticides are needed if a farmer is to get sufficient return on his money to pay back loans.

All this implies that the social impact of the Green Revolution has been very divisive, and that it widens the economic gap between rich and poor, to the disadvantage of the latter. This is often called the 'talents effect' after the parable of the talents in the Bible:

'For to every person who has something, even more will be given, and he will have more than enough; but the person who has nothing, even the little he has will be taken away from him.' Matthew 25, v 29

This effect may well be one that is important at the beginning of any process of change, but it gets less so with time (temporal change) as innovations spread to more remote areas (spatial change). See Figure 7.10 on page 168.

Environmentally two main problems stem from the high-tech approach. Firstly there is the overuse of agrochemicals. Pesticide runoff into drinking water sources (rivers or groundwater) is a concern but it is difficult to evaluate how widespread it is. A study in 1987 in the Philippines amongst smallholders showed a 27 per cent increase in the death rate since adopting pesticides (the statistics are thought to be especially reliable because where **double-cropping** took place two mortality peaks were recorded). Among the reasons were leaking knapsack sprayers, poor protection for farmers and poorly controlled crop spraying. A second issue is the increase in **salinisation** due to irrigation water raising the water table close to the surface so that salts can be concentrated in the root zone by surface evaporation. This can raise the alkalinity of soils to toxic levels when there is a marked dry period. Worldwide it is estimated that more land is lost through salinisation than is gained through irrigation each year. The problem can be eased by sprinkler irrigation rather than flood methods, by creating slow drainage by grading level areas into slopes, or by sinking wells which lower the water table; but all these are expensive.

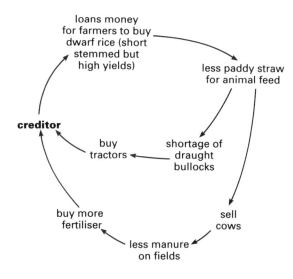

Figure 7.9 *The credit problem*

Such dependence is a particular problem for the smaller farmers as they have only their small plot as security against a loan. Should they default on this, they lose not only their land but also their livelihood.

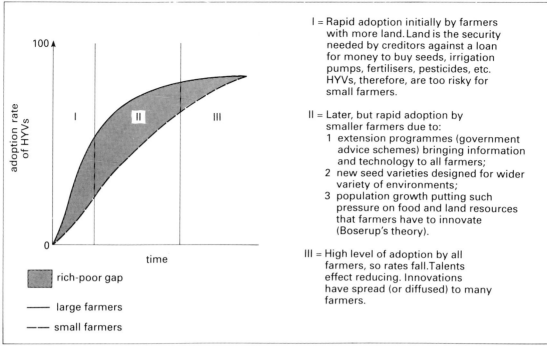

Figure 7.10 *Talents effect and diffusion of innovation*

Energy inputs			Energy outputs		
	Amount	% change due to Green Revolution		Amount	% change due to Green Revolution
Fertiliser			Sugar cane	165 425	+41
(1) Sugar cane	12 324	+138	Paddy rice	43 084	+190
(2) Rice	3 526	new	Subsistence food	1 530	−90
Pesticide	650	new		210 039	+57
Irrigation	146	0	Energy efficient decline	=	−25%
Human work			Energy yield increase	=	+57%
(1) Sugar	3 354	+26	Income to farmer (after inflation)	=	+20%
(2) Rice	1 440	+3	Casual labour employed	=	−66%
Bullock (draught)	294	0			
	21 734	+111	(Figures in energy units – megajoules)		

Figure 7.11 *Effect of Green Revolution on peasant farm in South India*

Q9. *Use Figure 7.11 to summarise the positive, negative, and neutral impacts of the Green Revolution. How successful has it been?*

The low-tech approach

The type of changes associated with the Green Revolution have left Africa (and to an extent South America) rather untouched because of the harsher environmental conditions of poorer soils and drought, and pests, especially locusts. There are now HYVs in crops such as sorghum, millet and cowpeas which are showing success. For example, in the Sudan a variety of sorghum doubled yields in the drought year of 1984 compared with yields in a normal rain year – but development has been slow and progress is perhaps 20 years behind Asia.

There has been a move towards technological innovations more appropriate to local conditions. Where these have given local people responsibility for management they have met with considerable success, albeit often small in scale. For example the one-ox plough has been developed in some of the Sahel countries. It is a metal skid attachment which takes the pressure off the yoke instead of a second ox and helps improve yields because farmers do not have to wait to borrow or hire a plough (a third of

168 FOOD FOR THOUGHT

farmers in Ethiopia do not have two oxen). It has proved a very popular innovation especially for the farmers who lost livestock in the droughts of the 1970s and 1980s.

Land use management

Technology is not the only (nor often the best) way of increasing food production. Indeed through the 1970s and 1980s taking more land into production (or **extensification**) was the main way of improving output. In the 1990s emphasis has been on the way farmers manage their land, and particularly towards **sustainable** practices. Two examples are alley-cropping and intercropping. Alley-cropping is a type of **agroforestry** in which nitrogen-fixing trees are grown in rows (or alleys) alternating with rows of crops. The trees provide nitrogen-rich leaves as a soil mulch for the crops as well as giving a supply of fuelwood and animal fodder. Crop yield increases of up to 80 per cent over non-fertilised soils have been experienced in Nigeria. The great advantages of this are that no extra technology is needed and it is not necessary to fallow land. Overall yields have increased, even though some land is lost from agriculture to tree growing. It is, however, very labour-intensive at a time when many rural areas are experiencing labour shortages. Intercropping is the growing of two or more different crops within the same field for mutual benefit. The substantial benefits are shown in Figure 7.12.

Agricultural change or rural change?

The thrust of reform is now moving away from simply changing agricultural systems towards trying to effect change in the whole of the rural economy and society. This should make it more sensitive to the needs of rural people and therefore more successful. Two examples from Africa demonstrate the approach.

In the tropical rain forest area of central Sierra Leone the traditional farming system of rotational bush fallowing is under threat because of declining fallows (between 1970 and 1985 the average fallow period fell from 20 to nine years), thus exhausting the poor soils. Forest (or bush) is cleared for cultivation in a year-on-year system.

> **Q10.** *Use a copy of the map in Figure 7.13 on page 170 to colour the cultivated land for next year (assuming all the present cropped land needs fallowing). In a separate colour add the extra land needed to cope with a ten per cent increase in food demand. What conclusions do you draw?*

With the possibilities of intensifying farming severely limited by the fragile ecosystem, the government has had some success in persuading farmers to adopt swamp rice cultivation in valleys (Figure 7.14, page 170) rather than upland or rain-fed rice. They have done this through an **integrated agricultural development project** in which wells are dug, new

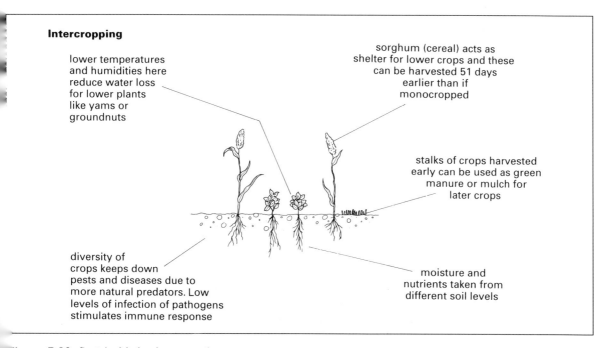

Figure 7.12 *Sustainable land use practices*

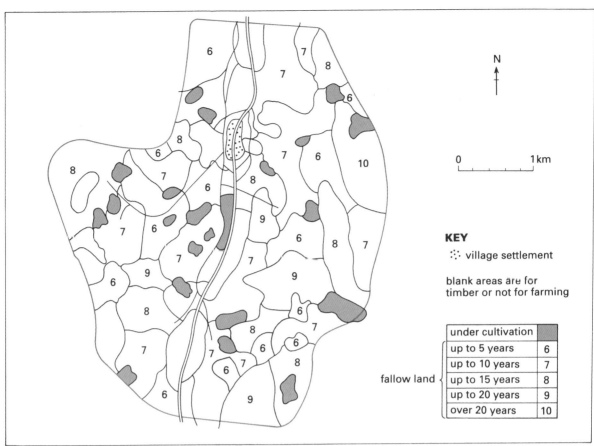

Figure 7.13 *Bush fallow system, Sierra Leone*

Figure 7.14 *Swamp rice farming, Sierra Leone*

access roads built for remoter villages, an advice service, a seed multiplication unit, and favourable credit facilities provided. Two particular problems remain: an increased disease risk from malaria and schistosomiasis whilst working in the swamps; and, ironically, by raising farmers' incomes, the project has stimulated a drift to the towns. This is because more money in rural areas has been used on education, raising young people's expectations in life which farming does not meet.

The second example is from southern Morocco in North Africa and is contained in the decision making exercise on page 187.

Organisational changes: agribusiness

Whilst land reforms and low technology changes have been concentrated in traditional farming areas (especially **marginal** ones), perhaps the most important change in modern farming areas, especially tropical lowlands, has been agribusiness. Here a firm, often a transnational company (**TNC**), controls most or all of the phases of agricultural production from the making of fertiliser through operating plantations, processing crops, to marketing and distributing the products. The agricultural system is run as a large business and is associated with three trends. First, there is a growing concentration on export crops for the highest profit. Sales are thus in EMDCs. This makes them vulnerable to fluctuations in world prices, the money markets being controlled in centres such as London and New York. For example in the last 20 years the price of cocoa has varied from between £800 to £3000 per tonne, and the trend is downwards.

One of the problems of export concentration is that it can take land from domestic food use, as in Brazil where increases in soya production were accompanied by a decline in the production of maize.

Secondly there has been an increase in the size of agricultural units, leading to a concentration of land and capital in the hands of increasingly few people and also leading to more landless peasants.

Thirdly there has been increased domination of agriculture by foreign firms and thus a loss of power over the economy. Some countries such as Guyana and Panama (which both grow sugar cane) have nationalised their plantations to overcome this problem but foreign companies can still dominate through marketing and trading powers.

> **Q11.** *Put these labels into the correct order to form a **positive feedback system:***
> *price rises; price falls; overproduction; concern over shortages; producers encouraged to increase output; producers cut back production.*
> *How do you think the following two measures should help the situation? Having a regulatory body like the International Cocoa Organisation which agrees production quotas for each country; alternatively stockpiling surplus cocoa.*

The case of bananas

Figure 7.15 shows that the bulk of the cost of buying a banana goes to the organising TNC. Only about ten per cent goes to the growers. The growers mainly

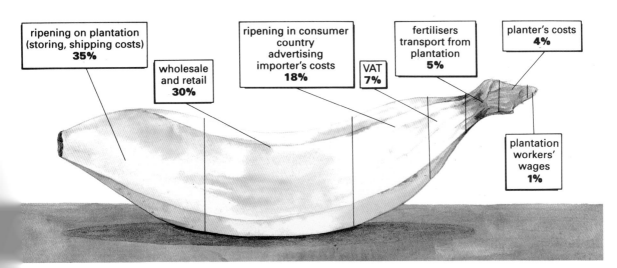

Figure 7.15 *Cost of Bananas, 23 March 1993*

come from the West Indies, Central and South America and are vulnerable to the market in different ways.

Q12. *Plot the data in Figure 7.16 on a scattergraph. Calculate the means of each variable and mark these as lines on the graph. This divides the countries into four groups. Match each group with the following descriptions: a powerful group with diverse economies; stable – unlikely to be affected drastically by changes; an extremely vulnerable group; powerful in the banana market but vulnerable due to heavy reliance on the product. Justify your answers.*

	Production of bananas (thousands of tonnes) 1991	Percentage share of market by volume
Brazil	5630	2.4
Ecuador	3520	25.0
Costa Rica	1550	25.0
Columbia	1520	1.0
Panama	1267	23.0
Honduras	1100	37.0
Guatemala	470	1.0
Jamaica	128	5.0
Nicaragua	104	7.0
St Vincent	33	27.0
St Lucia	92	41.0
Dominica	67	47.0

Figure 7.16 *Vulnerability to banana market*

The banana market is dominated by three American TNCs which control 70 per cent of the world trade (the dollar bananas): Del Monte; Chiquita; and Dole. Some countries deal exclusively with one of these, as for example Honduras with Chiquita, so that whole towns become dependent on 'the company' which provides not only jobs but schools, mortgages, and hospitals. Such bananas are plantation-grown to gain the economies of scale that come from bulk purchase of inputs like pesticides. So, a high level of technology causes a high production level, causing high profits which can be invested in even higher levels of inputs. However recent evidence suggests that there may be environmental costs to pay for this. In Costa Rica pesticide runoff has been blamed for pollution of coral reefs downstream of plantations perhaps because the latter go right up to the river edge even in National Parks.

Recently, too, a dispute has emerged about the preferential trade agreement (PTA) between some of the smaller producers in the Windward Islands (such as St Lucia) which have a guaranteed market organised by a smaller TNC, Geest Plc. This arrangement started after the 1939–45 war as a way of providing a regular supply of food that was in much demand and as a way of guaranteeing revenue to small, vulnerable, producer economies (see Figure 7.16).

Q13. *Read the tabulated arguments for and against the PTA and write a short essay setting out your views.*

FOR THE PTA:

- If this were to go, American TNCs would dominate the market even more and have an increasingly monopoly position.
- Small producers like the Windward Islands need protecting from market forces, especially those that rely heavily on export revenue.
- Small-scale growers on steeply sloping land may be relatively inefficient but provide much employment (e.g. Jamaica has 80 per cent of banana workers).
- A few extra pence on the price of bananas in EMDCs (where food is a low proportion of total spending) is a small price to pay for supporting the livelihoods of so many people for whom the only realistic alternative is the unreliable tourist industry.
- One plantation farm can produce as many bananas as the whole of St Lucia, surely this is unfair competition?
- Supported by Geest, Fyffes, Britain, Spain, France, small farmers, governments of Windward and Leeward Islands.

AGAINST THE PTA:

- It is a system developed between colonial powers and former colonies and has no relevance today.
- Bananas from small producers are grown inefficiently and are therefore expensive to buy. They are twice the price of plantation bananas. This is unfair on the consumer.
- The EU should stand for free trade and open competition and cannot have some of its members entering special liaisons. Germany, the largest consumer in the EU has no such agreement, and as such it should have the largest say.
- Supported by American TNCs, Germany, producer governments with close TNC links such as Honduras.

Figure 7.17 *Banana packing station in the Windward Islands*

This case study raises the issue of the extent to which agriculture should be opened up to market forces. If it is, it can be a problem between ELDCs: for example at the moment cheap, efficiently grown rice imported from Thailand is threatening Ghana's domestic production. For the present the EU has imposed a quota on dollar banana imports as well as a 170 per cent tax for any bananas imported over quota.

Changes in the ex-Soviet republics

Many countries in the ex-Soviet republics are experiencing organisational problems in farming similar to those in ELDCs. This is especially true of those like Estonia which are trying to significantly decrease economic links with Russia for political reasons. This has meant the shrinking of a large market for agricultural products and no large internal market to compensate. Exporting to the west remains a problem because the inefficiency of farming coupled with a lack of government financial support makes grain, potatoes, and dairy products uncompetitive with western prices.

In Estonia the result has been a fall in total agricultural production from 21 per cent in 1992 to 17 per cent in 1993. Some of the inefficiency of agriculture is due to outdated machinery and facilities, which is a legacy of the Soviet period. Some is also due to land reform. This means the privatisation of land and its return to the legal owners before Soviet rule started in 1939. This is a complex and lengthy process that leads to land being underfarmed or not farmed at all at present. Furthermore, some of the old, large, and often efficient collective farms have been sub-divided to private farmers, when a co-operative approach would have been more sensible economically.

Such rapid changes in economy and society are unusual, and Estonia should reach a low point in farm production in the mid to late 1990s. Agrarian reforms are likely to follow soon as farms look to new enterprises such as timber, farm tourism, fur farming, and soft fruits. But a major dilemma still lurks: the strong desire of individuals to own farmland leads to fragmented land and inefficiency, but encouraging land consolidation will cause an unwanted drift to the urban areas unless there are substantial developments in the rural economy.

FOOD ISSUES IN THE ECONOMICALLY MORE DEVELOPED COUNTRIES

Physical environment: constraint or opportunity?

Compared to ELDCs, the EMDCs tend to have fewer extremes in their natural environments because they are located in temperate regions of the world. The modest temperatures and amounts of rainfall, plus the lack of a continuous growing season restricts the range and type of crops that can be grown, but soil processes such as leaching and chemical weathering are also less vigorous, thus better preserving the soil resource. However there are physical constraints on farming and it is a reasonable hypothesis that the main way in which farmers adapt to this is by evolving a farm system that suits the ecosystem inputs. To test this idea, statistics for two climatically different areas of England are shown in Figure 7.18. Area 44 is the Exmoor area of North Devon and West Somerset, which is mostly livestock with rough grazing on higher land. Area 24 is in Essex and is dominated by cereals.

Q14. *Draw a climograph plotting temperature (x axis) against rainfall (y axis) for each area. Join the points in month order and add a vertical line at 6°C (minimum growing temperature). Secondly complete the P–T columns and display the results in some way. From this and the additional information, provide evidence to support the type of land use in each area.*

Such effects of the environment are even more marked at the individual farm scale. Swincombe Farm is a hill sheep farm of 230 ha in area 44. The better land is cropped for animal fodder (e.g. maize) or sown grass, whilst the lower quality land is used for rough grazing. About half the grass is cut (two or three times a year) for silage or hay for winter feed. A typical farm in Area 24 would be New Hall near Colchester growing mostly wheat and some oil seed rape. Here the farmer perceives that the following are important in deciding the farm system: the clay

	AREA 44 (latitude 51° N)				AREA 24 (latitude 52° N)			
Month	Air temperature	Precipitation (mm)	Transpiration (mm)	P – T	Air temperature	Precipitation (mm)	Transpiration (mm)	P – T
J	4.0	141	1		2.9	55	1	
F	4.1	99	8		5.4	45	10	
M	5.8	101	32		5.4	40	32	
A	8.1	71	51		8.4	40	57	
M	10.6	85	75		11.2	46	85	
J	13.4	68	85		14.4	49	95	
J	15.0	83	84		16.1	55	95	
A	15.0	99	70		16.0	66	78	
S	13.5	104	42		14.4	52	50	
O	10.8	121	20		11.0	54	22	
N	7.3	142	4		6.6	66	5	
D	5.0	151	1		4.3	55	0	
		1265	473			623	530	

Growing season:	267 days	248 days	
Day degrees over 10°C: (May to October)	640	770	
Winter degree days below 0°C:	95	145	
Irrigation need:	10 years in 20 need 50 mm rain equivalent	19 years in 20 need 140 mm rain equivalent	
Grass drought factor:	5 days	48 days	
average height of area:	208 m	93 m	
height range (slope):	0–420 m	0–77 m	

Figure 7.18 *Selected area climate statistics*

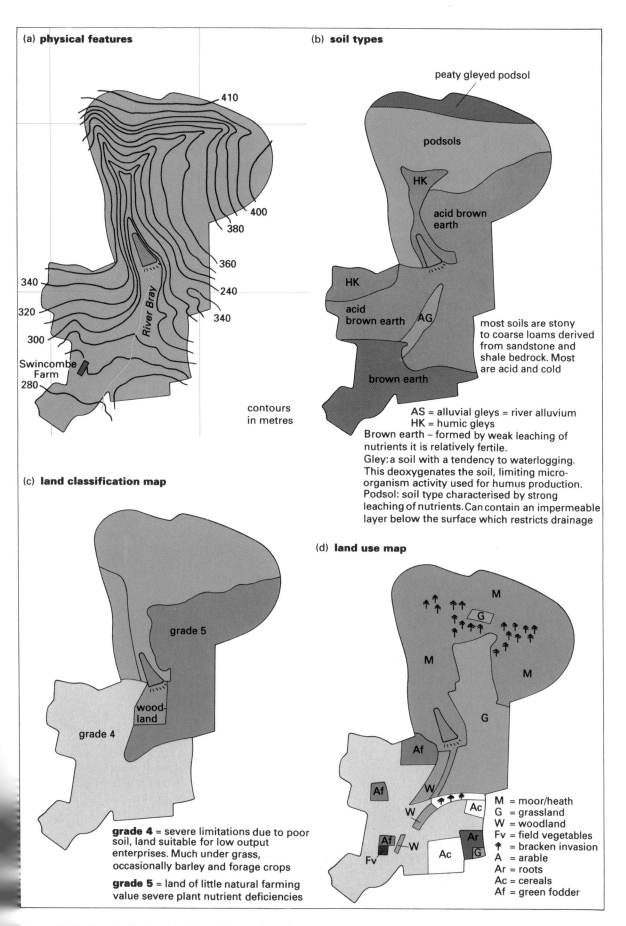

Figure 7.19 Data for Swincombe Farm (Exmoor) study

soil is wet and heavy in winter, ruling out many crops; animals left outside would cause **poaching** of the ground in winter – wintering them inside is too expensive; dry summers mean that grass production would be poor; and the relatively flat land allows the high level of mechanisation which cereal farming demands.

Q15. *Essential data for a farm in Area 44 is given in Figure 7.19. Use it to prepare a map showing areas of high, medium, and low agricultural potential. Write a justification of your map.*

However it would be wrong to view farmers as simply at the mercy of the elements. Such physical determinism is an over-simplification. In the real world farmers can modify physical constraints a good deal. Some examples of this are given in Figure 7.20 below.

Physical constraint	Possible solution	Problem with solution	Attempt to manage problems associated with solutions
1 Lack of soil moisture.	(a) Irrigation.	* Expensive * River abstraction can lower water tables in dry summers so that there is competition with supplies for industry and drinking water.	Regulation through water authorities.
	(b) Dry Farming – only cultivate when sufficient soil moisture built up.	Only low yields obtained.	Intensive irrigation.
	(c) Mulching using FYM or crop residues.	Crop residues often perceived as conserving diseases, so they are burnt.	Ban on burning in EU from June 1993.
2 Waterlogging of soils.	Drainage.	Loss of traditional landscape and ecology.	Incentive payments not to drain, or to revert to traditional wetlands – e.g. Countryside Stewardship Projects.
3 Slopes too steep for machinery. Slopes also promote erosion.	(a) Contour ploughing.	Good as long as not too steep.	
	(b) Develop machinery to cope with steepness.	Only economically viable up to 28° due to high cost.	Alternative land use for steep slopes – e.g. forest.
4 Too cold for good plant growth.	Glasshouse or plastic cultivation (using Greenhouse effect). Cultivate south-facing slopes (in northern hemisphere).	Expensive. Shiny texture can spoil landscape.	Planned growth outside areas of landscape value.
5 Infertile soil (usually caused by acidification).	(a) Hydroponics – i.e. growing crops like tomatoes in nutrient solution rather than soil.	Very expensive only justified in very intensive cultivation.	Government subsidies – e.g. Netherlands subsidise gas fuel to horticultural.
	(b) Lime the soil – calcium ions replace the hydrogen ions.	Can be costly to transport. Tackles symptoms, not causes, so only a temporary solution.	Policies to encourage farmers not to improve acid soils.

Figure 7.20 *Problems of physical environment for farmers in EMDCs*

The degree to which farmers adopt the solutions shown will depend firstly on their perception of their own needs. Broadly speaking optimisers will perceive more need than satisficers. It will also depend on the amount of economic gain involved and this has been heavily influenced by political pressures since the 1940s to maximise output. This started in the 1939–45 war with a drive to self-sufficiency triggered by the threat of a sea blockade on food trade.

Post-war economic and political changes in British farming

The processes of change

The pace and size of change in British (and EU) agriculture has led it to be called 'The Second Agricultural Revolution', ranking in significance with its predecessor in the late eighteenth century. More accurately it is described as the **industrialisation of agriculture**. It has involved seven processes.

1 Intensification: that is producing more output from the same amount of land. For example in just the short period from 1982 to 1991 sheep stocking rates rose by 20 per cent.

Q16. *Figure 7.21 shows wheat yield data. Calculate an Index for each year (1982 = 100). Index is the target year divided by base year, multiplied by 100. Produce a line graph to show the trend. Production controls on wheat were introduced in 1984, and in 1988 farmers were encouraged to take land out of arable farming. What effect do these controls appear to have had?*

2 Extensification: more land has been taken into production (see Figure 7.30, page 185) and this is reflected in the loss of certain natural habitats in the past 50 years (78 per cent of chalk downland and 58 per cent of lowland bog).

3 Farms have undergone specialisation focusing on one or two main products. For example New Hall has abandoned barley, beans, peas, cows and pigs in the last 45 years. Although such a move makes a farm more vulnerable to market price variations and to fungal disease, the advantages are great: the farmer can become an expert more easily, a smaller range of machinery is needed thus saving costs, and as there is a large area under crop large machinery can be used, so reducing labour costs.

4 Regional concentration: particular types of farming have become concentrated in an area to gain more benefit from environmental or market conditions. Changes in sheep numbers show this well (Figure 7.22). The effects of various subsidies for hill sheep farmers plus the arablisation of eastern England are clearly shown.

Year	Wheat yield	Year	Wheat yield
1982	6.4	1987	6.4
1983	6.6	1988	6.6
1984	8.1	1989	7.2
1985	6.7	1990	7.5
1986	7.3	1991	9.7

Figure 7.21 *Wheat yields in Britain (tonnes/ha)*

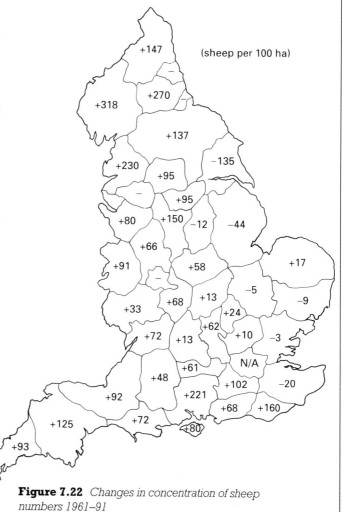

Figure 7.22 *Changes in concentration of sheep numbers 1961–91*

5 Structural concentration in farming is an organisational change towards fewer, larger, more efficient enterprises that lower costs through internal economies of scale. In the EU between 1950 and 1987 the number of farms over 100 ha grew by 113 per cent whilst those under 20 ha declined by 66 per cent. Farm amalgamation has been the main way this has happened as larger farms have bought up small ones when the farmers of the latter retire or leave farming. In areas like eastern England where land is more valuable the process is more pronounced.

6 Mechanisation has seen farmers using increasingly sophisticated equipment leading directly to a large decline in farm workers: in the 50 years up to 1990, farm workers in the EU fell from 30 000 to 9000.

7 Market manipulation by government consortia such as the Common Agricultural Policy (CAP). In this, 70 per cent of EU products including wheat, barley, beef, dairy products, and table wine benefit from a price support system in which the EU pays a common (i.e. guaranteed) price for each commodity and provides a guaranteed market for them. In this way farm incomes are protected. Figure 7.23 shows the mechanism.

Over-production and its management

The changes outlined above have been very successful in stimulating output to the extent that the EU is self-sufficient in many commodities such as milk, beef, and cereals. Whilst trade can be stimulated by surpluses, over-production has become a problem because first of all products may be 'dumped' on world markets undercutting local prices. (In the 1980s surplus EU beef arrived at the Ivory Coast in West Africa at a price 50 per cent lower than local beef from Burkina Faso). Secondly the cost of price support which gives rise to surplus, is very high: in 1991 it was Ecu 37 billion (1 Ecu = £1.30), or about £730 per year per average family of four. Ironically about 60 per cent of this figure goes to managing problems created by CAP in the first place – storage, export, and destruction of surpluses! This leads to the third problem, that of declining real incomes of farmers in the 1980s when support prices were level but import costs rose. The larger, well-supported cereal farmers, however, tended to increase incomes at the expense of the smaller, quota-hit dairy farmers. The 1990s have begun to reverse this as quotas have become valuable trading assets. Lastly there have been increasing worries about the environmental effects of the production drive.

Quotas have been a very successful way of managing production in milk and sugar. In milk the problem was created by an increase in yield per cow in the EU from 2400 kilos per year in 1950 to 4000 in 1980. This was caused by technological improvements to the cows (selective breeding and disease prevention), and to their feedstuff (new varieties of grass, imported concentrates), and by stable consumption with high guaranteed prices. In 1984 a limit was set on milk quantities bought by EU countries with heavy fines for exceeding quota. Production fell by 11 per cent from 1983 to 1989 and the scheme will last at least until the year 2000.

A second market mechanism is a stabiliser. If farmers collectively exceed a maximum guaranteed quantity at harvest, prices are reduced in the following year. The measure of stabilisation seems to have been insufficient and in the 1992 reform of the CAP they were scrapped in favour of the Area Payments Scheme where grain (and other product) prices will be cut and in return farmers with more than 39 ha of land can increase their earnings by taking at least 15 per cent of arable land out of production (i.e. set-aside). This is the first time that farmers, on a large scale, have been paid by the EU not to farm. Set-aside was introduced in Britain in

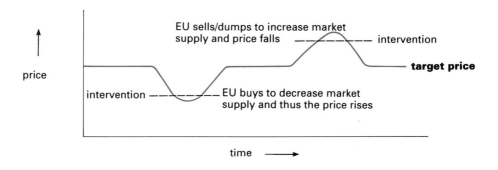

Figure 7.23 Intervention buying mechanism

1988 for arable farmers, and although by June 1991 it accounted for less than three per cent of arable land this was a 40 per cent increase over the previous year. It has, however, sparked a debate.

Q17. *Read the following notes for and against set-aside. Prepare an article for* Farmers' Weekly *entitled 'Should set-aside policy be set aside?'*

FOR SET-ASIDE
- Environmentally beneficial due to ban on pesticides, mowing only once a year to keep weeds down. Taller grasses encourage more animals, birds and butterflies.
- 15 per cent of land out of production will help limit over-production.
- Set-aside will add more variety to landscape, especially if rotated.
- Set-aside can be useful to a farmer especially as headland set-aside. This is land around the edges of fields that can be used to develop cross-country courses for horses and enable easier access to crops within the field for machinery.

AGAINST SET-ASIDE
- Abandoned land tends to encourage ragwort, blackgrass, and thistles which are of limited ecological value.
- 15 per cent less land cultivated will not necessarily lead to 15 per cent less production. From 1992 to 1993 there was a 13 per cent decrease in the area sown to cereals but yields increased by two per cent! Farmers can intensify their other land with more agrochemicals. Poorest land is often set-aside.
- Many farmers don't like it – they prefer to farm land rather than see it 'scruffy'.
- Five-year programmes do not allow sufficient time for imaginative schemes for using set-aside land.

Environmental impacts of agricultural industrialisation

Accelerated soil erosion

Soils with a good permanent cover of vegetation are unlikely to suffer soil erosion. The problem has occurred, especially under arable systems, where this is temporarily removed. About one-third of arable land in Britain is at risk. Sandy and **peaty** soils are especially prone to wind erosion because of their loosely compacted particles. The organic matter is acidic and often poorly decomposed so that polysaccharide gums (that cause particle cohesion and are produced in well-developed humus) are in short supply.

The Black Fens of East Anglia are a classic example: large, open fields and flat land allow a long fetch over which wind speeds build. Fifty years of intensive drainage of peat bogs allows good cropping but produces friable soils. Farmers have contained the erosion by planting rows of cereals between vegetables, but as the peat reduces in depth each year the acid clay sub-soil is ploughed in with the peat. This will resist erosion better, but is less fertile for high-value vegetable growing and can produce toxic acid-sulphate soils created by the oxidation of iron sulphides at the base of the peat when exposed by ploughing. Liming can help but treats the symptoms not the cause.

Water erosion is greatest where heavy rain falls on saturated soils, the resultant rill action taking humus, seeds and fertiliser away. Recent changes in farm practice have encouraged this process: ploughing up old permanent pasture for cropping; inorganic fertiliser replacing organic manure because of the decline in mixed farms; sowing crops in autumn for economic reasons leaves a smooth, fine seedbed with minimum ground cover at a time when rainfall is severe; ploughing up and down slopes made possible by modern machinery creates natural gullies for water movement (drill lines, furrows); compression by greater numbers of machines (that cover up to 90 per cent of a field area) increases surface runoff and the likelihood of **plough pans**. Overstocking in livestock systems can cause poaching of land especially on steeper or heavy land, or around feeding troughs.

Farm residues

Although amounts of fertiliser in the UK have stabilised since the mid-1980s, the level of application is very high as growth in use had been exponential prior to that date. Surplus soluble inorganic fertiliser which has not been taken up by the crop or fixed in the soil is highly prone to leaching out of the farm system. Nitrate is the main problem because it is so soluble. It can get washed into groundwater drinking supplies where there are fears that it could lead to oxygen starvation in babies or be linked to stomach cancers, or it can find its way into watercourses where it may cause damage through **eutrophication**.

> **Q18.** *Describe and explain the distribution of nitrate levels shown in Figure 7.24. Match levels to land use (grass/arable), areas of porous rock, and amounts of rainfall. This could be done as a series of overlays. N.B. 11.3 mg nitrate-nitrogen is the EU limit.*

Two sorts of action are being taken to reduce nitrate levels: water companies are blending high and low concentration waters, and Nitrate Sensitive Areas (NSAs) are being created. NSAs are voluntary agreements with farmers under which the latter agree to reduce nitrogen applications and to maintain a green cover.

Organic wastes have increased in livestock areas with the rising amounts of three things: animal excreta (urine and dung), which is now spread as slurry rather than manure, making runoff to streams more likely; silage effluent which can be very acidic; and effluent from washing down animals, milking parlours etc. Associated industries can add to the potential pollution problem: discharge from a creamery, for example, is thousands of times more toxic than raw sewage!

Pesticide residues mostly break up quickly into harmless substances, but they can affect non-target species if they persist and are concentrated further up the food chain either by temporary storage in animal tissues or longer storage in organs such as the liver. They will then be ingested by a predator. Before it was banned, DDT was found at concentrations of 0.4 ppm (parts per million) in small invertebrates but 3177 ppm in birds in the same ecosystem. Due to the effects of adult mortality and egg-shell thinning, birds of prey declined significantly between 1950 and 1980. Fish and butterflies have also been shown to be vulnerable. Such organo-chlorine compounds are now being

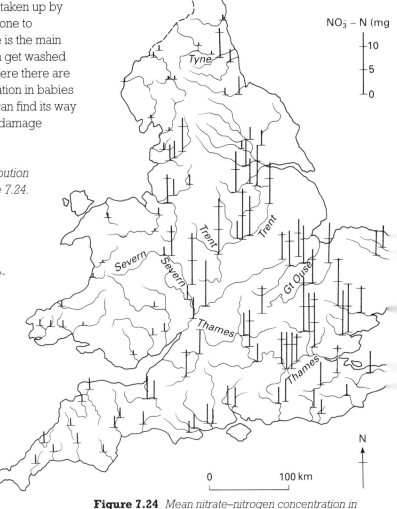

Figure 7.24 *Mean nitrate–nitrogen concentration in rivers in England and Wales, 1975–85*

replaced by organo-phosphorus ones which do not accumulate in invertebrates, but their use in sheep dip fluid is causing concern for human health, affecting the respiratory and nervous systems.

Habitat decline

Much of what has been discussed above has suggested that farm operations often contribute to ecological loss. Here, two particular issues will be investigated; hedges and drainage of lowland marsh.

About 25 per cent of hedgerows in Britain were lost between 1945 and 1985. This left about 400 000 km, and through the 1980s around 12 per cent of this was lost even though the rate of planting increased. About half of the decline has occurred on the cereal lands of eastern England. Neglect and mismanagement are the main causes of loss, not removal. In the past farmers were given government grants to grub-up hedges, but there are strong pressures now to get them to preserve the existing ones and to plant new ones.

FOR REMOVAL

- Costly for farmers to maintain (labour + equipment): up to £180/km/year.
- No longer needed on many farms and are therefore an attractive luxury. Many arable farmers no longer keep stock and many pastoral ones prefer moveable fencing.
- For every kilometre of hedge, one kilometre of cultivable land is lost.
- Field margins can be shaded, preventing grain from ripening.
- Provides a habitat for many weeds and crop pests (e.g. rabbits).
- Obstructs efficient use of large machinery.
- They can be replanted: they were man made in the first place (left after woodland clearance).

AGAINST REMOVAL

- Historically important. Some were created in Saxon times. Others are the result of the enclosures in the sixteenth to eighteenth centuries for stock control and as shelter for animals.
- Ecologically important. They are wildlife corridors which allow migration of insects, small mammals, and plants. Provide nesting sites for birds (e.g. barn owls). Some predators of crop pests live in hedges.
- In landscape terms they are important because they contribute to its variety (the 'patchwork quilt' idea).
- Older hedges are more diverse and richer in wildlife than newer ones.

In 1992 the Hedgerow Incentive Scheme was launched by the Countryside Commission, paying farmers between 50p and £1.75 per mile for hedgerow maintenance. Two further suggestions have been: a Hedge Preservation Order, as with some trees (but unlike trees, hedges need more management otherwise they become trees!); and a listed Hedgerow Scheme, like listed buildings, where a landowner would legally have to manage a hedge, if so listed.

Q19. *In groups, discuss all the information provided and then give feedback to a general session about your views on the following. Do you think the hedgerow loss is really a problem? What should be done for hedges in the future, if anything?*

Many valuable wetlands have been drained to increase productivity either to convert rough grazing to arable, as in the Norfolk Broads, or to allow grazing all year around and thus increase stock rates, as in the Somerset Levels (see Figure 7.25, page 182).

In the Somerset Levels and Moors three phases of development can be identified. Phase I, from the Iron Age to the eighteenth century, saw cattle grazing on the meadows that make up the low-lying 'levels' in summer, and moving to higher ground in winter as the water table rose with runoff from surrounding moors. This allowed wildfowl and wading birds to use the levels. In Phase II, up to 1981, there was increased drainage of the area beginning with the Enclosures of the eighteenth century which divided the area into fields separated by drainage ditches (acting as wet fences as well as drinking water for cattle). Since the 1939–45 war powerful diesel pumps have lowered the water table even further than the steam driven ones in an effort to prevent the area flooding. Phase III began a period of concern about ecological decline: breeding waders (lapwing, snipe etc.) fell by 50 per cent between 1977 and 1987 and wetland plants became rarer. The major cause was the falling water levels. Many of the Internal Drainage Boards controlling the pumps and sluices were composed of powerful landowners who wanted the lower water levels. But in 1981, under pressure from the Wildlife and Countryside Act, these bodies 'froze' water levels at existing settings. However species continued to decline because levels were still too low. Also, pumping seven days a week did not let soil moisture levels build up over weekends as they had previously. In 1986 large parts of the Somerset Levels were designated Environmentally Sensitive Areas (ESAs) to encourage traditional, low-intensity farming. Unfortunately high water levels were excluded from the conditions for getting grants. In 1991 Phase IV began with new ESA policies. Farmers who volunteer for the latter scheme can get up to £350/ha for maintaining their fields at conditions of 'surface splashing' in winter. The problem with the scheme is that all farmers with adjoining land have to enter a Raised Water Level Area together otherwise one farmer would have higher water levels than required! Figure 7.26 on page 182 shows the area and some of the success to date.

Figure 7.25 *Drainage on Somerset Levels – water kept low to allow cattle to graze*

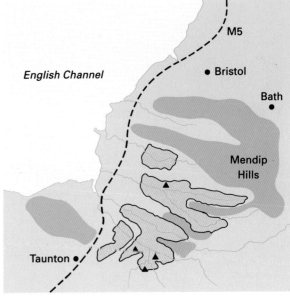

KEY
▲ raised water level scheme (600 ha to date)
▨ land over 100 m
▨ Somerset levels ESA
scale: 10 mm = 10 kms

Figure 7.26 *Somerset levels and moors area*

Landscape impacts of modern farming

Landscapes have tended to become less varied as modern human influences which are regular and angular become more prominent. The agricultural impacts on landscape are being seen as increasingly important in the planning of protected areas such as Areas of Outstanding Natural Beauty (AONBs), National Parks, heritage coasts, ESAs etc). This follows a move away from considering the countryside as being a unit of production with food supply as its only or main role, towards viewing it as a unit of consumption which outsiders consume as an amenity for walking, appreciating wildlife, playing golf, enjoying a view, and so on.

> **Q20.** *Figure 7.27 shows some graphics produced by the Yorkshire Dales National Park to try and gauge consumer opinions on the future planning of the area. Write down the differences you can spot in the landscape elements of each scene compared with today. Conduct a survey of people's opinions on the order of preference of the scenes, giving reasons.*

The third agricultural revolution?

As we have seen, more farmers are being paid not to farm land (set-aside) or to farm it in a traditional way, even if not already doing so (ESAs). There are large areas which are not under ESA rules, but many more are planned. Many farmers have entered Management Agreements as well: the farmer of Swincombe Farm (Area 44 in Figure 7.18) gets a grant from the Countryside Commission and National Park if he does not improve his land. The aim is to preserve the moorland for its landscape importance (Figure 7.28, page 184). The farmer is not allowed to plough or lime the area in question, and fences must blend with the contours. The economics of it might be something like this:

BENEFITS	With management agreement	If land improved
Gross margin/ewe	£47.00	£47.00
Number of ewes	55	164
Compensation paid	£937.00	Nil
COSTS		
Fertiliser	Nil	£6/ewe
Yearly costs	Nil	£1500
Miscellaneous	Nil	£2/ewe

(a) *Today's landscape*
Food production from livestock (yellow is mainly barley for fodder). Some tourism. Deciduous woodland suffering stock damage. Over-grazing of heather moorland. Many derelict buildings.

(b) *Food production landscape*
All subsidies taken away. Larger, modernised livestock ranches dominate. Grass dominates land use for grazing and silage.

(c) *Farm estate landscape*
Farming subsidies taken away. Leisure and game-shooting earn more money than farming. Nature trails, riding, rural industries. Subsidies for conservation.

(d) *Wildlife landscape*
No private ownership. Set-aside for wildlife. Some wildlife reserves. Otherwise open access for visitors. Some employed in outdoor recreation.

Figure 7.27 *Future landscapes*

Q21. Work out the overall costs and benefits for each strategy. Would you have entered the agreement? Why or why not?

Another trend is towards **farm diversification** where farmers turn to alternative uses for their land and buildings. Figure 7.29 on page 184 shows some of the possibilities. The growth of such ventures has been rapid so that in many areas they should be thought of as a normal reaction by farmers to market possibilities rather than as abnormal activity by a few farmers with financial difficulties. Distinct spatial patterns are emerging in the uptake of types of diversification: south west England is dominated by farm tourism, whilst recreation and horsiculture are more common in areas like the south east with a high population density and a relatively affluent market. Furthermore close to very large urban areas, direct marketing (Pick Your Own, farm shops) are very important. Economic motives are by far the most

Figure 7.28 *Swincombe Farm looking north towards moorland*

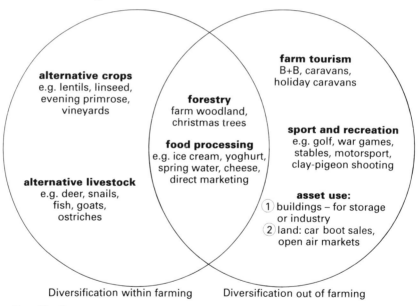

Figure 7.29 *Farm diversification*

important reason for applying for a diversification grant, together with an opportunity, such as a main road. The old, deterministic, model of land use of Von Thunen may well have a new application in looking at the spatial influence on land use of urban centres.

Diversification can in turn, however, create problems. For example the growth of golf courses and driving ranges (which suggests 70 new courses for the next ten years at least) has caused concern: they disrupt public rights of way, increase traffic on local roads at anti-social times, are a non-natural landscape feature and can stimulate semi-urban development (social clubs, restaurants). Those in favour emphasise the ecological importance of the rough areas, the prevention of total urban development (part of green wedges), and the provision of rural employment.

Diversification within farming is popular amongst farmers especially if it is something that attracts generous subsidies. Such is true of the pale lilac flowering crop of linseed seen in early summer. It is a useful break crop which at present gets a subsidy

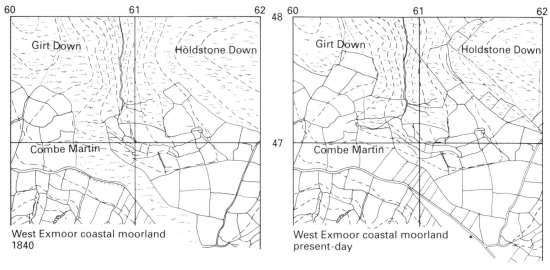

Figure 7.30 *West Exmoor coastal moorland*

of £571 per ha compared with wheat at £481. There is a large shortage in the EU and the main competitor, Canada, produces yields one-third less than the UK. Farmers who switch to the crop can harvest it with existing machinery.

Agriculture in the EMDCs, as typified by Britain, may well be entering a Third Revolution where its dominance in rural areas and as a food supplier are increasingly questioned. Agricultural land is a resource that can be used in many different ways: providing various services for an affluent, and ecologically-sensitive urban population is likely to become more valued in the near future. For a long time how to intensify farming has exercised the minds of researchers, farmers, and politicians. Now the opposite problems are facing these groups. For example, a good deal of the moorland has been reclaimed (extensification) in the past by ploughing, liming, and reseeding. How can the field on the right hand side of Figure 7.31 be reduced in fertility so that moorland vegetation with low intensity grazing will be returned to the landscape? Should the land simply be left, or would this result in tall grasses and high gorse? Should the humus layer be scraped away to provide the original poor soil on which moorland develops? Or will low intensity grazing open up the tall grasses to allow heather to develop sufficiently? These issues and others raised in this chapter can be brought together in the following exercise.

Figure 7.31 *Girt Down looking east from grid reference 604477*

FOOD FOR THOUGHT **185**

Role play exercise

What role should farming have in the future of the rural landscape of Britain?

Divide the class into four groups representing FARMERS, TOURISTS, GOVERNMENT (MAFF) and ENVIRONMENTALISTS.

One chair for each group is placed in the centre of the room facing each other. One person from each group sits on a chair, the rest of the group stands or sits behind them. The role play starts by one of the central four proposing a strategy for rural landscapes. The debate then begins.

The rules of the session are: anyone can speak provided they are sitting on the chair belonging to their group. To claim that right, a person touches the one occupying the chair on the shoulder. That person must give way immediately, even if in mid-sentence!

Not all people in the same group need have exactly the same opinion.

The session finishes when agreement has been reached or at the time limit. If the latter, a vote on the proposals that have arisen should take place. Follow up with a written account of the main points of the debate. Clearly the more thoroughly the roles have been researched, the better the debate goes!

DECISION MAKING EXERCISE

Rural planning in the Atlas Mountains, Morocco

This example concentrates on possible future directions for the Berber people who live in the valleys of the Atlas mountains. Use all of the information provided to complete your tasks and to present your answers in a report. Your tasks are to:

1 Outline the key problems of the environment and of the community of the Imlil area, giving supporting evidence where possible (see Figure 7.34, page 188).
    (10 marks)
2 Use appropriate graphical and/or statistical methods to provide an analysis of Figures 7.35 and 7.36. Comment on what the analysis reveals about the proposals for the area.
    (15 marks)
3 Evaluate the proposals for the area using all the information available. (15 marks)
4 Recommend and justify ONE of the proposals as your priority for development in the area.
    (10 marks)

The development proposals are:
Proposal A: Extensification: extend the upper limit of cultivation above 1900 m.
Proposal B: Intensification: by innovation (i.e. by starting to grow cash crops such as apples) and by improving water supply by lining irrigation channels with concrete or plastic piping.
Proposal C: Expand tourism: by improving the one access road, increasing facilities at Imlil (a new, medium-sized hotel), and bringing electricity to the area.
Proposal D: Afforestation: especially of the upper basin slopes.

Possible methods of analysis are:
a) Scattergraph and line of best fit (regression line): the independent variable (the measure you think causes the other one) should be the x axis. The regression line is best computer calculated as it may be a curved or straight line. If the latter, it can be found approximately by finding the means of both sets of data and plotting this point, and repeating this procedure for the data points to the left and to the right of this mean. The line should pass through all three points plotted.
b) Statistical **correlation** test. If the relationship is curved then the appropriate test is a Spearman's rank correlation. In this, both sets of data are ranked in order of size. Put the calculations in a table like the one shown opposite.

Sample point	rank for variable A	rank for variable B	difference in ranks (d)	difference squared (d^2)

The total of the last column is d^2. n is the number of pairs of data. Now put the results into this formula:

(correlation coefficient) $Rs = |-\dfrac{6(\Sigma d^2)}{n^3 - n}$

The closer the value is to 1, the higher the correlation. A positive value shows that as one variable increases so does the other. A negative value shows that as one variable increases the other decreases. To check how significant the relationship is, you need to put the correlation coefficient into a table of statistical significance.

x1 = remnants of climax dwarf-oak forest.
x2 = Berber village.
x3 = walnut trees.
x4 = floodplain.

Figure 7.32 *Natural environment of the High Atlas*

Figure 7.33 *Terrace farming in Imlil area*

FOOD FOR THOUGHT

The Imlil environment

The valleys have a humid, Mediterranean climate and are found between 1200–3000 m. Precipitation varies between 600–800 mm per year, with snow falling in the higher parts. Most rain and snow falls between November and April and consequently nearly half the discharge of rivers occurs between March and May. In the summer months when temperatures are at their maximum there is usually adequate water supply for irrigation via throughflow seepage which is taken off at high points in the catchment and gravity-fed to the fields (see Figure 7.33).

Just over one-third of the children under 14 are in education and 73 per cent of households own less than one-twentieth of a hectare of land. The population is increasing but only very slowly with out-migration acting as a safety valve. Tourism is on the increase, with a survey in 1990 finding that 40 per cent of services in Imlil were tourist-dominant. In response to a questionnaire in 1991, 60 per cent of visitors felt there were enough tourists in the area already, and half of those questioned said that there were inadequate facilities for tourists.

The energy situation around Imlil

Energy is in some crisis. Butane gas has been available since 1981, but even though the price is subsidised by the government, only those with a migrant worker or income from tourism can afford it. Many still do not cook with it because they believe it has a gaseous taste. Most families rely on wood collection for their main source of fuel: dead wood, bushes, and sometimes shrubs and live branches. Cutting with axes often disturbs the roots so that shrubs die. This makes the distance needed to travel to get wood even longer. Firewood collection is done mainly by women and takes between five and 12 hours a day, twice to seven times a week.

View of an environmental researcher, Marrakech University

Older Imlilis (people of the Imlil area) can remember that the slopes above the villages were once covered by natural vegetation up to the treeline of 2400 m (mainly red juniper forest), but there has been large scale deforestation due to increasing herd sizes and population growth which has led to more firewood collection. This has raised the potential for erosion. Loose scree threatens to damage upper level villages and can break through terrace walls and irrigation channels. Eroded material gets into the river courses and is transported and deposited on the wide flood plains of the Assif Ait Mizane. Such aggradation causes flooding in the winter (because of reduced channel capacity) which leads to two further problems: stones and boulders on the fields of the floodplain, and erosion of these fields by strong floodwater currents thus further reducing the amount of cultivable land.

View of a local farmer

As I see it there are two ways of improving my life. I could leave the area and go to Marrakech or Casablanca and look for work in industry or tourism. But it is a risk: I know many people who cannot find regular work like this. Or I could try to improve the way I farm. Government advisors have told us that the climate is suitable for growing fruits like apples and that people in the towns would buy them. It would also be a risk but might bring in more cash, now that we can graze fewer animals than we used to. The irrigation channels we depend on have been damaged in a few places so they need strengthening somehow. Some people have suggested plastic pipes to reduce water loss to a minimum, but I'm afraid they will be very expensive. Up until now our irrigation system has been very efficient but there is a worry that with more pressure on the land, farmers in the higher areas will take more than their allotted share of water.

Also, large government farms on the plains near Marrakech are becoming short of water and I fear the government may impose a limit on the amount we farmers in the mountains can have. This would then increase their underground supply.

View of a village elder

Although I own land in the area, making a living from farming is very difficult because of the shortage of land for pasture and crops. Quite a lot of tourists are attracted to the unspoilt atmosphere in the villages: eating by gaslight in the evenings, staying in a traditional home, and not having many western comforts. More tourist income could provide local people with things like the dispensary which has recently been built. A lot of erosion and flooding takes place in the valleys during the heavy rains and snow melt of autumn and winter because the steep upper slopes are bare of vegetation. The problem with planting conifer trees to help this is that the district government based in Marrakech (two hours away) see this as their project and don't consult local village councils. Locals might abuse the scheme therefore by taking young trees for wood before they are established.

Figure 7.34 *Resources for decision making exercise*

Site	Erosion Index	Height above sea level (m)	Slope angle
1	16	1983	39
2	04	1761	01
3	05	1840	07
4	13	1825	36
5	12	1875	30
6	01	1700	02
7	15	1975	36
8	11	1775	31
9	01	1710	01
10	10	1911	24
11	02	1734	04
12	07	1800	18
13	13	1953	32
14	09	1870	20
15	06	1920	05
16	19	1995	42
17	14	1962	38
18	03	1720	04
19	09	1950	13
20	20	2000	44

Erosion Index is a subjective scale from:
1 = no evidence of gulleying
20 = extremely severe erosion evident
A score of 10 represents a critical point between acceptable/unacceptable risk for farmers from erosion.

Figure 7.35 *Erosion levels at selected sites, 1989*

Income		
Sheep and goats	=	2070
Walnuts	=	1650
Total traditional economy	=	3720
Guiding	=	2400
Mule hire	=	720
Total tourism economy	=	3120
Relative sending money from Casablanca job	=	600
TOTAL	=	7440

Expenditure		
Barley and corn	=	2233
Wheat flour	=	1010
Local foods	=	4038
Total Food	=	7281
Taxes	=	78
Transport (market)	=	120
Clothes	=	300
Fuel (charcoal)		250
TOTAL	=	8009

Farm is modal in income. Figures in dirhams.
1 dirham = circa 8p

Figure 7.36 *Family budget for an agricultural year*

PROJECT SUGGESTIONS

1. **FARM STUDY** – visit a local farm, interview the owner/manager about farm changes: see how many of the seven processes involved (page 177) apply. To what extent is the case study typical of the area, and why?

2. **PHYSICAL FACTORS SURVEY** – measure and map soil variables (pH, infiltration, texture, organic content, bulk density, soil temperature, and nitrate level) and correlate with land use. Probably best done as a belt transect. What explains any unexpected results?

3. **AREA THEME SURVEY** – interview a number of farmers in an area to establish the degree of uptake of a recent trend in farming (e.g. set-aside, diversification). Try to include farmers who have and have not been involved. Try to explain the spatial variations.

4. **ORGANIC FARMING** – prepare a library-researched report on the organic farming debate using opinion and facts for both sides.

5. **HEDGE QUALITY SURVEY** – do a field survey of hedge quality, mapping well-managed ones, neglected ones (overgrown or gaps), and removed ones (you may need to interview the owner or go back to past OS maps for this). Include an ecological value survey (count the numbers of species of plants/visible animals) and correlate with different styles of management (flail-cut, layed).

6. **LANDSCAPE IMPACT ASSESSMENT OF AGRICULTURE**: can be done by finding or creating a list of landscape features and then scoring the contribution of farming (hedges, building, field size, crop colour etc.) to the landscape. Do this for different landscapes and views and try the assessment on people with different backgrounds. Can you explain your findings?

GLOSSARY

Agrochemicals general name for industrially manufactured pesticides and fertilisers.

Agroforestry growing trees in or around crops. The trees can be used for timber or for fruit or nuts.

Correlation statistically finding the level of match between two sets of data.

Double-cropping two harvests per crop per year. Implies a very intensive approach.

Eutrophication pollution of water by algal growth due to over-feeding with nutrients.

Extensification taking more land into production.

Farm diversification farmers going into a variety of non-traditional economic practices.

Integrated agricultural development project a project in which farming changes are part of a package of rural change.

Intensification increasing the output of a farm, usually by raising yields or stocking rates.

HYVs High Yielding Varieties of crops formed by cross-breeding different strains.

Marginal farming farming in physically harsh, and therefore economically marginal, areas.

Peat poorly decomposed organic matter due to cold and/or wet conditions. Slow release of nutrients to plants results from this.

Plough pan compacted layer in a soil immediately below the plough limit.

Poaching the compaction and puddling of ground by animal hooves. Can start erosion.

Positive feedback system a loop in a system that increases the original effect.

Salinisation build-up of salt in a soil to a level that damages plant growth.

Sustainable agriculture where an increase in inputs results in an increase in outputs with no deterioration in the soil resource.

TNCs Transnational Companies which operate across national boundaries.

Trophic level feeding level within a food chain within an ecosystem

REFERENCES

T Bayliss-Smith, *The Ecology of Agricultural Systems*

R Bradnock, *Agricultural Change in South Asia*

Bull, Daniel and Hopkinson, *The Geography of Rural Resources*

Education Europe 2000: Module 301, The CAP

P Harrison, *The Greening of Africa*

P Harrison, *The Third Revolution*

HMSO, *Agriculture, Fisheries, and Forestry*

B Ilbery, *Agricultural Change in Great Britain*

Robinson, *Conflict and Change in the Countryside*

Williams, *The Diversification Guide*

Winchester and Ilbery, *Agricultural Change in France and the EEC*

The Countryside Commission, John Dower House, Crescent Place, Cheltenham, Gloucestershire, GL50 3RA

Ministry of Agriculture, Fisheries and Food (MAFF), MAFF Publications, London, SE99 7TP

Farming and Conservation in 1990s, computer package and database from Centre for Rural Studies, Royal Agricultural College, Cirencester, Gloucestershire, GL7 6JS

SECTION 8

Life in the city

by Keith Grimwade

KEY IDEAS

- What are towns and cities like?
- Why do so many people live in towns and cities?
- How are towns and cities structured?
- What goes on inside towns and cities?
- How do we manage the urban environment?

WHAT DOES 'URBAN SETTLEMENT' MEAN?

The United Nations, for statistical purposes, defines an urban settlement as one with a population of over 20 000. This is a straightforward definition but it raises a number of problems.

- National censuses. Different countries use a wide range of figures to distinguish between urban and rural settlements: for example, in Denmark it is 250 people; in the USA 2500; and in India it is 5000.
Boundaries. Cities are usually an administrative region and their boundaries often include an area of surrounding countryside which contains hamlets, villages and even small towns; for example, it is estimated that Shanghai's true urban population is six million rather than its official 12 million.
Functions. Urban settlements are associated with 'high order' functions; for example, government offices, department stores, main cinemas and bus termini. However, in countries with small and scattered populations, e.g. Iceland, some or all of these functions can be found in settlements with populations of little more than 300.

- Occupations. Rural settlements have traditionally had a greater percentage of their workforce employed in farming but in recent years many villages in Economically More Developed Countries (EMDCs) have become **dormitory settlements** with their populations commuting to work in a nearby city.
- Attitudes. Rural areas used to be behind the times in matters such as fashion and ideas. However, better communications – particularly radio and television – mean that people in rural areas are now as well informed as people in towns and cities.
- Environments. The quality of urban environments varies a great deal; for example, Figure 8.1 shows different parts of London. A short, simple definition of an urban settlement could leave out these important differences.

Q1. *Think of nine characteristics of urban settlements. Arrange these into a 'diamond nine' and justify your top three.*

Q2. *What criteria could you use to compare the quality of urban environments in Figure 8.1, page 192?*

i) Residential suburb, Islington

ii) Lambeth House

iii) Hyde Park

iv) Oxford Street

Figure 8.1 *London*

> **Q3.** Carry out a bi-polar analysis of people's opinions of towns and cities using the framework provided in Figure 8.2. Compare the views of different groups; for example, the young, students and the elderly. Present, and comment on, your results.
>
> **Q4.** Write down what you think would be a good definition of urban settlement in no more than 50 words.

	1	2	3	4	5	
noisy						quiet
polluted air						fresh air
dirty streets						clean streets
crowded						spacious
dangerous						safe
boring						interesting
poor facilities						good facilities
unwelcoming						friendly
poor						wealthy

Figure 8.2 *People's opinions of towns and cities*

Who invented the city?

The earliest archaeological evidence of urban settlement dates back to 4000 BC and is found in Mesopotamia – the region between the rivers Tigris and Euphrates in present-day Iraq. By 2000 BC cities were also found in the valleys of the Nile and Indus, and in China: these may have developed independently, or the idea may have spread (Figure 8.3).

Area	Date	Cities
Mesopotamia	4000–3500 BC	Lagash, Ur, Uruk
Egypt	3000 BC	Memphis, Thebes
Indus	3000–2250 BC	Mohenjo-dara, Harappa
China	2000 BC	Cheng-Chon, An-Yang

Figure 8.3 *The first cities*

Urban settlement could not have developed without a food surplus to support a non-agricultural population. Settled agriculture, which had developed in Mesopotamia around 8000 BC, provided this surplus and improvements in transport meant that food could be delivered to the cities while it was still fresh. However, why people decided to build the first cities is not known, although it is likely that economic, political and social factors all played a part.

Mesopotamia, the Nile and the Indus offered great potential for farming but they were not self-sufficient – for example, they needed flints for farm equipment and timber for building – and this encouraged trade. In turn, this could have led to the growth of cities as centres of commerce.

Irrigation was important to agriculture in all of these regions but it required careful organisation: this could have encouraged the development of cities as centres of political control.

It was a period of great religious activity and temples occupied important sites in all early cities: thus, it is possible that social developments, such as new religious customs, could have been the stimulus for urban growth.

The largest of these early cities was probably Babylon with a population not more than 80 000. It is likely that up to 80 per cent of its working population were farmers who lived inside the city walls and 'commuted' to their fields in the surrounding countryside.

Urban settlement did not spread more widely until the eighth and seventh centuries BC when the Greeks began to settle along the shores of the Mediterranean. They set up 'city states': an urban settlement with a surrounding area of farmland. Athens had a population of between 100 000 and 200 000 but most city states had populations of between 5000 and 20 000.

Greek cities were laid out according to a grid plan with straight streets at right angles to each other. This had the advantage of being easy to set out, and its use has continued through to the twentieth century; for example, look at street plans of New York and Los Angeles. The grid plan was popularised by Hippodamos, the first recorded town planner, who rebuilt Miletus in the fifth century BC after it had been destroyed by the Persians (Figure 8.4).

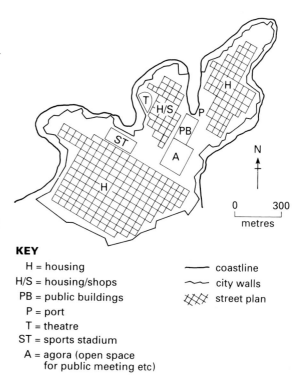

KEY
H = housing
H/S = housing/shops
PB = public buildings
P = port
T = theatre
ST = sports stadium
A = agora (open space for public meeting etc)

— coastline
~ city walls
XXX street plan

Figure 8.4 *Miletus*

The Romans took the idea of the city with them when they conquered north and west Europe. Their empire gave them access to large quantities of food, and their skill at building harbours for transport, and aqueducts for water supply, allowed Rome itself to reach a population of 200 000. However, most cities remained small; for example, London, founded in the first century AD, probably had a maximum population of 30 000. Again, the grid plan dominated the design of Roman settlements.

The collapse of the Roman Empire in the fifth century AD led to many towns in Europe being abandoned. However, as trade increased in the twelfth and thirteenth centuries hundreds of small market towns grew up, usually at the focus of routeways, and by the sixteenth century the pattern of urban settlement in Europe was established. These medieval towns are often depicted with narrow, winding lanes but in fact many were still laid out using a grid plan; for example, Monpazier in France which was built in the thirteenth century (Figure 8.5, page 194).

From the sixteenth century onwards there was a

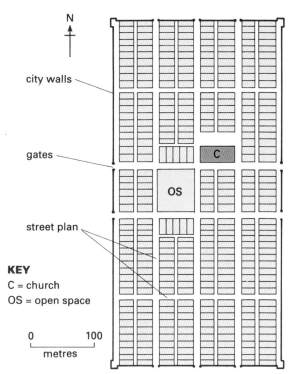

Figure 8.5 Monpazier

great deal of interest in developing new styles of town plan. An example is Bath, the site of a Roman town built in the first century AD, which became the most fashionable spa town in eighteenth century England. It was designed with crescents, radial avenues and squares (Figure 8.6).

The nineteenth century in western Europe was a period of rapid urban expansion stimulated mainly by the Industrial Revolution. The often unsatisfactory housing which was built for the rural-urban migrants was set out using the grid plan because it was easy to organise and the basic design of a block of houses could be repeated quickly and cheaply. Consequently, large parts of the inner city are still dominated by back to back terraced housing and blocks of flats with narrow roads set at right angles to each other. It was only away from the city centre, in some of the better off suburbs, that the more elegant plans of the eighteenth century continued to be used.

There were, however, some exceptions. A number of industrialists who were concerned about the quality of the urban environment and its effect on the welfare of their workforce, planned and built their own settlements. An example is Port Sunlight on Merseyside, built by the Lever Brothers in 1886: it was designed with low density housing, parks and recreational areas for children.

Towards the end of the nineteenth century there was also a great deal of discussion about what towns and cities should be like. In 1898 Ebenezer Howard published his thoughts about the garden city and these were particularly influential: they led directly to the building of Letchworth and Welwyn Garden City and provided many of the ideas for the UK's New Towns.

Figure 8.6 Bath: aerial view

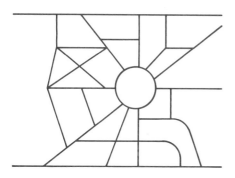

(a) geometric pattern
1900–1950s

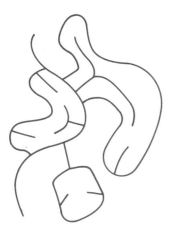

(b) irregular pattern
1960s–1970s

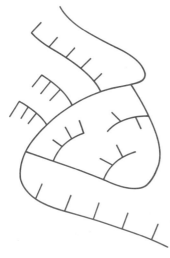

(c) cul-de-sacs
1980s

Figure 8.7 *Twentieth century street patterns*

The slower rate of urban growth, increasing prosperity, improved transport and communications and discussions about the 'ideal' city have meant that the grid plan has been less of a feature of western towns and cities in the twentieth century (Figure 8.7). At the turn of the century a variety of geometric patterns were used in the suburbs. The irregular pattern of the 1950s 1960s and 1970s was an attempt at making the urban landscape more varied and interesting. The 1980s saw a return to a more regular street plan dominated by cul-de-sacs; the idea was to make roads and pavements safer by eliminating through traffic.

> **Q5.** Mark onto a map of Europe, Asia and Africa: the Rivers Tigris, Euphrates, Indus and Nile; and the cities Ur, Memphis, Athens, Rome and London. Add dates and labels to show the spread of urban settlement.
>
> **Q6.** Draw and label a simple sketch of the aerial photograph of Bath (Figure 8.6) to show the main features of its urban plan.
>
> **Q7.** What are the advantages and disadvantages of a) the grid plan and b) the plans shown in Figure 8.7?
>
> **Q8.** Look at the OS map of Peterborough (Figure 8.8). Sketch the street plans in squares 1899, 1900, 1901 and 1601. Describe the main characteristics of each area and suggest when it might have been developed.

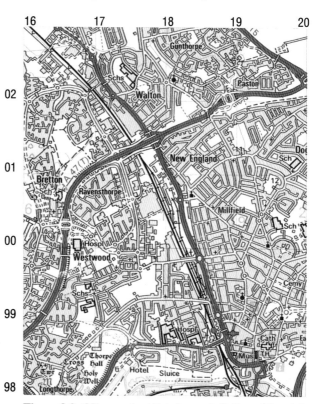

Figure 8.8 *OS map extract, Peterborough,* © *Crown Copyright*

LIFE IN THE CITY

WHAT IS THE WORLD PATTERN OF URBANISATION?

Urbanisation takes place if **rural-urban migration** is greater than urban-rural migration and/or life expectancy is greater in urban than in rural areas. Until the nineteenth century the proportion of the world's population living in towns and cities remained very small. However, the last 200 years have seen an urban revolution: it has been one of the biggest ever changes in the way people live (Figure 8.9).

The world pattern of urbanisation shows significant variations (Figure 8.10). Generally, EMDCs have a higher percentage of their population living in towns and cities than Economically Less Developed Countries (ELDCs) and there is a positive relationship between level of urbanisation and Gross National Product (GNP). It is therefore tempting to conclude that urbanisation is the result of economic development.

However, the rate of urbanisation also shows significant variations (Figure 8.11) and it can be seen that there are high rates of urbanisation in some of the poorest countries where very little economic development is taking place. In these countries urbanisation is being stimulated by other factors, many of which are to do with social and economic conditions in the countryside (see page 200).

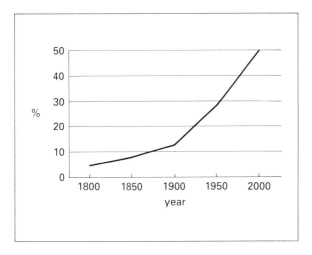

Figure 8.9 *World urbanisation*

Q9. Use the statistics given in Figure 8.10 to construct a choropleth map (see the technique box opposite) to show the world pattern of urbanisation.

Q10. Describe and explain the pattern shown on your map.

region		urban population %	region		urban population %
Africa			Latin America		
	1	43		11	64
	2	23		12	58
	3	19		13	74
	4	38		14	
	5	52	Europe		
Asia				15	83
	6	62		16	82
	7	26		17	63
	8	29		18	68
	9	34	USSR (former)	19	66
North America	10	75	Oceania	20	71

Figure 8.10 *World pattern of urbanisation*

Q11. *Calculate the degree of correlation between a) urban population and GNP and b) urban growth and GNP. Compare your two results. What conclusions can be drawn from them?*

Country	Urban population (% 1987)	Rate of urbanisation (% increase, 1950–87)	GNP (1990, US$ per capita)
Egypt	48	16	600
Ghana	32	17	390
Zaire	38	19	230
Kenya	22	16	370
Pakistan	31	13	380
India	27	9	350
China	38	17	370
Thailand	21	10	1 420
USA	74	10	21 700
Mexico	71	28	2 490
Jamaica	51	21	1 510
Peru	69	33	1 160
Argentina	85	20	2 370
UK	92	8	16 070
France	74	18	19 480
Poland	61	22	1 700
Italy	68	14	16 850
Romania	49	23	1 640
Australia	85	10	17 080

Figure 8.11 *Urbanisation and development*

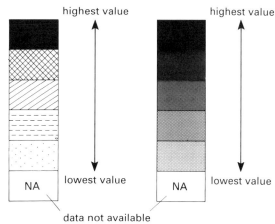

Figure 8.12 *Shading techniques*

TECHNIQUE BOX
A choropleth map uses shading to show the relationship between quantities and area in an administrative region: densities, ratios, averages and percentages can often be presented in this way.

1 Draw boundaries onto the map if necessary.

2 Divide the values into groups. Use an arithmetic progression (0–10, 11–20, 21–30 etc.) if the range of values is small. Use a geometric progression (0–10, 11–30, 31–70 etc.) if the range of values is great. Avoid having too few groups or the map will appear too uniform but also avoid having too many groups or the map will appear confusing: between three and seven groups is usual.

3 Decide on your shading method. This is very important because you are trying to give a clear visual impression of the different values and their rank order from high to low. Therefore, if you use one colour a graded pattern is necessary while if you use different colours you must choose shades which show a gradual transition (Figure 8.12).

4 Complete the map by adding title, key, scale, compass, source of statistics and any appropriate labels.

Urbanisation in England and Wales

Urbanisation in England and Wales can be divided into four main stages (Figure 8.13, page 198). In stage one the percentage of the population living in towns and cities went up sharply. Natural increase accounted for only a small part of this rise because the extremely poor conditions found in the working class districts of most urban areas meant that the death rate was much higher than the birth rate (Figure 8.14, page 198). Most of this rise was because of rural-urban migration with people being attracted by jobs in the growing number of factories in the urban areas and 'pushed' from the rural areas by unemployment which was mainly the result of mechanisation.

In the second half of the nineteenth century natural increase became as important to the rate of urbanisation as rural-urban migration. This was mainly because improvements in the quality of water supply, in sanitary provision and in medical knowledge significantly reduced the death rate (Figure 8.15, page 198).

Between 1900 and 1950 the rate of urbanisation levelled off. Farming was still relatively labour-intensive and this guaranteed a rural population; also, many towns and cities were reaching saturation point.

Since 1950 the trend has been in the other direction with an increasing percentage of people living in the countryside. This has been described as **counterurbanisation** and it has been the result of a number of interrelated factors:

- transport improvements have made long-distance commuting possible;

- people have left the cities in search of a more pleasant living environment;
- the cost of housing in rural areas has been lower than in the cities;
- footloose industry has become increasingly important.

Counterurbanisation was at its greatest in the 1970s with the main **conurbations** losing 6.7 per cent of their population. Change in the 1980s is shown in Figure 8.16; although the rate of counterurbanisation slowed down, the main conurbations lost a further 1.8 per cent of their population.

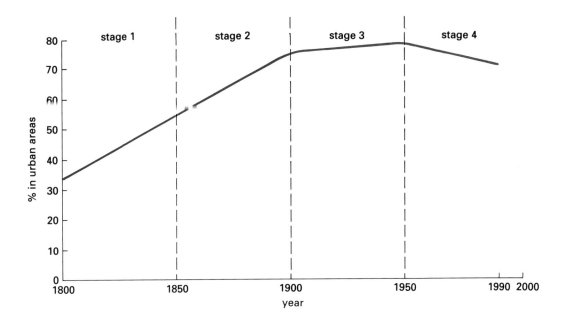

Figure 8.13 Urbanisation in England and Wales

- **The River Medlock district of Manchester in 1832**
'An unhealthy district which lies so low that the chimneys of its houses, some of them three storeys high, are little above the level of the ground. (It is) surrounded on every side by the largest factories of the town, whose chimneys vaunt forth dense clouds of smoke, which hang heavily over this insalubrious region.'
- **Wolverhampton in 1840**
'A dense population is congregated in these places (and) in the formation of the buildings everything has been sacrificed to secure a large (financial) return. (The buildings) are often of the worst construction, and in immediate contact with extensive receptacles of manure and rubbish. Many have only one privy for several families. Should any epidemic occur, its victims can be scarcely otherwise than numerous. Even in the new buildings in the town, regard for health of the public does not appear to exist, particularly as respects drainage and the facility of removing refuse. Fever is constantly present.'
- There were major cholera epidemics in 1831, 1848 and 1866 resulting from polluted water.
- Life expectancy in London in 1841 was 36 years; in Manchester it was 26 years.
- In Liverpool in 1840–41, 259 out of every 1000 children died before their first birthday compared with a national average of 148.

Figure 8.14 Urban conditions in England and Wales, 1800–1850

- 1848, the first Public Health Act.
- 1866, the first Sanitary Act.
- 1868, local authorities were given powers to force owners to repair or destroy sub-standard housing.
- 1875, Artisans' Dwelling Act: this gave local authorities the right of compulsory purchase for the demolition and/or redevelopment of sub-standard housing.
- 1875, the second Public Health Act: this divided the country into urban and rural sanitary districts, supervised by central government.

Figure 8.15 Beginning to manage the urban environment, 1850–1900

Q12. What do you think were the main problems with living conditions in the growing towns and cities of England and Wales in the first half of the nineteenth century?

Q13. What do you think were the main causes of these problems?

Q14. To what extent is the counterurbanisation trend reflected in Figure 8.16?

Q15. Consider the advantages and disadvantages of counterurbanisation for a) urban areas and b) rural areas. The information in Figures 8.1 and 8.18 will help you with this answer.

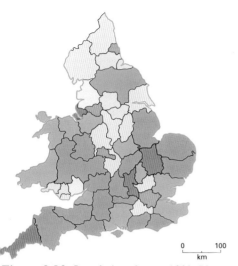

annual percentage rate of population change 1981–1991

■ increase 1.50 and over
■ 0.90 to 1.49
■ 0.45 to 0.89
□ 0.15 to 0.44
□ 0.00 to 0.14
□ –0.15 to –0.01
□ –0.45 to –0.16
■ –0.90 to –0.46
■ –1.50 to –0.91

England and Wales: counties

change in population present on census night 1981–91

Cambridgeshire = 1.16
Buckinghamshire = 0.95
England and Wales = 0.04
Greater Manchester = –0.55
Merseyside = –0.91

Figure 8.16 *Population change 1981–91*

Rural poor 'lose out to affluent arrivals'

Poverty and deprivation are rife in country areas as well as inner cities because of "population substitution" — people forced to move out of villages while affluent people move in — a report of the Church of England Synod says this week.

An underprivileged remnant were left with "poor housing, poor education, poor health, few jobs and low incomes."

The report on the Church's rural strategy follows the Archbishop of Canterbury's controversial report last year, Faith in the City. The Bishop of Norwich, the Rt Rev Peter Nott, co-author, said yesterday that the document was "not a reply, but a natural consequence" of the cities report.

Nott urges a full-scale archbishop's report on rural areas to match the cities report. This project has already been approved in principal by the synod, but funds are lacking for a start to be made before 1988.

The report says that "the new villagers — the retired, the commuters and occasional residents" — allow local services vital for the under privileged to disappear.

A much higher proportion of council housing had been privatised in villages than in cities. "Unequal competition" for such housing had created a "hidden need in rural areas which results in many young people moving reluctantly from the village in order to find housing elsewhere," the report says.

Many small villages "have no services of any sort and elderly residents are trapped at a considerable distance from shops, social services, doctors practices, and banks."

The report says: "the sharp decline in the availability of public transport in rural areas (and the cost of what remains) is one of the principal features of deprivation in rural areas."

The closure of village schools resulted in the bussing of children many miles from home.

To help meet the challenge, the church needed special training for country parsons, the report urged. Most now came from "urban or suburban backgrounds" and had no special training.

The Bishop of Norwich commented yesterday that there was "a positive side, in that the rural challenge is attracting younger and abler men to tackle it."

The Guardian, 25 September 1986

Figure 8.17 *Counterurbanisation and rural areas*

Cutbacks fuel urban sprawl

Support for a campaign to prevent a "swathe of urban development across much of lowland England" as farmers are forced to cut back food production, is being sought from thousands of local amenity societies and naturalist groups.

The plea is made today in a new year message from the president of the Council for the Protection of Rural England, the film maker Mr David Puttnam. The crisis in farming combined with government plans to abolish the strategic planning role of county councils "could mean new and chaotic urban development pressures over much of England's unspoilt countryside," he says.

Mr Puttnam says pressure to cut food production could make up to 20 per cent of England's land surface available for new uses. This in turn would create the conditions for an inevitable swathe of new urban development across much of the lowland England which we have kept green until now."

While such options are being vigorously discussed in Whitehall, Westminster, and Brussels the interests of conservation and public enjoyment of the countryside are not being kept at the centre of the debate "because public opinion is not yet fully aware of what is at stake."

The Government's proposed abolition of county structure plans—leaving plan making to district councils—may be necessary, but it will weaken strategic control over the location of necessary new development and the protection of natural heritage, he argues.

Organisations responding to the Government's consultation paper on structure plan abolition should note the connection with the farming crisis.

The Guardian, 29 December 1986

Figure 8.18 *Counterurbanisation and urban sprawl*

LIFE IN THE CITY

Urbanisation in India

It can be seen from Figure 8.19 that both the absolute number and the relative percentage of people living in towns and cities in India has increased, particularly in the last 30 years.

Year	Total population (millions)	Urban (millions)	% of total	Rural (millions)	% of total
1901	238.3	25.9	10.8	212.4	89.2
1911	252.0	25.9	10.3	226.1	89.7
1921	251.2	28.1	11.2	223.1	88.8
1931	278.9	33.5	12.0	245.4	88.0
1941	318.5	44.1	13.9	264.9	86.1
1951	361.0	62.4	17.3	287.4	82.7
1961	439.1	78.9	18.0	345.9	82.0
1971	547.9	109.1	19.9	419.8	80.1
1981	684.0	159.0	23.2	525.0	76.8
1991	845.1	217.2	25.7	627.9	74.3

Figure 8.19 *India's urban population*

Natural increase has been an important part of this process. In the period 1971–81 it accounted for 45 per cent and between 1981–91, 60 per cent of the absolute increase. It has also contributed to the relative increase because infant mortality rates have been lower in the cities than in the rural areas at 60 deaths per 1000 births compared with 139 deaths per 1000 births.

However, over the last 30 years rural-urban migration has been the main process. There have been a number of push factors forcing people out of the rural areas. Some migrants have moved away because of drought. The **Green Revolution**, although it has increased food production, has put many small farmers out of business because they cannot afford the fertilisers and pesticides required by the new seeds and this has added to the rural exodus. A greater use of machinery has also increased rural unemployment.

There have also been pull factors attracting migrants to the cities. Many perceive job opportunities to be better in the cities. However, few find regular work in a factory or office and the majority have to make a living in the **informal sector** of the economy, picking up casual work such as street selling or shoe-shining; this can still produce a better income than they would earn in the countryside (see page 207). The cities also offer greater access to health care, schools and entertainments, and they represent freedom from traditional village customs.

Some migrants move straight to the city. Others – step migrants – go to a small town first and then move to a bigger town when they are more used to the urban way of life. More recently there has been a growth in circular migration which involves leaving the village for a number of months a year to work in the city but returning home when labour is needed, for example, during the harvest. This type of migration poses particular problems for urban authorities because it means that the city's population, and therefore the demand on services, is variable.

In the past, most rural-urban migrants were young men (Figure 8.20). They joined friends and relations already living in the city until they could find work and somewhere to live on their own, and they then brought their families to live with them. However, in the last ten years this pattern has begun to change with an increasing number of young women moving to the cities because of a demand for cheap female labour in the growing number of factories. There are now more female than male migrants in the 16–25 age group in ELDCs as a whole, although it would appear that many women go back to live in their villages when they get married.

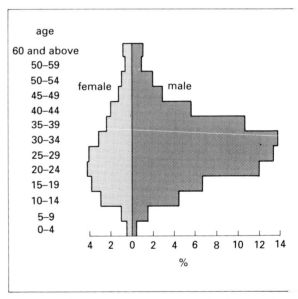

Figure 8.20 *Population structure of a city distorted by rural–urban migration*

The rate of urbanisation has been so great that India's towns and cities have had problems coping. Most of the migrants end up living in illegal squatter settlements, in very poor conditions, and the consequences of this are explored further in the case study of Bombay on pages 214 to 216.

However, it can be seen from Figure 8.19 that the rate of urbanisation between 1981–91 slowed down compared with the previous ten years and this may offer some hope for India's urban authorities. Why this has happened is unclear but a number of ideas have been suggested:

- some of India's largest cities have reached saturation point, with little suitable land left for building even a squatter settlement;
- rural development schemes have kept people in the countryside;
- harvests in the late 1980s were good;
- road and rail improvements have made commuting easier.

These developments have placed pressure on the urban-rural fringe. It is often the only place left for building; it is a convenient location for industries which are trying to get away from the crowded central areas of cities; and it is the ideal location for dormitory settlements. This could add sprawl to India's urban problems.

Q16. *Draw appropriate graphs to show the absolute and relative increase in India's urban population since 1901.*

Q17. *Describe and explain the trends shown on your graphs.*

Q18. *Draw a diagram to show the different types of migration described in this section.*

Q19. *Comment on the population pyramid shown in Figure 8.20 and explain how and why the situation represented by this graph is changing.*

Q20. *How and why might India's urban planning priorities change in the next ten years?*

CAN MODELS HELP US TO EXPLAIN URBAN STRUCTURE?

There are a great many similarities between towns and cities; for example, the main business district is usually found in the town centre and it is often surrounded by an area of run down housing, while the best housing is usually found on the outskirts. Geographers have used models – theories of urban structure – to explain these similarities. These models are very different to each other; however, the main reason for studying them is not to decide which is right and which is wrong but to learn about the factors and processes which are at work in shaping any city.

The concentric model (Figure 8.21) was published by Burgess in 1925 and was based on Chicago. His theoretical city has a circular structure. The main area of shops and offices is in the town centre and is known as the Central Business District (CBD). This is surrounded by the transition or twilight zone which is now more commonly referred to as the inner city. It is a mixture of run down houses; areas of redevelopment and areas of **gentrification**; declining traditional industries and new light industrial estates; congested city roads and urban motorways; close-knit working class communities and immigrants. As you move away from the transition zone the quality of housing improves.

A number of factors and processes help to explain this model. **Accessibility** explains the location of the CBD because the town centre is the easiest place for everyone to get to, wherever they live. Concentric growth explains the circular structure with the town growing outwards from a central point like the

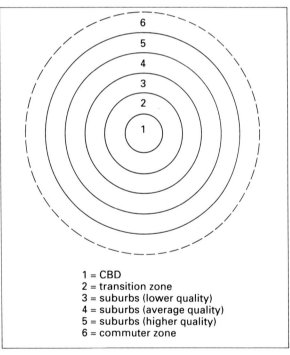

Figure 8.21 *Concentric model of urban structure*

ripples on a pond. Concentric decay, with all of the buildings in a zone getting older at the same rate, explains the variation in housing quality between the older inner suburbs and the newer outer suburbs. Attraction means that some types of land use are found near each other, like banks and offices. Repulsion means that other types of land use try to avoid each other, like industry and upper class housing.

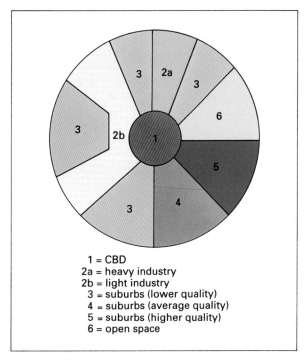

Figure 8.22 *Sector model of urban structure*

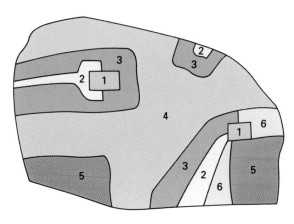

Figure 8.23 *Multiple nuclei model of urban structure*

In reality, there are many reasons why cities do not have a perfect concentric structure. Physical factors such as hills or lakes can disrupt the pattern directly, or indirectly by attracting particular types of land use. Economic factors are also significant; for example, the need for good transport links can lead to industrial development along major routeways. Social trends can make, or keep, parts of a city fashionable, such as Mayfair in central London. Political decisions can also disrupt Burgess's theoretical pattern; for example, the location of overspill estates on the outskirts of Glasgow.

In 1939 the sector model of urban structure was put forward by Hoyt (Figure 8.22). It retains a CBD but the other land use zones grow out from the centre in a radial pattern. Accessibility, attraction and repulsion are important to this model but sector growth is the main feature: different types of land use become established in the centre and then develop outwards because they attract more of the same type. This growth often takes place alongside main routeways.

The sector model pays more attention to economic factors but it is affected by the same physical, social and political factors as the concentric model. Also, the age of buildings in a city is likely to show a concentric pattern, whatever their function, so as these buildings get older, concentric decay could disrupt a radial pattern.

A third theory of urban structure, the multiple nuclei model, was published in 1945 by Harris and Ullman (Figure 8.23). It allows a city to have more than one growth point, unlike the concentric and the sector models which assume that a town or city develops around a single core. Consequently, it is more useful when analysing the structure of large urban settlements, such as conurbations, which are the result of smaller urban settlements merging together. The processes at work in the concentric and sector models still operate in the multiple nuclei model, so a variety of city forms are possible.

Recent attempts at analysing urban structure have concentrated on identifying and mapping groups of social, economic and political factors; an approach known as factor analysis. It has been discovered that these demonstrate concentric, sectoral and multiple nuclei patterns and this reinforces the point that a wide range of processes can operate in a city simultaneously. For example, family status, migration and the **urban-rural fringe** display concentric patterns; economic activity displays sectoral trends; and ethnicity and poverty show a multiple nuclei pattern (Figure 8.24). The structure of a city is the result of these trends and patterns being superimposed one upon the other.

The models described so far are appropriate for cities in Europe and North America but elsewhere the history of urban growth has been very different and this has affected urban structure; consequently, different models are needed. Figure 8.25 shows one for South East Asia. Most of the cities in this region were first developed as ports by European traders. During the **colonial period** they took on an administrative function with the European sector

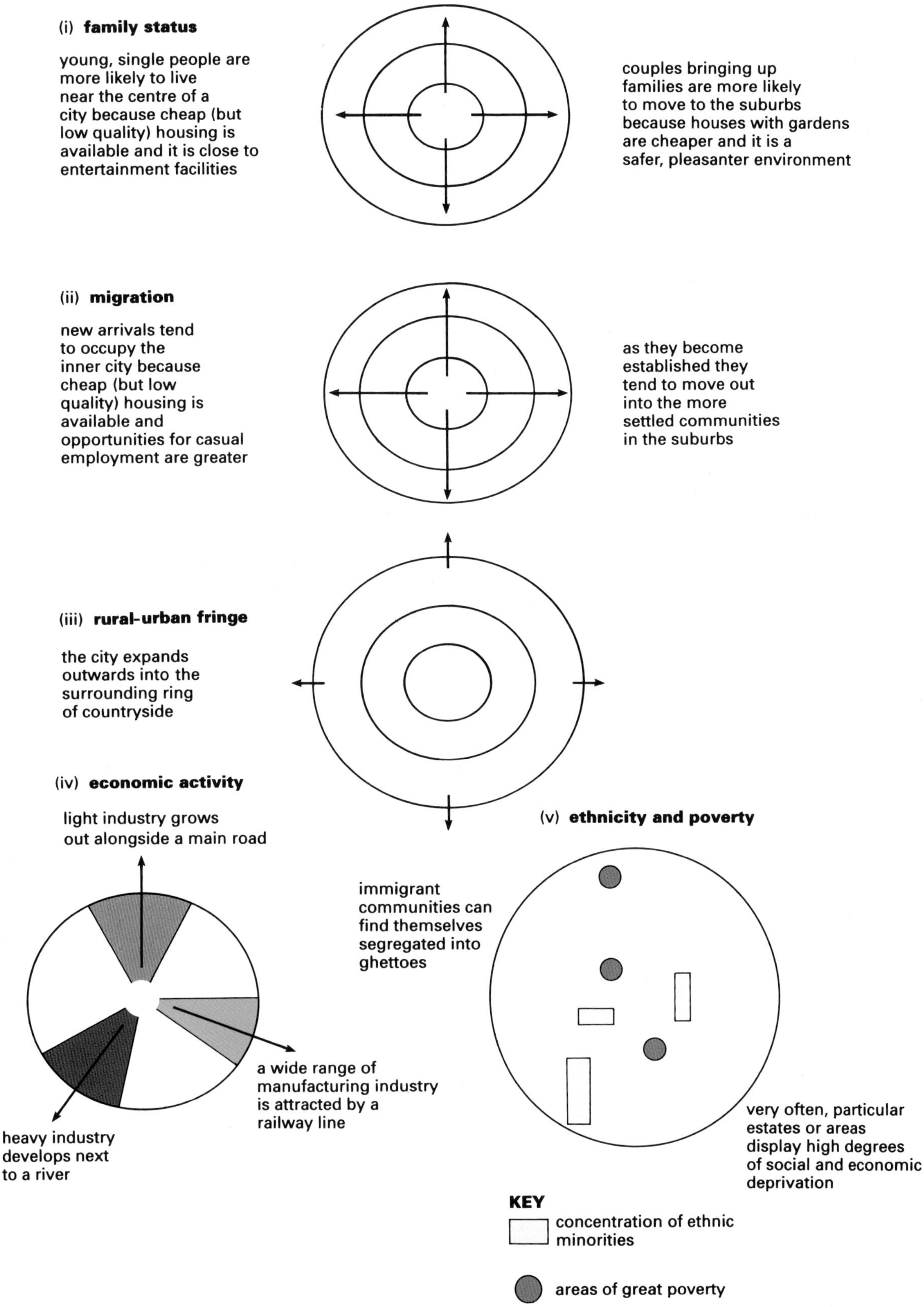

(i) family status

young, single people are more likely to live near the centre of a city because cheap (but low quality) housing is available and it is close to entertainment facilities

couples bringing up families are more likely to move to the suburbs because houses with gardens are cheaper and it is a safer, pleasanter environment

(ii) migration

new arrivals tend to occupy the inner city because cheap (but low quality) housing is available and opportunities for casual employment are greater

as they become established they tend to move out into the more settled communities in the suburbs

(iii) rural-urban fringe

the city expands outwards into the surrounding ring of countryside

(iv) economic activity

light industry grows out alongside a main road

a wide range of manufacturing industry is attracted by a railway line

heavy industry develops next to a river

(v) ethnicity and poverty

immigrant communities can find themselves segregated into ghettoes

very often, particular estates or areas display high degrees of social and economic deprivation

KEY

☐ concentration of ethnic minorities

● areas of great poverty

Figure 8.24 Factor analysis of urban structure

LIFE IN THE CITY

forming a distinct enclave. Expansion has tended to take place outwards from the port zone. Squatter settlements have occupied locations on the edge of the settlement where land is available for building, and in the central area where opportunities for casual work exist. Such models are, of course, affected by physical, social and economic factors in the same way as those generated for Europe and North America.

Q24. *Draw and label a diagram to show the main processes at work in the model for cities in South East Asia.*

Q25. *Which of the models of urban structure help to explain the land use zones of Glasgow (Figure 8.39, page 211), and why?*

Q26. *To what extent does the model for cities in South East Asia help to explain Bombay's land use zones (Figure 8.45, page 214).*

Where do people live, and why?

Above all else, towns and cities are places where people live. However, it is striking that even in a small town, there are variations in location, type and ownership of housing. There can be luxury penthouse flats in the CBD; run down deck-access housing in the inner city; and exclusive 'garden suburbs'. There are one bedroom flats, renovated terraces and detached mansions. Some housing is owner-occupied; some is owned by local councils or housing associations; and there are illegal **squats**. These different types of housing can be very close to each other and this helps to give cities a unique fascination and interest, not only to geography, but also to other disciplines such as sociology and economics.

The character and location of many of these different types of housing are described and explained in other parts of this section. However, one aspect of housing in EMDCs which generally receives little attention is the growing number of homeless people. In Britain it is estimated that there are now more than two million people in this category (Figure 8.26). It is not just a problem in the big towns and cities – at least 40 per cent of this total is to be found in small towns and rural areas – but with an estimated 75 000 homeless in London, such totals are equivalent to the shanty town populations of cities in ELDCs.

People sleeping rough on the streets are the tip of the iceberg. The conditions that squatters live in can be nearly as bad because the buildings they occupy rarely have gas, water or electricity. Those in bed and breakfast accommodation find themselves living in cramped, unsatisfactory conditions and sharing facilities with many other families. The number of 'hidden homeless' is a more contentious figure: these are people who are living with friends and family and some think that they should not be counted as 'homeless' in the same way as those living on the streets. However, studies have shown that the majority of these people are living in overcrowded

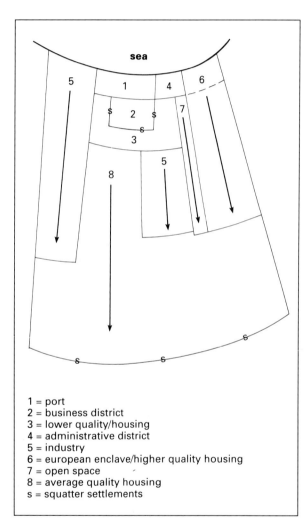

1 = port
2 = business district
3 = lower quality/housing
4 = administrative district
5 = industry
6 = european enclave/higher quality housing
7 = open space
8 = average quality housing
s = squatter settlements

Figure 8.25 *Model of South East Asian cities*

Q21. *Draw and label a diagram to show the main processes at work in the concentric model of urban structure.*

Q22. *In what ways are a) the sector model and b) the multiple nuclei model different to the concentric model?*

Q23. *Describe and try to explain the trends shown in Figure 8.24.*

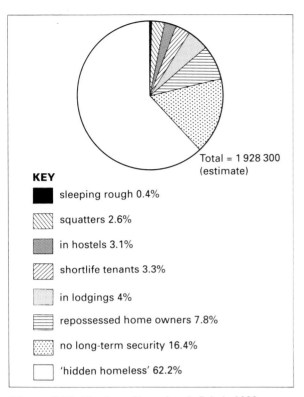

Figure 8.26 *Number of homeless in Britain 1992*

and unsatisfactory conditions very much against their will.

Those sleeping rough are usually found in the CBD. This is no coincidence: this area is busy during the day but is deserted at night, so there is the possibility of bedding down in shop and office doorways without causing too much inconvenience or being moved on or arrested (Figure 8.27). Bed and breakfast accommodation tends to be concentrated in run down hotels on the edge of the CBD, such as Earls Court in London. Most squatters are found in the inner city because this is where the greatest amount of empty housing is to be found. The 'hidden homeless' can be spread throughout the residential areas of the city but the greatest concentration tends to be in the inner city and the inner suburbs because these are the poorest neighbourhoods.

There are several reasons for the increase in the number of homeless. The decrease in state involvement in housing provision is one of the main ones (Figure 8.28, page 206). The government has stopped local authorities from building new council houses, which used to be the major source of cheap accommodation, but measures to expand the role of the housing associations and the private sector have

Figure 8.27 *Sleeping rough, London*

LIFE IN THE CITY **205**

	1968	1990
Local authority	133 145	13 434
New Towns	6 301	—
Housing associations	5 538	11 743
Government departments	4 236	—
Private	203 324	130 132

Figure 8.28 *Houses completed in England 1968 and 1990*

failed to bridge the gap. The economic recession of the late 1980s saw many people getting into difficulties with their mortgage repayments; for example, there were 70 000 repossessions in 1991. The 'hidden homeless' also includes people who want their own house but perhaps would not have expected it in the past, such as young people and single-parent families.

The homeless have a very different experience of the city environment to the rest of the population. Sleeping on the streets is dangerous (Figure 8.29). There is a direct correlation between unsatisfactory living conditions and poor health. Getting full time employment when you do not have a permanent address is extremely difficult. Consequently, these people find themselves in a situation which it is difficult to get out of.

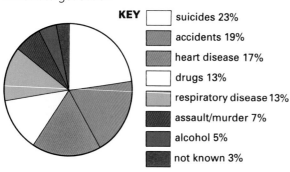

Figure 8.29 *Causes of death of people sleeping rough, London 1992*

Q27. *Comment on the scale; the character; the geographical distribution; the causes and the consequences of homelessness.*

Q28. *What are your own feelings about Britain's housing crisis?*

Where do people work, and why

As well as being places where people live, towns and cities are also places where people work. They have the main concentrations of a country's secondary, tertiary and quaternary activities and their location reveals some interesting geographical patterns.

In the growing towns and cities of nineteenth century Europe and North America, manufacturing industry, heavy and light, was found mainly in the inner city, close to its source of labour. In the twentieth century, as towns and cities have become more congested, and as bulk transport has become larger, these types of activity have moved out of the inner city, first to industrial estates on the main roads leading into the town (Figure 8.30), and more recently to sites on the urban-rural fringe. In ELDCs manufacturing is still to be found in the inner areas of towns and cities, often in cramped and unsatisfactory conditions which would not be tolerated by planning legislation in EMDCs.

Figure 8.30 *The Hoover factory on Western Avenue, London*

Tertiary and quaternary activity has been, and still is, concentrated in the CBD. The accessibility of the CBD was its original advantage and this has been reinforced by the advantages of agglomeration, with shops, banks, insurance companies, for example, finding it useful to be near each other. Competition for land has seen dramatic high-rise developments, many of which have a lot to do with prestige, i.e. their height and design is arguably a symbol of how important the company is, and not just a function of how much space is needed (Figure 8.31). This approach to development has been copied in many ELDCs (see page 209).

This concentration of activity in the CBD has brought with it the problems of traffic congestion, environmental pollution and commuter stress, which are at their worst during the rush hour (Figure 8.32). However, whilst some decentralisation of service activities has taken place in EMDCs, especially of

Figure 8.31 *National Westminster Building, London*

Figure 8.32 *Traffic congestion, London*

shopping facilities with the growth of out-of-town superstores, the pre-eminence of the CBD largely remains intact.

For some time it has been recognised that the traditional classification of industry is of limited value in describing work in the cities of ELDCs because such a large percentage of their populations do not have officially recognised full time employment. Rather, they have unofficial casual employment with little, if any, security. This sector of the economy is described as 'informal' and includes small-scale manufacturing industries such as recycling paper, and service activities such as street-vending and shoe-shining (Figure 8.33). It is unskilled, labour intensive and easy to get into: which helps to explain the lure of the city to rural-urban migrants.

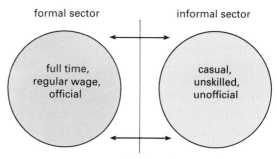

Figure 8.33 *Formal and informal sectors of economy*

Much of this activity takes place within the squatter settlements themselves, such as cooking, selling food and building houses. However, there are also links between this and the more formal sector of the economy; for example, the street-vendors buy their goods from the wholesale markets. Also, while wages are generally lower and more unreliable in the informal sector this is not necessarily the case; for example, officially employed cleaners can have a lower wage than the successful street-vendor who works 12 hours a day.

Q29. Draw and label a diagram to locate and explain the categories of industry in a typical city in an EMDC. Show how and why this pattern has changed since the nineteenth century.

Q30. Modern technology (computers, modems, fax machines etc.) makes it theoretically possible for many service activities to be carried out from home. What benefits could this style of working bring, and why do you think it is so slow to catch on?

Q31. Complete a copy of Figure 8.33 by labelling the correct sector with the following characteristics: small-scale; fixed prices; government help a possibility; simple technology; credit available from banks; no advertising costs; stock bought from day to day; capital intensive.

Q32. In what ways does the formal sector benefit from the informal sector?

Q33. How would the experience of the urban poor in ELDCs be improved if the formal sector of the economy was expanded?

HOW ARE URBAN AREAS MANAGED IN THE UK?

Strategy

The government has been involved with urban planning since the middle of the nineteenth century (see Figure 8.15, page 198), but concern about the urban and industrial framework came to a head in the wake of the 1930s depression. The initial response was a series of Royal Commissions, e.g. the Barlow Commission on the Geographical Distribution of the Industrial Population. Many of the recommendations made by these Commissions were incorporated into Acts of Parliament in the years immediately after the 1939–45 war. The government's main aims were to stop urban sprawl and to deal with slum housing which was found mainly in the inner cities. There was also bomb damage to repair in towns and cities.

Of the most important Acts, the 1946 New Towns Act set up the mechanism for the planned overspill of many thousands of people from the country's main conurbations to new settlements in the regions; and the 1947 Town and Country Planning Act nationalised the right to develop land, i.e. it made it compulsory for developers to apply to the local authority for planning permission before starting to build.

The government's broad strategy had three main elements.

- **Comprehensive redevelopment.** This involved demolishing the run down areas of the inner city and building new, often high-rise, housing.
- **Green belts.** These were set up as rings of land completely surrounding a town or city upon which no new building was allowed.
- **New Towns** and **overspill estates.** One of the aims of comprehensive redevelopment was to reduce overcrowding so these new settlements were necessary to accommodate people who could not be rehoused in the inner city.

This policy was pursued until the 1970s when a shift in emphasis came about for a number of reasons. Firstly, the country's population had begun to stabilise so some of the pressure to provide new housing had eased. Secondly, much of the worst housing had been dealt with. Thirdly, although comprehensive redevelopment at first seemed to be a success, it had run into major problems (see the case study of Glasgow on pages 211 to 213). Fourthly, 1979 saw the election of a Conservative Government with a less interventionist ethos.

As a consequence, the strategy since the 1970s has involved an emphasis on **urban renewal** (renovating rather than demolishing buildings); a wide range of initiatives which have established **partnerships** between the government, private industry and the public sector; and **spatial targeting** (concentrating government assistance into specific areas, such as Enterprise Zones, e.g. the Isle of Dogs in London).

Process

The government sets the main aims and objectives of planning policy through Acts of Parliament and the Department of the Environment (DoE) deals with this aspect of its work. Major projects, such as the building of a New Town, usually have their own development corporation or are run by the DoE itself. In the counties it is left to the County Councils to interpret government policy and to set a framework for development through a structure plan. This leaves the District Councils with responsibility for most day to day planning decisions, such as the development and location of an out-of-town superstore. It draws up a Local Plan against which it judges planning applications, all of which have to go through a clearly defined procedure (Figure 8.34). In the metropolitan areas, the Metropolitan District Councils are required to produce unitary plans consisting of a Part 1, which deals with the overall picture (like the County Structure Plan), and a Part 2, which deals with specific issues (like the District Council Local Plan).

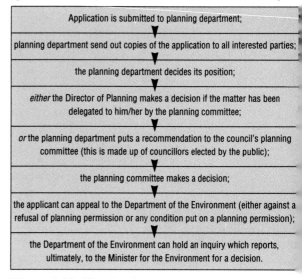

Figure 8.34 *The planning process*

Q34. *Identify and explain the main changes in urban planning strategy since 1945.*

Q35. *Explain the involvement of a) national government and b) local government in the planning process.*

WHAT ARE THE ALTERNATIVE APPROACHES TO URBAN DEVELOPMENT?

The character of any town or city is the result of many processes. Decisions taken by urban planners are very important but there are many different ideas about the most appropriate type of urban development. These ideas can be grouped together into three 'models' of urban planning. However, it should be remembered that some planning strategies are difficult to classify, and that a city may contain more than one of these planning models.

Western model

In the western model, economic growth is the top priority, the idea being that as the city becomes richer wealth 'trickles down' to the poor. There is an emphasis on building offices, roads and industrial estates. Planning controls are kept to a minimum and the city is allowed to develop according to 'natural' urban processes such as suburbanisation and gentrification.

Most EMDCs follow the western model but, compared with the nineteenth century, they now operate a great many planning controls. In ELDCs the western model tends to be followed with very few controls. ELDCs justify this approach on the grounds that it has been successful in the past; and western-style developments also impress investors. However, there is little evidence of the trickle down of wealth taking place.

Self help model

In the self help model the needs of the poorer sections of the urban community are put first. There is an emphasis on housing, basic services and employment. In EMDCs a variety of schemes have been developed in recent years; for example, the joint initiatives launched in Glasgow in the 1980s gave residents a major say in the renewal plans for their districts, and enabled them to carry out much of the work themselves by providing grants and cheap loans. In ELDCs self help projects range from supplying squatters with subsidised building materials so that they can improve the structure of their dwellings, to providing the shell of a house with water, sewerage and electricity laid on, and allowing the squatters to improve it as and when they are able.

Self help schemes have enjoyed some success but because they are by definition small in scale they are of limited value in dealing with large-scale urban problems. Also, they need to be part of an overall plan, or they do not fit in with, for example, the development of the city's infrastructure. In ELDCs they have the advantages of being relatively cheap and popular but they contribute very little to the economic growth of the city and they are less impressive to investors than is, say, an urban motorway.

Radical model

The radical model is also concerned with reducing the quality of life gap but it is very critical of self help schemes which it sees as continuing the existence of rich and poor, even if the poor are marginally better off than they were before. This approach can only be put into operation if the existing power structure is overthrown and complete control is taken of all aspects of urban society.

The planning strategy of the former Soviet Union was an example of the radical model. It reduced the quality of life gap but generated vast areas of uniform blocks of flats. Their design was basic: 9–12 m^2 was allocated per person and the average flat had two rooms with a kitchen and bathroom. Residents were given little choice about where they could live and this led to many problems, such as families being separated and people having long journeys to work. The government ran a flat swap agency but it was notoriously slow and 'unofficial' transfers became increasingly common towards the end of the communist era.

Figure 8.35 *Self help building project, Sri Lanka*

Figure 8.36 *Moscow's suburbs*

Figure 8.37 *Central Bombay*

Figure 8.38 *The World Bank's view of urban development*

The World Bank yesterday called for a more comprehensive approach to urban problems which would move beyond slum improvement to reforms making urban centres more productive.

Traffic jams in Bangkok, the lack of telephones in Sao Paulo and building codes that inhibit construction in Kuala Lumpur not only affect the quality of life for millions, but also, says the bank, hit productivity and therefore economic growth.

The call for a new approach to urban policy, contained in a World Bank policy paper, marks a shift in bank thinking. Since the early 1970s, government and donor efforts have tinkered with low-cost investment projects affecting shelter, water supply and urban transport. While many were reasonably successful, the bank says, they did not affect the broader issues of managing urban centres and how their economic performance linked with national development.

"Experts working in the urban sector did not appreciate how their activities could effect macroeconomic performance," asserts the bank's Urban Policy and Economic Development document.

The call for a shift in thinking towards urban development comes when population growth will add 600 million to cities and towns in developing countries during the 1990s.

As the bank notes, urban poverty is evident in all cities in developing countries: vast neighbourhoods of squatters living outside the legal framework, lacking water, sanitation, transport, adequate shelter and deprived of social services. All this takes place against a backdrop of a deteriorating environment.

The bank identifies four key constraints on urban productivity. The first is infrastructure deficiencies, which restrict productivity of private investment in most cities in developing countries. Firms in Lagos, Nigeria, must invest 10–35 per cent of their capital in power generation. Congestion in Bangkok, Cairo and Mexico City impedes movement. Second is inappropriate regulations. In Lima, 11 permits are required to establish a small textile plant.

Third is the dominant role of government in planning and financing urban infrastructure, starving local governments of financial resources. Finally, poorly-developed financial sectors hinder investment in infrastructure, housing and other urban activities.

The bank calls for reforms in several areas. They include: the regulatory framework governing land and housing markets and use of the private sector to provide infrastructure services; fiscal, financial and administrative relations between central and local government; and policies affecting savings and investment in housing and infrastructure.

The bank also calls for a sustained research effort to tackle the continuing environmental degradation of the cities. There is almost no data on air pollution in any African city.

By calling for a new approach towards urban problems, the bank has set itself an ambitious task. Unfortunately, many of the steps it calls for involve hard cash. The Group of 24 developing countries noted this week the industrialised world and the banks have not borne their side of the bargain by providing new resources.

The Guardian, 30 April 1991

Q36. Summarise each of the three models of urban development under the following headings: main characteristics; advantages; and disadvantages.

Q37. Match Figures 8.35 (page 209), 8.36 and 8.37 with their model of urban development. Justify your choice.

Q38. Read Figure 8.38. Which of the models of urban development is the World Bank in favour of? Which models does it criticise and why? Comment on its main proposals.

Case study: Glasgow

Glasgow is the largest city in Scotland and is the centre of a conurbation with a population of 1.04 million (1991). The original settlement was built on areas of slightly higher ground above the floodplain of the River Clyde, including remnants of raised beach, river terraces and drumlins. It was the **lowest bridging point** on the Clyde and therefore a focus for routeways. Its contemporary land use zones (Figure 8.39) are the product of many years of often dramatic change.

In the twelfth century the city developed as a religious centre, with the building of the cathedral, and as a trading centre. In the eighteenth century trade with North America stimulated the expansion of the city and provided the raw material (cotton) for its first important manufacturing industry (textiles).

However, its major period of growth began in the first part of the nineteenth century because Glasgow was in the ideal situation to take advantage of the Industrial Revolution; there were deposits of coal and iron ore nearby; the Clyde provided a safe and easy means of bulk transport with the rest of the world; merchants had accumulated great wealth from trade so money was available for investment; and rural unemployment meant that there was a ready supply

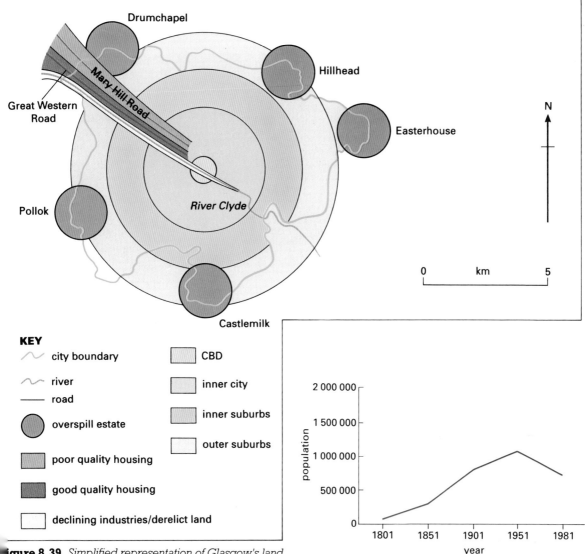

Figure 8.39 Simplified representation of Glasgow's land use zones

Figure 8.40 Glasgow's population growth

LIFE IN THE CITY 211

of migrant labour. Iron, steel, shipbuilding and engineering became the main industries; in 1914, 50 per cent of the world's shipping tonnage was built on the Clyde.

Throughout the 1800s Glasgow's population expanded rapidly (Figure 8.40, page 211). Between 1850 and 1900 over 100 000 dwellings were built in the inner city: most of these were one or two bedroom flats in four or five storey blocks known as tenements. However, the pace of building could not keep up with the increase in population and this led to overcrowding and poor conditions.

In the twentieth century Glasgow has had to face a series of economic problems. Its narrow range of heavy industries were badly affected in the 1930s' depression with over 30 per cent of its workforce being unemployed. Since the 1950s, its geographical advantage has changed completely. The increasing size of ships has meant that port functions have transferred downstream to Greenock and Port Glasgow. The coal and iron ore deposits have been worked out. Its inland location, the development of new materials, such as plastics, and foreign competition have contributed to the decline of the iron and steel industry. Its last integrated works, Ravenscraig, closed in 1992. The limits of the relatively shallow and narrow river to the size of vessel which can be manufactured, and foreign competition, have seen an end to shipbuilding.

The combination of a poor housing stock and a fragile economic base meant that Glasgow in the 1950s was faced with serious socio-economic problems. In parts of the inner city, population density was as high as 96 100 per km^2. One-third of the city's entire housing stock was considered to be 'at or near the end of its structural life' and living conditions were amongst the worst in the country (Figure 8.41). It is against this background that the planning initiatives of the last 50 years must be assessed.

% of households	Central Clydeside	West Midlands	Greater London
without own lavatory	29.2	8.9	2.6
living in one or two rooms	42.7	2.3	5.6
with more than three people per room	7.0	0.5	0.2

N.B. Population: Central Clydeside = 1.1 million; West Midlands = 2.4 million; Greater London = 8.3 million.

Figure 8.41 *Socio-economic conditions in Glasgow, West Midlands and Greater London 1956*

The first strategy adopted by the city was comprehensive redevelopment. This involved the demolition of tenement blocks in the inner city; the rehousing of people in mainly high-rise flats, overspill estates on the edge of the town, or in New Towns, such as Glenrothes; and the building of urban motorways. By 1975, 95 000 demolitions had been completed.

At first it appeared as if this strategy was going to be a success. Overcrowding was significantly reduced (Figure 8.42), the new dwellings had proper facilities and the motorways made the city an easy place to get to. However, within only a few years major problems surfaced. Many of the new buildings had been badly constructed: they were damp and had structural problems. High-rise living was unpopular with tenants. Communities were split up.

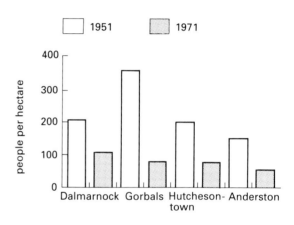

Figure 8.42 *Inner city densities before and after comprehensive redevelopment*

Lack of money meant that many areas were not rebuilt, leaving gaps in the urban landscape. The motorways separated districts and, rather than attracting economic activity, made it easier to go straight through the city without stopping. In some areas the problems with the new buildings were so bad that they had to be abandoned only a few years after they had been first occupied. For example, Hutchesontown 'Area E' was built in 1968, abandoned in 1982 and demolished in 1987.

The shortcomings of comprehensive redevelopment led to a change of strategy in the 1970s with the emphasis being placed on urban renewal, i.e. renovating, rather than demolishing, existing buildings. Glasgow Eastern Area Renewal (GEAR) was a major urban renewal scheme which ran for ten years from 1976. GEAR covered an area of 1600 ha which had seen a dramatic fall in its population but, despite comprehensive redevelopment, still had great socio-economic problems (Figure 8.43). Particular emphasis was given to renovating the remaining tenements, although this could only be done where they were structurally sound and it required knocking two into one to create an acceptable housing density. In an attempt to improve employment opportunities an industrial estate was set up and linked to the motorway network and firms were given grants and other forms of assistance to encourage them to set up.

Indicator	GEAR	Strathclyde
Households with pensioners	41	29
Households including handicapped persons	29	21
Households with children	30	40
Households with no car	84	61
Male unemployment rate	20	13
Female unemployment rate	17	11
Income less than £1750	55	39

Figure 8.43 *Socio-economic indicators 1976 (%)*

In total, around £300 million was spent on a variety of programmes (Figure 8.44). The decline in the area's population was halted, 15 000 dwellings were built or fully modernised, and 250 new factories, workshops and offices were created. However, not all of its aims were fully met: Glasgow's Deputy Director of Planning, Michael Evans, concluded that *'ten years is nowhere near long enough for a major urban renewal project'*.

Recent initiatives have been guided by two principles: partnership, which has meant working with the local community; and diversity, which has meant a wide range of solutions to urban problems, rather than just a single strategy. On the Barrowfield estate in the inner city, the planning department helped the tenants to form an association to control the spending of the money allocated for the renewal of the estate. Barrowfield Community Business was set up to carry out the work and, whenever possible, it had to employ local labour. On the overspill estates, such as Easterhouse, an alternative strategy known as 'homesteading' has been used. Derelict dwellings are made structurally sound and sold at minimal cost to private owners who are then eligible for a range of self-improvement grants. Even some of the high-rise flats have been renovated and, because of more careful management, are proving to be very popular. For example, each block is given a caretaker and security cameras; young families and the elderly are allocated flats at or near the ground floor; and an association is set up so that the tenants can have a say in how the block is managed.

Glasgow has done much to improve its image in the 1980s, although much still remains to be done. The overspill estates require a great deal of attention and there is a concern that unless more money is available for renewal schemes, there will be a return to comprehensive redevelopment. Dwellings successfully renovated in the 1970s are now in need of maintenance. Also, its unemployment is still well above the national average at one in seven of the workforce. However, the city can take confidence from the fact that it is in a better position than it was 50 years ago.

Q39. Describe Glasgow's **site** and **situation** and account for its rapid expansion in the nineteenth century.

Q40. Use an appropriate technique to present the statistics in Figure 8.41. Comment on what they show.

Q41. Compare Glasgow's three main post-1945 planning strategies – comprehensive redevelopment, urban renewal and partnership/diversity – under the following headings:
- the period when used;
- main features;
- the degree of community involvement;
- advantages/successes;
- disadvantages/failures.

You could set this answer out in the form of a table.

Q42. What do you think are Glasgow's planning priorities for the 1990s? How important do you think adequate funding is to meeting the city's needs?

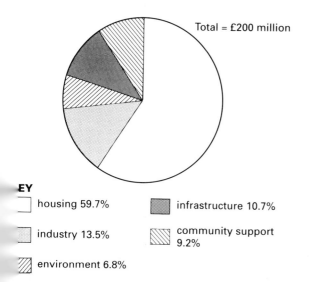

Total = £200 million

KEY
- housing 59.7%
- industry 13.5%
- environment 6.8%
- infrastructure 10.7%
- community support 9.2%

Figure 8.44 *Expenditure in GEAR 1976–84*

Case study: Bombay

Bombay is India's second largest city with a population of 11.2 million (1991). Its site was originally a number of poorly drained islands but it was developed by the East India Company in the seventeenth and eighteenth centuries because it had a good natural harbour and because it was ideally situated to serve Britain's increasing involvement with the Indian sub-continent. By 1900 a great deal of land had been reclaimed and the process has continued this century so that Bombay now forms a continuous peninsula to the west of Thana Creek (Figure 8.45).

In the nineteenth century it developed not only as a centre of trade and commerce but also as a centre of textile manufacture. The cotton was grown locally and the machinery was imported from Britain. In 1901 the city had 102 000 textile workers (94.4 per cent of the manufacturing workforce) and this is still a major employer today.

The Bombay Metropolitan Authority (BMA) recognises four main components of the housing stock. *Bungalows* are detached buildings with good facilities and they are usually occupied by the upper classes: many are found in the former European enclave. *Flats* are classified as self-contained dwellings with two or three bedrooms, good facilities and are found throughout the city. They can be occupied by the moderately well off to the very rich: it is an interesting aspect of Indian cities that even the wealthy often live in dwellings of modest quality, in the company of many members of their extended family. *Chawls* are four or five storey blocks of flats, similar in design to Glasgow's tenements; many were built in the nineteenth century to house the textile workers and most of them are now in a very poor state of repair. Finally, it is estimated that over half of Bombay's population live in huts in **squatter settlements** known as *zopadpattis*.

The zopadpattis are found in four main locations: on vacant land in the business district because of the opportunities for casual employment; on the pavements next to main roads, and on land next to railway lines, not because of good transport links but because there is space for huts to be built; near to certain industrial sites, not because of employment opportunities but because the risk of pollution makes the land unattractive for residential development; and on low-lying land which has yet to be properly reclaimed. Very few zopadpattis are found on Bombay's periphery because it is so far from the business district, and in this way it is different to most other cities in ELDCs.

Conditions in the zopadpattis can be very poor.

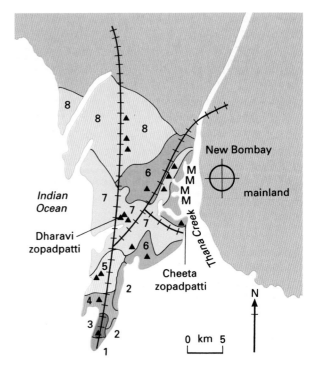

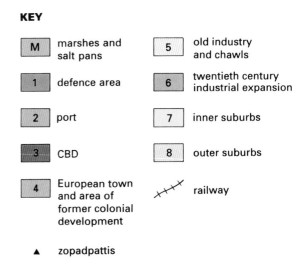

Figure 8.45 *Simplified representation of Bombay's land use zones*

They are very overcrowded with, on average, four o five people per room and population densities as high as 250 000 per km². The huts themselves are usually made of scrap materials and cannot stand up to the monsoon rains which give 1800 mm of rain in three months. The streets are unpaved and unlit. At best, water is available at stand pipes and has to be queued for. Drains are often open and public latrine: if they exist, are in such a poor state of repair that they are unusable. Cheeta colony, built on marshlan in the north west of the city is an example of a zopadpatti where many thousands of people live in these extremely poor conditions. The site was occupied in the late 1970s when the well-established 30-year old Janata colony was cleared by the authorities at gun point because it had become an embarrassment for the Bhaba Atomic Research Centre which it was next to.

However, some zopadpattis have seen improvement. For example, an increasing number of people in Bombay's biggest squatter settlement, Dharavi, have been able to improve their own homes with help from the authorities; some have even been able to build new ones with proper materials. Water supply and sanitation have also been improved. Better off people have even been moving into Dharavi simply because the cost of 'official' accommodation in Bombay is so high.

Conditions may be poor but the rate of rural–urban migration remains high and this is the main reason for the dramatic increase in the city's population (it was only five million in 1971). Rural poverty is the main push factor and this has been made worse by the increase in the number of peasants made landless by debt (see page 200). At least Bombay offers the hope of casual employment and very few of the migrants intend to return to the countryside.

The transport system presents a further challenge for the BMA. Bombay's business district is in the south of the peninsula, near the historic core of the city, but the suburbs stretch 40 km to the north. Most commuter journeys are concentrated into this north–south corridor and this results in major congestion. The problem is made worse by the mixture of cars, buses, bikes, motorcycles and animal carts found on Bombay's roads.

Urban planning in Bombay has a long history, stretching back to the City of Bombay improvement Act of 1898. The first plan for Greater Bombay was drawn up in 1947 and this was followed by the Master Plan for Greater Bombay in 1964. This was superseded by the Bombay Metropolitan Regional Plan in 1973 which is still in force. Its current aims are:

- a freeze on office and commercial employment in central Bombay;
- a limit on new industrial development in Greater Bombay;
- population stabilisation at its current level;
- improvements to urban transport;
- improvements to the water supply;
- improvements in the zopadpattis; including water supply, sewerage, pavements and street lights; house building and repairs, particularly in the area of chawls;
- the financing of self help schemes;
- the building of a new town on the mainland, New Bombay, to take pressure off the city;

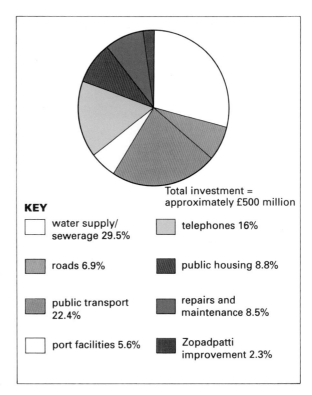

Figure 8.46 *Invesment in Bombay Metropolitan region in 1970s*

- the development of urban growth points in the region surrounding Bombay.

The task facing the BMA is an enormous one and as the city's population continues to rise it becomes even more difficult. This makes it hard to judge the authority's attempts to meet the challenge, although an analysis of how they have allocated their funds (Figure 8.46) gives an indication of their priorities.

Q43. Describe Bombay's site and situation and briefly account for its growth and development since the seventeenth century.

Q44. Describe and explain the distribution of the four main components of Bombay's housing stock.

Q45. Why is security of tenure likely to lead to improvements in living conditions in zopadpattis?

Q46. Draw up a list of Bombay's main urban problems. To what extent are these the result of its physical site?

Q47. To what extent does the BMA's allocation of funds reflect the aims of its regional plan?

Q48. What does CIDCO consider to be the main advantages of New Bombay's CBD (Figure 8.47)?

Q49. What do you think the BMA's two most important priorities should be and why?

Figure 8.47 The Guardian, 28 May 1993

DECISION MAKING EXERCISE

An out-of-town superstore for Huntingdon?

Huntingdon is a market town in Cambridgeshire. It is located at a bridging point on the River Great Ouse; it is close to the A1; and it is one hour from London by train (Figure 8.48). Its population has risen from 5282 in 1951 to 15 434 in 1991 mainly because of the Town Expansion Scheme which lasted from 1957 until 1981. This was an agreement with the Greater London Council to accommodate overspill – those people that moved out from the run down areas of London where comprehensive redevelopment was taking place.

Huntingdon's growth has been accompanied by transport developments and improvements in its shopping facilities. In the 1960s a ring road was built to ease traffic congestion in the town centre and the town itself is now bypassed to the north and south. Two shopping precincts have been built and the high street has been pedestrianised. It has many major stores, such as Waitrose, Boots and W H Smith.

However, the town centre is hemmed in by the river, the railway and housing, leaving very little room for expansion. Consequently, despite the improvements which had been carried out, traffic congestion and parking remained as major problems in the 1980s. Many people were choosing to drive 17 km to the nearest out-of-town superstore at Bar Hill, or even further to Peterborough, Bedford and Cambridge, rather than shop in Huntingdon.

Then, in 1989, both Tesco and Sainsbury put forward proposals to build out-of-town superstores on the newly completed Northern Bypass. The District Council's planning department took the view that one superstore would be of great benefit to the town but that two would take too much trade away from the existing town centre. They decided which of the proposals they thought was the best but the unsuccessful applicant appealed to the Department of the Environment and a public inquiry was held to examine the whole issue.

You have two tasks: firstly, to decide whether or not you agree with the planning department that a superstore would be of benefit to the town; and secondly, to decide which of the two proposals is the best. You should begin by studying the information presented in Figures 8.48 to 8.53. Consider the social, economic and environmental impact of the proposals on their immediate area, on the town centre and on the urban-rural fringe. Set out your answer as a report (which you can illustrate with appropriate maps, graphs, diagrams and tables) using the following structure:

- an introduction which should include necessary background information and a clear statement of the issues involved;
- an analysis and evaluation of the advantages and disadvantages of Huntingdon having an out-of-town superstore;
- an analysis and evaluation of the advantages and disadvantages of the two proposals;
- your personal recommendations, each with a brief justification.

The result of the real public inquiry is given at the end of this section, but do not look at it until you have completed your report!

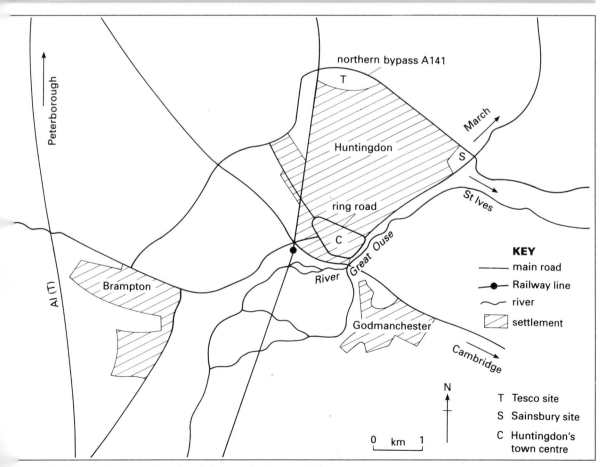

Figure 8.48 *Location map: Tesco or Sainsbury for Huntingdon*

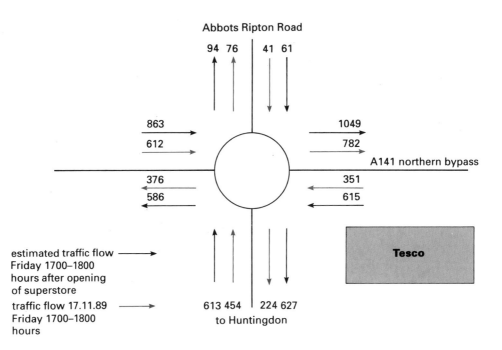

Figure 8.49 *Traffic flows: before and after the opening of the superstore*

Miss Rampakash
I am in favour of a superstore in Huntingdon. At the moment, the shops in the town centre give us very little choice and it is very difficult to park. The nearest out-of-town superstore is at Bar Hill which is 17 km away: this is an inconvenience and we must remember that car journeys have an impact on the environment.

Mr Williams
I am against the Sainsbury proposal. I live in Hartford which until quite recently was a small village separated from Huntingdon by farmland. Although it has been swallowed up by the growth of Huntingdon it has retained its village character. However, a big supermarket on our door step will spoil the area completely and ruin the trade of the local shops.

Mr Low
I own and farm the land where Tesco want to build their supermarket. I'm in favour of this sort of development. The government is encouraging farmers to take land out of production and here is a good use for it, and one which earns us money. I would prefer to get rid of my land on the edge of the town anyway because it is difficult crossing the bypass in slow moving farm vehicles, people use your fields as a rubbish dump and it's too near the road to keep animals.

Mrs West
I don't want any out-of-town superstores in Huntingdon. I cannot drive and I am too old to carry heavy loads of shopping. The High Street is fine for me: there is a regular bus service and I can potter from shop to shop and meet old friends. I am worried that if either of the proposals goes ahead, shops in the town centre will have to close.

Mr Wright
I run a craft shop in the High Street. I am very worried about the proposals. Most people come into my shop because they have seen something in the window as they go past on their way to one of the supermarkets or chain stores. A lot of these people will go to the out-of-town superstore and never see my shop at all.

Mrs Boon
I'm in favour of an out-of-town superstore. I've got a full time job and two young children at school. I don't have time to queue up for a space in a car park and then to walk from one end of the High Street to the other. I want to do all my weekly shopping in one place.

Mrs Fullbright
I am completely against both proposals. As Huntingdon has expanded it has swallowed up farmland, trees, hedges and open space. It is wrong to cover land with concrete just for the convenience of people who want to do their weekly shopping in as short a time as possible. There are sites in the town centre which could be redeveloped but they don't want to do this because it would be more expensive. I value the environment but they don't!

Mr Street
I am in favour of the Sainsbury proposal. It is the more accessible and more imaginative of the two plans. I like the idea of the Manor Farm building being restored for community use because it is the sort of facility which Huntingdon lacks. A bit of competition might make the town centre smarten up its act as well!

Figure 8.50 *Resident's opinions*

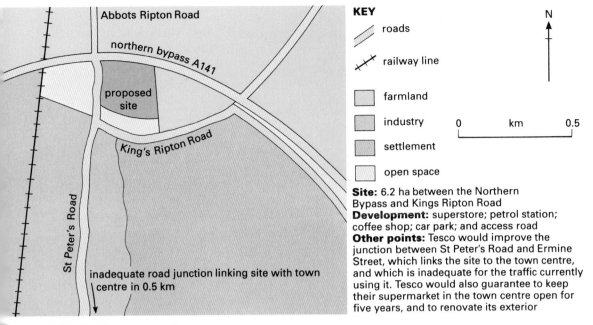

Site: 6.2 ha between the Northern Bypass and Kings Ripton Road
Development: superstore; petrol station; coffee shop; car park; and access road
Other points: Tesco would improve the junction between St Peter's Road and Ermine Street, which links the site to the town centre, and which is inadequate for the traffic currently using it. Tesco would also guarantee to keep their supermarket in the town centre open for five years, and to renovate its exterior

Figure 8.51 *The Tesco proposal*

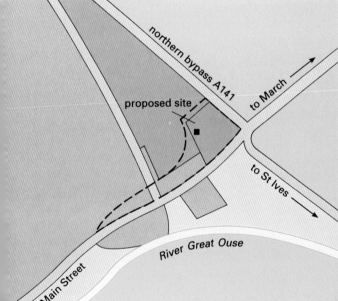

Site: 4.4 ha. at Manor Farm between the Northern Bypass and Main Street, next to the St Ives roundabout
Development: superstore; petrol station; car park; and coffee shop
Other points: The site lies within the Hartford Conservation Area. This means that developments should be sympathetic to the 'village character' of the area. Sainsbury would landscape their supermarket so that it would be hidden from view. Also, they would restore the Manor Farm building and offer it for community use

Figure 8.52 *The Sainsbury proposal*

About the Tesco proposal
Its commitment to keeping its town centre store open for five years would give the High Street time to adjust;
the store is on a main access road for the industrial estate and this could lead to traffic problems.

About the Sainsbury proposal
It will create employment opportunities close to a main residential area of the town;
the noise of its night-time deliveries will disturb the nearby houses.

Figure 8.53 *Comments made by the Inspector at the public inquiry*

PROJECT SUGGESTIONS

Urban areas offer many opportunities for fieldwork. The following piece allows three hypotheses to be tested: that the oldest buildings are found in the centre of a town; that different types of land use are found in distinct zones; and that environmental quality is best on the outskirts of a town.

One problem posed by urban areas is their size: it is simply not possible to record information for every plot of land. Sampling strategy must therefore be one of your first decisions (Figure 8.54). You could choose places by using random numbers to select co-ordinates on a map: this would prevent sample bias but unless you are going to visit a great many sites you could miss out important parts of the town. A transect drawn from one side of the town to the other, and which passes through the town centre, will make testing the above hypotheses easier, especially if you are able to divide the task up among several groups so that all parts of the town are covered. It is perfectly legitimate to use your existing knowledge of the town to ensure that all of its land use zones are covered: this is known as stratified sampling.

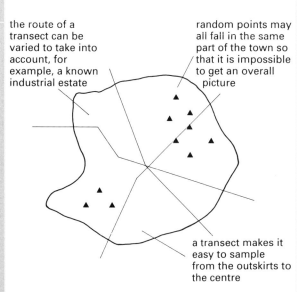

Figure 8.54 *Urban fieldwork: sampling techniques*

Another problem is finding out the age of a building. Some have their age carved above the lintel. However, it is usually necessary to estimate age from architectural style and from the materials used. You should begin by looking at a general text (see the references on page 222) but styles and materials vary from one part of the country to another so you will need to follow this up with research in your local library, and perhaps by contacting the building control officer at your local planning department. You can then make your own identification chart using estate agents' pictures from the local newspapers.

The next stage is to prepare a results table like the one in Figure 8.55 to be used in conjunction with the key and score chart (Figure 8.56). If the town is divided up between several groups you will need to discuss into which category certain types of land use should be put; for example, should a bank be classified as a shop or as an office? You should also discuss your ideas of environmental quality so that each group's opinions are similar. You must also agree on the 'ground rules'; for example, are you going to survey every fifth, tenth or twentieth plot? The size of the town and the time available are two important considerations but you must end up with a meaningful set of data.

Having collected your data, your next task is to present it. This can be done in a variety of ways, only one of which is shown in Figure 8.57. A good idea is to show the three different sets of data – age, land use and environmental quality – on separate sheets of tracing paper so that they can be used as overlays.

There are many ways of analysing the data. You could begin by trying to produce generalised maps of age of building, land use and environmental quality. You could then look for correlations between these maps using your overlays. It is also possible to carry out statistical analysis of the data; for example, if you divide the town up into a series of zones, Chi-Squared (see page 125) could be used to test the hypothesis that 'there are no differences in environmental quality in the town studied'.

It is important that you try to explain all of your findings. Judging them against the models of urban structure is a good way forward. For example: are the oldest buildings found in the centre because this was the first part of the town to be built, or are the newest buildings in the centre because of comprehensive redevelopment? Does land use follow the sector model because of the influence of a main road? Is environmental quality best on the outskirts because it is the area of most recent suburban growth or is it worst because of the location of overspill estates?

You should also mention the limitations which affect the significance of your results; for example, renovations and extensions could affect your decision about the age of a building, while your assessment of noise could be affected by the time of day.

Finally, you should return to the three hypotheses and, by way of a conclusion, state briefly your main findings.

TRANSECT/ZONE NUMBER					DATE						RECORDED BY					
Characteristic / Location	Age	Ground floor land use	First floor land use	Second floor land use	Land use on other floors	Building condition	Gardens, walls, fences	Street and pavement	Street furniture	Traffic safety	Parking	Noise	Litter	Air pollution	Wirescape	Landscape	TOTAL EQA SCORE
1																	
2																	

EQA = Environmental Quality Assessment

Figure 8.55 *Urban fieldwork: record sheet*

(i) Key for age of building and land use

Age of building
1 = 17th Century or earlier (e.g. Tudor)
2 = 18th Century (e.g. Georgian)
3 = 19th/early 20th Century (e.g. Victorian)
4 = 1920s–1940s
5 = 1950s–1970s
6 = 1980s to the present day

Land use
R = residential (house, flat etc.)
S = shop
OS = open space
PB = public building e.g. police station, church
I = industry (manufacturing or distribution)
O = office
E = entertainment
X = other (state type)

(ii) Score chart for Environmental Quality

Building condition
very good repair — 20
satisfactory repair — 10
poor state of repair — 0

Gardens, walls, fences
very good condition — 10
satisfactory condition — 5
very poor condition — 0

Street and pavement
in very good repair — 15
satisfactory but with some pot holes/cracks — 7.5
very poor state of repair — 0

Street furniture
(street lights, post boxes, litter bins etc.)
In very good condition — 5
In need of some maintenance — 2.5
In need of a lot of maintenance — 0

Traffic safety
cul de sac — 5
some residential through traffic — 2.5
main road — 0

Parking
all off-street parking — 5
some on-street parking — 2.5
all on-street parking — 0

Noise
very quiet — 10
some traffic noise etc. — 5
very noisy — 0

Litter
no litter — 10
some litter — 5
a lot of litter — 0

Wirescape
no telephone wires/satellite dishes etc. — 5
one wire and one aerial per house — 2.5
more than one wire and one aerial per house — 0

Landscape
many trees and green spaces — 10
some trees/green spaces — 5
no trees/green spaces — 0

Air pollution
no traffic/industrial fumes or smoke — 5
some smell of traffic etc. — 2.5
noticeably smelly/smoky — 0

Figure 8.56 *Urban fieldwork: key and score chart*

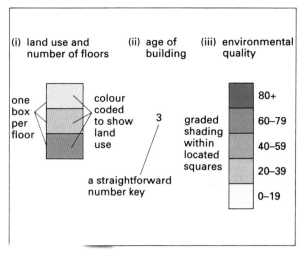

Figure 8.57 *Urban fieldwork: presenting the data*

GLOSSARY

accessibility how easy a place is to get to.

agglomeration a grouping together of activities which benefit from being near each other.

colonial period at its height in the nineteenth and early twentieth centuries when mainly European powers ruled countries in Africa, Asia and South America.

comprehensive redevelopment knocking down large areas of slum housing and rebuilding, usually at a lower density.

conurbations large urban settlements which are the result of towns and cities merging together.

counterurbanisation movement of population in developed countries away from large conurbations and cities to small towns and remote rural areas.

deck-access housing blocks of flats, access to which is via stairs and walkways which go past people's front doors.

dormitory settlements ones with a large percentage of commuters who sleep there but travel to work elsewhere.

gentrification part of a town is renovated and then occupied by a group of better off people.

green belts rings of land surrounding towns or cities upon which no new building is allowed.

Green Revolution a scientific approach to improving crop production, the term is particularly associated with the use of High Yielding Varieties of seeds in ELDCs.

Informal sector casual, irregular and unofficial work.

REFERENCES

Alternative approaches to development, Longman, 1985
M Carr, *Patterns: Process and Change in Human Geography*, Macmillan, 1987
D Drakakis-Smith, *The Third World City*, Routledge, 1987
W K D Davies and D T Herbert, *Communities Within Cities: An Urban Social Geography*, Belhaven Press, 1993
J Frew, *Advanced Geography Fieldwork*, Nelson, 1993
N Pevsner, *A History of Building Types*, Thames and Hudson, 1976
J Short, *An Introduction to Urban Geography*, Routledge, 1984
F Slater, *Societies, Choices and Environments*, Collins, 1991
Urban planning for a Third World City, Longman, 1985

DECISION MAKING EXERCISE: THE RESULT OF THE PUBLIC INQUIRY

The planning department made a decision in favour of Tesco. The Inspector at the public inquiry agreed that Huntingdon would benefit from an out-of-town superstore and also decided on Tesco, mainly because it would have less of an impact on existing housing.

lowest bridging point on a river the nearest place to the sea where a bridge can be built.

natural increase population increase because the birth rate is higher than the death rate.

New Towns urban settlements built from scratch since 1945.

overspill estates areas, usually on the outskirts of a town or city, where people are moved to when their homes have been demolished as part of a comprehensive redevelopment programme.

partnerships a planning strategy which combines public and private money.

rural-urban migration the movement of people from the countryside to the town.

site the land a settlement is built on.

situation a settlement in relation to the area around it; for example, roads, rivers, land use.

spatial targeting a planning strategy which concentrates assistance on those who need it most.

squats housing which has been illegally occupied

squatter settlements areas of makeshift housing built on what is often illegally occupied land.

urbanisation the process by which an increasing percentage of a country's population comes to live in towns and cities.

urban renewal renovating slum housing.

urban-rural fringe the zone where a town or city meets the countryside.

SECTION 9

All the people

by Steve Burton

> **KEY IDEAS**
>
> - **Population growth is the world's greatest environmental threat.**
> - **Population growth is out of control in some parts of the world.**
> - **Population growth rates vary between countries.**
> - **Some countries are concerned about their ageing populations.**
> - **There are different approaches to managing populations.**
> - **Population policies need to encourage sustainability.**
> - **Migration threatens the stability of cities and the countryside.**
> - **Women need to be given greater control over their lives.**

GLOBAL POPULATION CHANGE

Slowly but surely the global time bomb is ticking away. It isn't the risk of nuclear holocaust or extreme natural events that threatens the planet. It is the pressure of population. The evidence for this is startlingly bleak:

Over 100 million acts of sexual intercourse take place each day. These result in 910 000 conceptions and 356 000 sexually transmitted bacterial and viral infections. About 50 per cent of the conceptions are unplanned, and about 25 per cent are unwanted.

Three people are born every second; more than 250 000 every day. At the beginning of the 1990s the number of people being added to the planet each year was 93 million; by the end of the decade the annual increase will be more than 100 million.

- About 150 000 unwanted pregnancies are terminated every day by induced abortions. One-third of these abortions are performed under unsafe conditions and in an adverse social and legal climate, resulting in some 500 deaths each day.
- Every day, 1370 women die in the course of pregnancy and childbirth.
- Some 25 000 infants and 14 000 children aged one to four years old die each day. One in 12 infants born this year will not see their first birthday, and one in eight will not see their fifth birthday.
- Over 300 million couples do not have access to family planning services.
- Aids will have infected 40 million people by the end of the decade with the possibility of a million deaths per year.

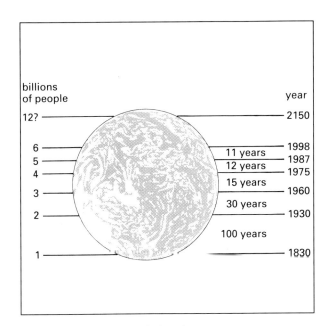

Figure 9.1 World population change

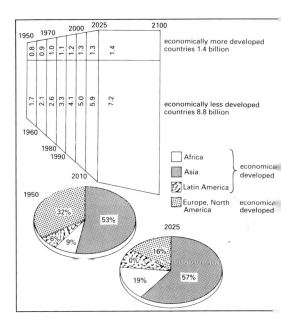

Figure 9.2 Growth of world population 1950–2100

Population projections

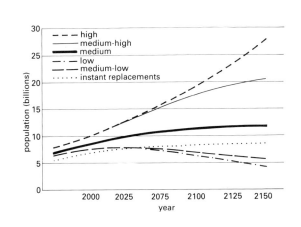

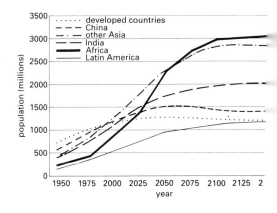

Population projections bring the issues of population growth into focus by illustrating the possible paths along which population can grow. In 1992 the UN released a set of long range global and regional population projections up to the year 2150. These complimented the medium range country specific projections which are prepared each year up to to the year 2125. The projections use a wide range of fertility scenarious to present five possible population sizes. Population projections are important in providing planners with a realistic framework and in assessing the results of today's population policies. For example, they help to create an understanding of how many women and children will need health care services, the number of young people who will need schooling, how many workers will be looking for new jobs, and how many new families will require housing. Some countries use population projections for target setting in their 'five year plans'.

Preparing population projections is actually a much more detailed process than it may appear. Rather than simply projecting a total population figure forward by using an assumed growth rate, individual age cohorts, or age groups, are 'aged' forward by using survival rates from life tables. Thus, if 2.3 per cent of women ages 40–44 in a country die before reaching the ages of 45 to 49, one multiplies the number of women in the 40–44 age group by 97.7 to carr the survivors forward to the next age group. If there were 1 201 019 women ages 40 to 44, then there would be 1 201 019 x .977 = 1 173 396 women ages 45 to 49 five y later.

Mortality is one of three components of the projection, th other two being fertility and migration. To calculate the number of births, the number of women in the childbeari age cohorts (ages 15–49) is multiplied by age-specific bi rates. Thus, the number of births is dependent not only o the birth rate, but also upon the number of women in the childbearing years. Simiarly, the number of immigrants c emigrants is added to or subtracted from each age group migration is a factor.

One of the factors important in population projections is 'replacement level'. This is the rate that will eventually le a no-growth situation. Looking at the future, the year 20 example, population size does not vary much under the quite different, yet plausible, fertility scenarios presented above. But under these same scenarios, the projections that the world could have an eventual population size of anywhere between 4.3 and 28 billion!

Figure 9.3 Population projections

Each of the facts on page 223, are alarming. But the urgency of the problem can only be realised when the rate of population change is considered (see Figure 9.1). The world population, which reached 5.5 billion in mid-1993 has not actually been growing any faster than the 1975 rate – at about 1.7 per cent each year. In fact the average number of children per family has fallen slightly: it was 3.8 in 1975–80 and 3.3 in 1990–5. However, because of past growth the number of people added each year is still rising very quickly: a population equivalent to that of Southampton or Manchester is added daily and by the year 2001 the increase will be the equivalent of a new China, with nearly a billion extra people in a single decade. Even a pandemic like Aids cannot significantly affect the growth. And most of this growth, approximately 95 per cent, is occurring in Economically Less Developed Countries (ELDCs) (see Figure 9.2). In Economically More Developed Countries (EMDCs) the problems come from gradually ageing populations. Rapid population growth, projected into the future, is still the dominant feature of global demographics and will be for the next 30 years (see Figure 9.3).

But what do these statistics mean for the planet and us individually? About 10 000 years ago the world population was around five million, about half the present population of London. By the time of Christ the population had slowly grown to around 200 million, a result of the move from hunting and gathering to settled agriculture along rivers like the Nile and Ganges. Over the next 1650 years the global population crept up to 500 million. The first billion was not reached until 1830. It then took 100 years to reach the next billion, 30 years to reach the next and 15 years to reach four billion. This final sprint meant that population growth was exponential (a logarithmic or regular doubling), and frighteningly fast! Growth of the world's population is not expected to stop for about another century, when the population may stabilise at approximately 10.2 billion – nearly twice its present level. Even this figure is a guestimate – no one is really sure what the total may be.

More people means more mouths to feed, more land to find for farms and homes, more minerals and wood for shelter, tools, transport and basic goods, and more medicines, hospitals and doctors. It also means more waste: every European leaves a lifetime's waste of 1000 times his or her body weight! However, the coming population crisis is not a resource crisis, although individual countries will face these with growing severity as they come up against limits of land, water and food. There could be a health crisis as countries have to face the cost of an increasing health burden and tackle pandemics like Aids, but there is most likely to be a pollution crisis, the impacts of the solid, liquid and gaseous wastes that pour into the world. Ten years ago the fear was that population growth and increasing consumption would mean that we would run out of the non-renewable resources like oil, gold or tin. Today it is the **renewable resources** which are under attack as population growth and consumption threaten the planet's **bio-diversity**.

Q1. a) *Identify the ways in which global population change may affect the world's resource base. Rank the effects in order of importance justifying your choice.*
b) *Study Figure 9.4. For each diagram describe what has happened to world population growth. What are the main differences and similarities? Which diagram, in your view, is the best method of illustrating population growth?*

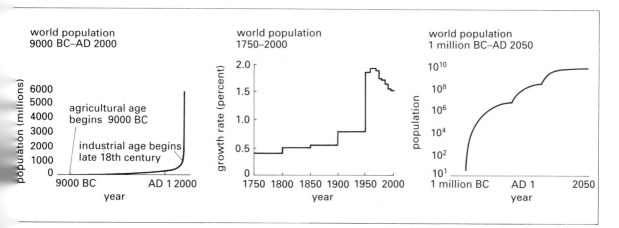

Figure 9.4 *Different perspectives on population change*

Demographic disaster

Demographers are population experts and **demography** is the study of population. Pure demography is concerned with the collection, evaluation, analysis and projection of population data, while social demography explains demographic patterns and processes (see Figure 9.3). In the last section some elements of social demography were identified: in some parts of the world there are too many people using a deteriorating resource base; in other countries zero or little population growth means nations age slowly. The population paradox, too many or too old, is one view of demographic issues.

Too old?

Although the Japanese live with some of the highest population densities in the world the problem here is not too many people, but too few. Like other nations in the EMDCs, Japan's population is 'greying'. The ageing is currently happening more rapidly than previously expected, with the proportion of elderly predicted to reach the critical level of 20 per cent in 2007, according to the Nihon University Population Research Institute (NUPRI). At this time Japan will have 26.3 million people aged 65 or older. This is a dramatic increase from the ten per cent of elderly in the population in 1985. The effect of population ageing is already being felt in many parts of Japan: in some industries acute labour shortages exist; funeral ceremonies in some towns outnumber live births; seasonal festivals are being cancelled because there are not enough young people to carry the *mikoshi* (portable shrines); and women bear the increased burden of caring for the bedridden or senile elderly. NUPRI has warned that Japan will need an 'ingenious policy' to cope with its ageing population.

The Japanese problem is mirrored in the countries which make up the EU (see Figure 9.5). Germany will have the greatest proportion of the elderly in 2030, closely followed by Holland and Denmark. Students leaving their classrooms to find work in the 1990s will need all the earning power they can secure to pay for the pensions that today's workers have promised themselves. Across Europe the rise in the number of pensioners to those of working age is striking. In 1950 the average dependency for Europe was under 20 per cent; by 2040 it will be around 30 per cent. Paying pensioners is just one of the issues: when people cannot draw pensions they often sign on for disability benefits. In Finland, France and Holland fewer than half the men aged between 55 and 64 are in work. Many receive disability payments. Older people are more likely to fall ill, require hospital treatment or specialist care. The future cost for Europe's health and welfare services is enormous.

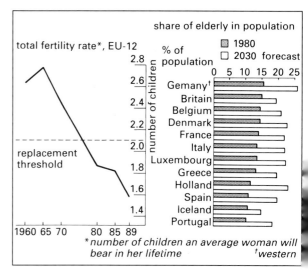

Figure 9.5 *From here to eternity*

Too many?

Too many people create pressures on economic, social, political and physical systems (see Figure 9.6). Although average family size has decreased from 6.1 children in the late 1960s to 3.3 in the early 1990s, and population growth rates have also fallen, the absolute numbers continue to increase. These increases will move forward like a demographic wave into the next century. More than half the world's population will be under the age of 25 by the year 2000. Many of these will be young children, dependent on their families and the state for survival. Most will be living in countries in ELDCs: those countries least equipped to provide for them.

The idea that there are too many people in ELDCs is linked to the concepts of **carrying capacity** and **overpopulation**. Carrying capacity has two aspects: those of productive and waste. The productive carrying capacity of land, for example, is the ability of the land to sustain people by providing food or minerals without a deterioration in the land itself. The waste carrying capacity is the ability of the land to absorb a certain level of pollution or degradation without significant change. When significant change does occur then the carrying capacity has been exceeded, as is graphically described in Figure 9.6. Carrying capacity is closely linked to changes in technology. As populations have grown people have intervened to raise productive carrying capacity, by sowing grain, irrigating or adding nutrients. Most changes in agricultural technology have been designed to raise carrying capacity. However the earth's waste carrying capacity is far less flexible – the limits have been exceeded in many cases. Paradoxically, attempts to increase productive carrying capacity have led to waste carrying capacity being overstretched.

Population growth and deforestation in Madagascar

Madagascar's forests have been reduced to a narrowing strip along the eastern escarpment. Of the original forest cover of 11.2 million hectares only 7.6 million hectares remained in 1950. Today this has been halved to 3.8 million hectares—which means the habitat for the island's unique wildlife has been halved, in just forty years. Every year some three per cent of the remaining forest is cleared, almost all of that to provide land for populations expanding at 3.2 per cent a year.

The story of one village, Ambodiaviavy, near Ranomafana, shows the process at work. Fifty years ago the whole area was dense forest. Eight families, 32 people in all, came here in 1947, after French colonials burned down their old village.

At first they farmed only the valley bottoms, easily irrigated by the stream running down from the hilltops. There was no shortage of land. Each family took as much as they were capable of working.

Over the next 43 years, the village population swelled ten times over, to 320, and the number of families grew to 36. Natural growth was supplemented by immigration from the overcrowded plateaux, where all cultivable land was occupied.

The valley bottom lands had filled up completely by the 1950s. New couples started to clear forest on the sloping valley sides. They moved gradually uphill until, today, they are two thirds of the way to the hilltops.

There was a parallel decline in the size of each family's paddy holding—also fuelled by population growth. When children marry, parents have to subdivide their own land and give them a plot. So holdings in the irrigated valley bottoms have dwindled. Today only a few are big enough to feed a family.

The more children in a family, the smaller their share as adults will be. The village chief lives in a small mud hut, looking out over a valley which he once owned entirely. Since then he has had ten children, and given parcels away to each. Though he is the wealthiest man in the village in cattle, his sons are among the poorest. They have only half a hectare of paddy each. They moved from prosperity to pauperdom in a single generation.

Zafindraibe's small paddy field feeds his family of five for only four months of the year. In 1990 he felled and burned two hectares of steep forest land to plant hill rice. The next year cassava would take over. After that the plot should be left fallow for at least six or seven years.

Now population growth is forcing farmers to cut back the fallow cycle. As land shortage increases, a growing number of families can no longer afford to leave the hillsides fallow long enough to restore their fertility. They return more and more often. Each year it is cultivated, the hillside plot loses more topsoil, organic matter, nutrients. No-one here uses fertiliser.

I came upon Marie Rasoaninna burning the stumps of last year's pineapple plants on a quarter-hectare field. This year she would plant beans and cassava. Then she would leave it fallow for two years, instead of the six it needs.

Figure 9.6 *Population pressures in Madagascar*

Overpopulation of a country occurs when the population exceeds the ideal or optimum (this may, for example, be linked to the most efficient use of a country's resources). The consequences of overpopulation may include unemployment, underemployment, environmental degradation and loss of wildlife, pollution, housing shortages, low living standards, war and racism. Hitler used the concept of overpopulation to invade other countries during the 1939–45 war and to justify attacks on the Jews. Carrying capacity is fundamental to the environment while overpopulation depends to a large extent on the definition of the optimum or ideal population level. The acceptance of a lower standard of living may mean that a country is no longer overpopulated.

> **Q2.** *How might attempts to increase productive carrying capacity affect waste carrying capacity? How could these effects be reduced or managed?*

In general terms it is possible to identify four resource impacts that arise from having too many people.

Mineral resources

These are the fixed resources that in the 1960s and 1970s most people thought would run out. In reality the world reserves of many minerals has actually increased, despite rising use. For example, copper reserves have 'grown' from 91 million tonnes in 1950 to 500 million tonnes in 1985. The paradox is explained by advances in technology, exploitation and exploration techniques, and changes in the commodity prices which influence demand. But such a process cannot continue indefinitely – a growing population means greater demands on the resource base, far greater than the demands of the 1970s and 1980s. The demand for non-renewable resources could easily double in the next two decades.

Renewable resources

These are the most threatened by population growth and consumption. Land, for example, is being lost through desertification in semi-arid areas,

Figure 9.7 *Desertification in Burkina Faso, Africa*

salanisation of irrigated areas, intensive farming in agribusiness zones and by soil erosion and urban sprawl everywhere. Estimates for the loss of land of productive potential vary across the globe. By the mid-1980s Asia had lost ten per cent of productive potential on nearly 50 per cent of the land area. The proportion affected in Africa was 40 per cent and in South America it was 27 per cent. Severe degradation – where more than 50 per cent of the yield potential is lost – affected 17 per cent of the land area of Africa, 16 per cent in Asia and ten per cent in South America. In economically more developed regions less than seven per cent was severely degraded. Despite these changes, yields have increased almost everywhere although food production per person, a more realistic measure of the effect of population growth, has gone down in all regions of the developing world except Asia. In South America cereal production fell by five per cent per person from 1970–1989, in West Asia by 18 per cent, and in Africa by 20 per cent.

Population growth inevitably results in a shrinking land base. In 1990 the world arable area per person was 0.27 ha: in ELDCs it was 0.20 ha, while in EMDCs it was 0.53 ha. If these figures are projected to the year 2050 (medium projection) then the average person will have 0.165 ha or 41 m^2. In ELDCs each person will have only 0.11 ha, or 33 m^2. Projecting future land needs also raises some important issues: an extra 1.76 million km^2 will be needed for crops worldwide by the year 2050, with 1.6 million km^2 needed in ELDCs. At the same time non-agricultural demands (land needed for houses, roads, factories and offices) will increase land pressure by 2.75 million km^2 by 2050. The effects of this will be to use up more agricultural land around towns and cities which will have to expand by 2.75 million km^2 to compensate. This expansion is impractical in large parts of Asia and Africa where the limits to land have either been approached or exceeded.

> **Q3.** *Water is another renewable resource under threat. How does population growth affect water supply? Using your library and CD-ROM carry out some research into areas of water shortage. Are these locations also experiencing overpopulation?*

Climatic change

This is evident in many parts of the globe as the waste carrying capacity has been exceeded. Natural systems are increasingly overloaded and beginning to collapse. 'Forest Death', the massive death of trees first noticed in central Europe, indicates excess dioxide and nitrogen oxides. The ozone 'hole' is the result of excess CFCs (chlorofluorocarbons). Eutrophication of lakes and streams through dissolved nitrogen is slowly killing aquatic life. Polluted air, water and land directly increase health risks. Population growth is a major contributor to global warming and in sub-Saharan Africa for example, carbon dioxide output between 1980–8 rose by 68 per cent. In Brazil the figure was 76 per cent and in Indonesia it was 42 per cent. In these countries some of the increases in Greenhouse gases come from methane which is produced by livestock and rice paddies. Slower population growth would help to slow down the Greenhouse effect.

Bio-diversity loss

Population growth plays an important role in species variety and survival. The list of species in danger lengthens every year. One in every 13 species of plants is threatened while one in every ten mammals and birds are at risk. Estimates put species loss at between two and 11 per cent of all species per decade. Like humans, species have limits on their minimum size. The greatest risks to bio-diversity come from habitat loss rather than overhunting or pollution and wildlife habitats are under threat all over the world. Since 1900 the world has lost half its wetlands to drainage for agriculture, clearances for forestry, urban and tourist development. Africa has lost nearly 30 per cent and Asia as much as 60 per cent. Twenty per cent of the world forest area was lost in the last decade: of this loss 79 per cent can be attributed to population growth; the rest to increased consumption for agriculture.

> **Q4.** *Using information in this section draw a flow diagram to show links between population and resource change. What other resources, apart from those mentioned, may be affected? Try and add these to your diagram.*

Causes of population change

In order to manage the demographic disaster it is necessary to understand world population changes. There must be more births than deaths for an overall **natural increase** or more deaths than births for a natural decrease. However, when different parts of the world are considered, population may increase or decrease as a result of natural change (the difference between the number of people being born and dying in a community) and as a result of people moving into or away from a community. This relationship can be expressed as the formula:

$P = B - D +/- M$

where P = population, B = births, D = deaths, M = migration.

The relative importance of the two components of population change can vary over time and space. For example, during the early nineteenth century much of the USA's population growth came from immigration as settlers arrived from a large number of European countries. Later, in the 1940s, natural change became much more important in the country's population growth. Most of the British New Towns experienced significant population increases due to immigration during the early phases of their growth; later growth phases have occurred because of natural increase as the towns move towards their planned target sizes.

THE MISSING 20M

Yoruba chiefs in Lagos have declared total disbelief, election officials have been left stuttering, politicians are flabbergasted. For a nation whose enormous ego is linked to its enormous size, the 1991 census figures released on March 19th came as a shock. If the numbers are correct, and UN observers say they are, then there are only 88.5m Nigerians—at least 20m fewer than expected.

Overnight Nigeria became richer. Assuming an income per person of $250, the World Bank had ranked Nigeria the 13th-poorest country in the world. Now the figure is more like $360, putting Nigeria closer to China and India. Nor is the country's population explosion as severe as some had thought: the population will double in 30 years, not 20. The new numbers suggest an adult population of about 35m, which makes it curious that the National Election Commission has 50m names on its electoral rolls—even after 20m fraudulent names had already been struck off.

By showing that mainly Muslim northerners are more numerous than mainly Christian southerners, the census may help the northerners' campaign to have a northern president. About 5.6m people were counted in Lagos, the same number as in the northern state of Kano. People in Lagos grumble that Kano does not seem so heavily populated. They may be worrying that Kano and Lagos will now receive equal shares of federal revenues.

Nigeria's previous attempts to count itself foundered on these kinds of problems. Since independence, three censuses have produced impossible results, often because the figures were the result of negotiation between regional leaders rather than enumeration. The 1962 results were thought too controversial to publish. In 1963 the north alleged that the southern population was too high and in 1973 the south said the north had inflated its numbers.

This time UN monitors and the census-takers' avoidance of sensitive questions about ethnicity and religion ensured better results. Nigerians spent three days under curfew last November, while 500,000 census-takers paddled up creeks and forded rivers to count them all. It would be a pity to disbelieve their results.

The Economist, 28 March 1992

Figure 9.8 *The missing 20 million in Nigeria*

Demographers refer to the births and deaths in a population as the **vital rates** of **fertility** and **mortality**. These are usually given as units per thousand of the population, although other measures are also used. These statistics and the many factors which may help to change vital rates are described in more detail below. Information on population change comes from the **census** but we must be careful how we interpret this information (see Figure 9.8).

Q5. *There are often problems of data collection when any sort of census is attempted. For example people may refuse to fill in their census forms or register births, deaths or marriages. Demographers depend on using accurate, trustworthy data. Data from ELDCs is often distorted by various factors.*
  a) *Try and identify these factors.*
  b) *Which of these factors may help to explain the missing millions in Nigeria?*
  c) *How can some of the difficulties of carrying out a census be overcome?*
  d) *Suggest ways in which the reliability of census data could be measured.*

Changing fertility

Fertility in population studies, means the occurrence of live births. The simplest measure of fertility is the **crude birth rate**. National crude birth rates in the world today normally fall between the range of 15–50 per 1000.

One method of classifying the factors which influence fertility rates is shown in Figure 9.9, page 230. One of the most common reasons for identifying a fall in fertility is the link between economic growth and the crude birth rate. In the United Kingdom, for example, the increasing wealth and living standards after the Industrial Revolution reduced fertility, as the need for the **extended family** was replaced by the **nuclear** ideal. Added to this were the considerable cost factors of bringing up children in a consumer-oriented society. In other countries, particularly ELDCs, fertility patterns have not shown the same fall, even with increasing living standards and development. Crude birth rates have remained high, sometimes to supply family labour to ensure support during old age or to help run the family farm. Often high levels of fertility exist for other reasons such as the desire to have a male to continue the family name or the need to have boys because girls cannot inherit land or property. In some Muslim societies high fertility levels are one of the ways in which males dominate women and are able to show off their sexual prowess. We know that in global terms fertility

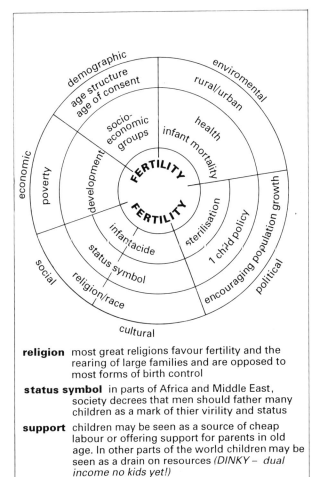

religion most great religions favour fertility and the rearing of large families and are opposed to most forms of birth control

status symbol in parts of Africa and Middle East, society decrees that men should father many children as a mark of thier virility and status

support children may be seen as a source of cheap labour or offering support for parents in old age. In other parts of the world children may be seen as a drain on resources (DINKY – dual income no kids yet!)

politics encouraging population growth e.g. Germany, Italy, Japan before the second World War

Figure 9.9 *Factors influencing fertility*

rates have slowly been falling, however the rate of change varies enormously and is dependent on a wide range of factors, such as the availability of effective programmes of birth control.

Changing mortality

Mortality in demography means deaths. Like fertility the simplest measure is the **crude death rate**, and this statistic also has its limitations. For example, in 1992 the crude death rate for the UK was 12 per 1000 and for Kenya it was 13 per 1000; however a person can expect to live some 21 years longer in the UK than in Kenya. This is because the age structures of the two populations are very different: the UK has 15 per cent of its population over 65 years while Kenya has only two per cent in that age group. Unlike fertility, mortality rates do not vary as much over space, although there have been significant variations over time. One reason for this is because the global trend in mortality rates over the last 300 years has been downwards. Many factors have been identified to explain this retreat from death: the influence of health and hygiene programmes; vaccinations and inoculations against 'killer' diseases; the enormous technological advances in medicine and health care; improved social and welfare provision and improvements in food production and distribution.

One of the automatic consequences of reducing mortality has been the raising of **life expectancy**. This statistic is often calculated separately for each sex because of significant differences in mortality between the sexes. In Japan, for example, life expectancy was 47 years for males and 50 years for females during 1933–7; from 1985–90 life expectancy had increased to 76 and 82 years respectively. Sierra Leone currently has the lowest life expectancy in the world at 35 years. This does not mean that all people live to 35. High mortality in the early age groups lowers the average life expectancy, so having survived their childhood a person could expect to live to well over 35.

Another important statistic is **infant mortality**. Children are particularly vulnerable to disease, especially the new-born, so infant mortality and child mortality are often used as general indicators of a country's state of health. For example a high infant mortality rate would be 159 per 1000 live births in Mali (1990–5) compared to five per 1000 live births in Finland (1990–5). In some countries women have a higher risk of dying during childbirth often as a result of complications in pregnancy and childbirth. This population indicator is measured by the maternal mortality rate and is expressed as deaths per 1000 live births in a given year. An African woman, for example, is 200 times more likely to die from these complications than a woman in western Europe.

The five models

There have been a number of attempts to 'model' and account for population change (see Figure 9.10). In population studies, modelling population change has involved the population projection (see page 224). It is possible to have a number of alternative population projections and therefore a number of different future situations when it comes to the resource/population balance. The resource limit in these different scenarios can be thought of as the carrying capacity described earlier. All the models, apart from the Demographic Transition Model, try to show how population will affect and respond to the resource limits.

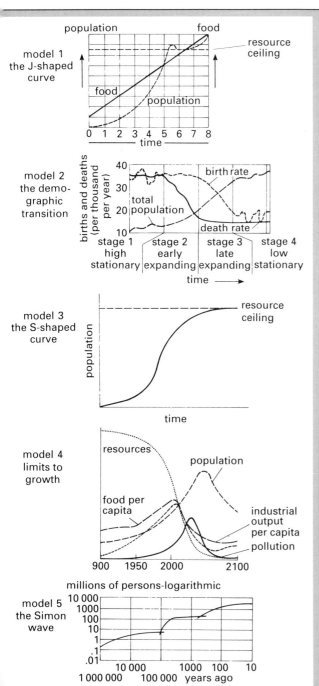

MODEL 2 – THE DEMOGRAPHIC TRANSITION
The Demographic Transition Model, different to the other models, was first developed in the 1940s. It uses changes in the relative balance of the population components, fertility and mortality, rather than linking population growth to a resource limit. The model is development driven, arguing that changes in the vital rates only occur because of changes in the economy. The net result of all countries following the model will be increasingly lower population growth, or zero growth. Research has shown significant differences between EMDCs when they are compared to ELDCs with the latter failing to show the same falls in fertility. In these countries the result of falling mortality has been a large net increase in total population. The recent evidence of falling fertility rates may suggest that some of these nations are now moving towards stage 3. The real issue here is whether they can achieve progress quickly enough before the resource limits are met or exceeded.

MODEL 3 – S-SHAPED CURVE
Contradicting the Malthusian view was the S-shaped curve of Esther Boserup. The essence of Boserup's argument was that *'necessity is the mother of invention'*: population growth encouraged agricultural development which in turn stimulated further population growth, increased development, more growth, and so on. Boserup pointed out that as the resource limit was neared societies regulated their population sizes so that it was not exceeded. These ideas were later developed further by Simon's wave model. Both these models saw population growth as having a positive rather than negative effect on society.

MODEL 4 – THE LIMITS TO GROWTH
This model was developed by the Club of Rome in 1972. The Club, which involved ten countries, produced a computer model to show what would happen to food supply, industrial output, resources and pollution if population growth continued. The computer simulation showed that the limits to growth would be reached in the next 100 years; the most probable result being a sudden and uncontrollable decline in population and industrial capacity. This model had a far reaching impact on thinking in the 1970s and 1980s.

MODEL 5 – THE SIMON WAVE
The Simon Wave was developed by Julian Simon, a USA economist in 1981. Simon believed that moderate population growth was good for development. He did not believe that population growth was geometric (the Malthus view); instead he saw sudden increases in population growth as the result of major economic, technical or health improvements. Population growth was gradually moderated as the advances were assimilated. Simon felt that more people could cause problems in the short term, but at the same time there could be more people to solve the problems!

MODEL 1 – J-SHAPED CURVE
The earliest model to be developed was the J-shaped curve of Thomas Malthus. Malthus was an English clergyman and economist who lived from 1766 to 1834. While writers before him had warned of the dangers of overpopulation, Malthus's views of the problem in 1798 created a debate among social scientists which has not yet subsided. It may be hard to agree with Malthus's 'moral labels' but it is easier to accept that one or more of his checks must operate in every society, keeping growth rates below the earth's theoretical maximum. According to Malthus no amount of technology could invalidate this 'principal of population'. Later in the 1970s the 'Limits to growth Model' provided computerised support for the Malthus idea.

Figure 9.10 *The five models of population change*

DECISION MAKING EXERCISE

Investigating population change in three countries

Your task is to prepare a report to the *United Nations Population Fund* (UNFPA) about the changes in population in the three countries shown in Figure 9.11. Your report should include:

a) an assessment of the current population situation;
b) an analysis of the changes in population from 1990–2025;
c) an evaluation of the problems that may result from these changes;
d) a discussion of the possible solutions that may result from c).

Background information

USA

The USA has shown quite remarkable changes in its population over the last 200 years, taking it from insignificance in global politics to the position of a superpower. When the first census of population was held in the USA in 1790 the population stood at four million. At the time the country was restricted to the 13 original states on the eastern seaboard and the 'black country'. The next 80 years saw a westward expansion, encouraged by the development of river, canal and railroad networks. The new transport systems assisted the spread of native born Americans from east to west and also helped bring in more European immigrants. Between 1820 and 1880, 7.3 million immigrants arrived from three major sources: Germany (nine million), Ireland (2.8 million) and GB (1.5 million). In the south a plantation system based on slave labour saw the additional arrival of 3.5 million blacks by 1850. Between 1820 and 1880 the number of immigrants quadrupled to 26 million. Population in 1865 was 35 million. In 1950 it had reached 151 million. Since the 1950s immigration has been significantly restricted and recent population growth is attributable to increasing fertility and the 'yuppie factor'.

BRAZIL

Brazil, the largest nation in South America, has a rapidly expanding population. Rather like the USA a large proportion of Brazil's early population growth can be attributed to the country's colonial past and its association with the slave trade. Portuguese occupation has been the dominant colonial impact, imparting the Portuguese language, religion and culture to the country. There have also been Spanish, Dutch, African and British influences creating a cultural and ethnic melting pot. Brazil's population growth has been rapid over the last 120 years. In 1872 the population stood at 9.9 million. By 1940 it had reached 30 million and in the 1950s and 1960s very rapid growth was experienced as mortality rates showed a significant decline. High rates of natural increase have been matched with enormous progress in industrialisation, the building of cities, roads, and power schemes. Fertility on the other hand has tended to remain high in Brazil. One factor explaining this trend is the fact that Brazil, the largest catholic country in the world, has great difficulty implementing effective programmes of birth control. This is because of the 1968, and recently reaffirmed 1993, Papal Encyclicals which express disapproval of the use of mechanical methods of birth control.

BANGLADESH

Bangladesh is a much younger nation than the USA or Brazil. Roughly half the size of the UK and with about twice the number of people Bangladesh's population problems are acute and worsening rapidly. Created in 1973 after the division of Pakistan, the country is one of the poorest in the world. With an estimated population of 113.9 million (1993) living within a geographical area of 143 998km^2, Bangladesh is considered to be the most densely populated country on the globe. Limited resources, rapid population growth and recurrent natural disasters have kept per capita incomes to around $170. It has been estimated that around 55 per cent of the population is below the poverty line while the literacy rate is less than 30 per cent. Infant and maternal mortality rates are very high indicating the poor development of health facilities. The economy of Bangladesh is predominantly agricultural with agriculture accounting for over 50 per cent of GDP and involving 82 per cent of the total population. Jute, or the 'golden fibre', is still considered to be the main cash crop and the major source of foreign exchange earnings. Population pressures on the small land area are intense. Many farmers cultivate tiny fragments at barely subsistence levels. Even the introduction of new technologies such as fertiliser and better seed can do little to raise surplus production which trails behind population increase.

Figure 9.11a *Population data for the USA, Brazil and Bangladesh*

Bangladesh

Indicator	1990	2000	2025
Population total (/000)			
Total	115593	150589	234987
Males	59560	77523	120596
Females	56033	73066	114391
Sex ratio(/100 females)	106.3	106.1	105.4
Urban	19005	34548	99078
Rural	96589	116041	135908
Per cent urban	16.4	22.9	42.2
Functional age groups (0/0)			
Young child 0–4	16.5	15.4	8.9
Child 5–14	27.4	25.2	17.9
Youth 15–24	20.6	20.6	19.5
Elderly 60+	4.6	4.6	7.6
65+	2.9	2.9	4.7
women 15–49	22.7	24.2	27.5
Median age (years)	17.7	19.3	27.0
Dependency ratio: Total (/000)	87.9	76.9	46.0
Aged 0–14	82.5	71.8	39.2
Aged 65+	5.5	5.1	6.8
Population density(/sq km)	803	1046	1832

Indicator	1985–90	1995–00	2020–25
Average annual change (000)			
Population increase	2889	3674	2974
Births	4569	5399	4562
Deaths	1679	1725	1588
Net migration	0	0	0
Annual population growth rate (0/0): Total	2.67	2.60	1.31
Urban	6.33	5.81	3.31
Rural	2.02	1.74	-0.03
Crude birth rate(/1000)	42.2	38.2	20.0
Crude death rate(/1000)	15.5	12.2	7.0
Net migration rate(/1000)	0.0	0.0	0.0
Total fertility rate(/woman)	5.53	4.71	2.30
Gross reproduction rate(/woman)	2.70	2.30	1.12
Net reproduction rate(/woman)	1.95	1.78	0.99
Infant mortality rate(/1000 births)	119	96	53
Life expectancy Males	51.1	55.1	65.1
at birth Females	50.4	54.9	66.1
(years): Both sexes	50.7	55.0	65.6

USA

Indicator	1990	2000	2025
Population total (/000)			
Total	249224	266096	299884
Males	121561	129992	146755
Females	127663	136104	153129
Sex ratio(/100 females)	95.2	95.5	95.8
Urban	186835	205002	253632
Rural	62389	61094	46251
Per cent urban	75.0	77.0	84.0
Functional age groups (0/0)			
Young child 0–4	7.3	6.5	5.9
Child 5–14	14.1	13.7	12.0
Youth 15–24	14.5	13.5	11.9
Elderly 60+	18.9	16.8	20.5
65+	12.6	12.8	19.8
women 15–49	26.4	25.5	21.3
Median age (years)	33.1	36.5	40.0
Dependency ratio: Total (/000)	51.6	49.3	60.4
Aged 0–14	32.5	30.1	28.7
Aged 65+	19.1	19.1	31.7
Population density(/sq km)	27	28	32

Indicator	1985–90	1995–00	2020–25
Average annual change (000)			
Population increase	1988	1567	1027
Births	3665	3453	3520
Deaths	2147	2317	2943
Net migration	450	450	450
Annual population growth rate (0/0): Total	0.81	0.60	0.35
Urban	0.98	0.91	0.68
Rural	0.33	-0.39	-1.39
Crude birth rate(/1000)	15.1	13.2	11.8
Crude death rate(/1000)	8.8	8.8	9.9
Net migration rate(/1000)	1.8	1.7	1.5
Total fertility rate(/woman)	1.83	1.88	1.95
Gross reproduction rate(/woman)	0.89	0.92	0.95
Net reproduction rate(/woman)	0.88	0.90	0.94
Infant mortality rate(/1000 births)	10	7	5
Life expectancy Males	71.9	73.7	76.7
at birth Females	79.0	80.4	82.8
(years): Both sexes	75.5	77.1	79.8

Brazil

Indicator	1990	2000	2025
Population total (/000)			
Total	150368	179487	245809
Males	74992	89323	121590
Females	75376	90164	124219
Sex ratio(/100 females)	99.5	99.1	97.9
Urban	112643	144721	215478
Rural	37725	34766	30331
Per cent urban	74.9	80.6	87.7
Functional age groups (0/0)			
Young child 0–4	12.6	10.8	8.3
Child 5–14	22.6	20.9	16.3
Youth 15–24	19.1	18.8	15.8
Elderly 60+	7.1	8.0	13.8
65+	4.7	5.4	9.3
women 15–49	25.7	26.6	25.6
Median age (years)	22.7	24.7	31.3
Dependency ratio: Total (/000)	66.4	59.1	51.4
Aged 0–14	58.6	50.5	37.3
Aged 65+	7.7	8.6	14.1
Population density(/sq km)	18	21	28

Indicator	1985–90	1995–00	2020–25
Average annual change (000)			
Population increase	2961	2881	2398
Births	4086	4121	4222
Deaths	1125	1241	1824
Net migration	0	0	0
Annual population growth rate (0/0): Total	2.07	1.67	1.00
Urban	3.15	2.31	1.25
Rural	-0.85	-0.78	-0.72
Crude birth rate(/1000)	28.6	23.9	17.6
Crude death rate(/1000)	7.9	7.2	7.6
Net migration rate(/1000)	0.0	0.0	0.0
Total fertility rate(/woman)	3.46	2.91	2.28
Gross reproduction rate(/woman)	1.69	1.42	1.11
Net reproduction rate(/woman)	1.52	1.31	1.06
Infant mortality rate(/1000 births)	63	51	30
Life expectancy Males	62.3	64.7	69.1
at birth Females	67.5	70.4	75.3
(years): Both sexes	64.9	67.5	72.1

Figure 9.11b *Selected population data: medium variant*

POPULATION STRUCTURE – THE DEMOGRAPHIC WEDGE

Every country in the world has a demographic fingerprint. Like human fingerprints these are unique and allow different countries to be identified demographically. In reality the fingerprint is made up of a number of wedges (see Figure 9.12) which lie on top of each other and sometimes form the shape of a population pyramid.

Population pyramids

Population pyramids are graphic methods of representing the age and sex structure of a population at one point in time. The shape of a population pyramid can tell us much about its population history at a glance. It will reflect demographic and socio-economic changes within the population over the period of approximately a century. The shape, or arrangement of wedges, is the result of past fertility, mortality and migration. Pyramids are very sensitive to wars (see Figure 9.12), epidemics, economic conditions, baby booms and population planning policies.

In some population pyramids each age group is represented by percentages rather than by absolute numbers, so the pyramids all have the same area, regardless of the actual numerical size of the population. Comparisons in the age-sex ratios between countries can be made but not comparisons of population size.

One of the most common measures which can be worked out from population pyramids is the **dependency ratio** (see Figure 9.12). For this calculation the age structure must be divided into three broad age groups: elderly dependants aged 65 and over; youthful dependants aged 0–14; working population aged 15–64. To work out the dependency ratio the combined proportion of dependants, young and old, is divided by the working population. This statistic is very useful in assessing the population crisis. For example, if the ratio is high then the dependant group will be a relatively large proportion of the total population who are not economically active. These groups consume resources without producing any in return.

Dependency ratios are generally higher for ELDCs, where the bulk of dependants are young, with relatively few old people. However, this does not mean that young people are dependant in similar ways to those in EMDCs. Quite often children are economically active at a very young age (around four to five years) and the older generation will often continue to work past the age of 65.

It is possible to identify four modifying factors which are evident in most pyramids:

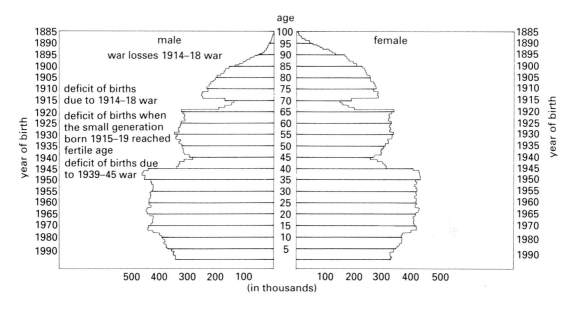

Figure 9.12 *Effects of war on population (France)*

Migration

Population pyramids often reflect the fact that migration is age and sex selective. In parts of Africa and India, for example, it is common to find that males migrate from rural to urban areas creating an imbalance in the destination and origin areas. In European countries migration is common during retirement and as a result the coastal areas of Britain, Spain and France often have an unusually large proportion of the elderly.

Population policies

When these are rigorously pursued then it is possible to see the effects on population. In China the 1980s' one child policy has altered the base of the Chinese pyramid, narrowing it in the early age groups. In France, where population growth is encouraged, the wedges have widened according to the rigour of the campaign and the number of incentives.

Famine, disease and war

These conditions are most likely to have the greatest effects on children and the elderly because they have a lower immunity and are often immobile. The effects of the civil war in the former Yugoslavia in the 1990s had a devastating effect on the young Muslim population. Aids, on the other hand, can have a serious impact on working adults (21–41) and also on infants, through placental transmission of the HIV virus. Researchers have estimated that this could remove up to ten per cent of the population from these age groups.

Social factors

In economically developed societies there is often a higher death rate amongst middle-age men compared to women. The causes of this difference are less clear although some scientists have pointed to the fact that men tend to have higher cholesterol intakes, consume large quantities of alcohol and relieve stress at work by smoking. Recent evidence suggests that some of this difference is due to increasing male suicide, often a response to the difficulties imposed by unemployment.

Demographic wedge and population structure

Population pyramids, like population projections are invaluable in the preparation of planning strategies for the future. It is not just governments that are interested in the demographic wedge. Businesses also need to plan for changes in demand which may come about by the momentum of a baby boom or a decline in fertility: motorcycle manufacturers, for example, look closely at the number of males in the 16–31 groups and adjust their motorcycle production accordingly. In England and Wales, the 1960s' baby boom will result in an 'echo' peak of older people beginning in 2020. This will place a considerable burden on the future working labour force which, in relative terms, will be diminishing each year. In order to put off the effects of an inadequate state pension provision the British Government has offered incentives for those people who opt out of the State Earnings Related Pension Scheme (SERPS). The USA is planning to raise retirement age from 65 to 67 (male) in the year 2000 to reduce the dependency burden and increase the working population. The UK raised the female retirement age from 60 to 65 years in 1993.

Population structure and the census

The census provides information on a number of population aspects. Its reliability has been looked at in Question 5. In spite of the difficulties of obtaining information a census remains the most important source of demographic data. Censuses are carried out periodically in almost all countries. The aim of any census is to provide statistics on the size, distribution and characteristics of all persons in a population. The United Nations (UN) has tried to achieve some sort of international standardisation by encouraging all countries to collect the same types of information at regular intervals. The list includes: place of birth, citizenship and married status; residential location (urban or rural); total population and it's distribution in the country; sex and age; mother tongue, literacy and education; fertility and economic characteristics such as education.

In Great Britain the first nationwide statistical survey was carried out by William the Conqueror in the eleventh century and resulted in the famous Domesday Book of 1086. It took over 700 years for another national survey to be conducted. The first official census occurred in 1801. Although this was a large time gap there is some historical population data available from the Poll Tax of 1377, the Hearth Tax of 1662–74, Parish Registers and Bills of Mortality. Today a full census is carried out in Britain every ten years, with the last one occurring in 1991. This is in line with UN recommendations.

There is also a smaller census conducted every five years inbetween the decade counts which attempts to provide information for a representative sample. Surprisingly there are still a few countries in the world where a census has not been conducted: Mali has not had one and Ethiopia and Afghanistan have only had a single census.

In order for a census to successfully record information, heads of household are usually required by law to complete a form containing questions about all members of the household. The topics covered include age, sex, birthplace, nationality, **migration**, education and economic activity. There may also be additional questions on housing tenure (ownership), amenities, car ownership etc. Data collection is made easier if the country or region is divided into areas known as 'enumeration districts'. Someone, known as an enumerator, is then made responsible for the district and must ensure that all questions on a census form are answered. Where people are unable to read or write, then the enumerator must interview them. Data is published by the Office of Population Census and Surveys (OPCS) on microfiche, microfilm, floppy disk or a hard copy computer print out. A number of enumeration districts are often combined to create a ward.

As well as the census, Vital Registration and Special Surveys are also used to collect population data. In EMDCs there is a legal requirement to register vital rates, births and deaths, and also marriages, with an official of the registry office. This is increasingly the trend in other countries. In Britain these statistics are published every year. Special surveys can be carried out for various reasons: governments in poor countries need more data or a more reliable sample; other issues not covered by the census may be required. For example the British General Household Survey (GHS) is carried out every year covering topics like leisure, smoking and education.

Collecting data for the census in large countries can be a logistical nightmare. In other countries, like the USA, errors occur during the collecting and processing stages. These can have significant financial implications. For example, the 1990 census probably underestimated the population by five million people, costing local authorities where such people lived some $US17 billion in federal aid, passing the burden onto the local tax payer.

Q6. *This question is about **population structure** and makes use of Figure 9.14. It is a good idea to work in groups on the tasks.*
   a) *Using information technology (IT) or drawing the charts by hand, produce three population pyramids for 1995, 2010 and 2025 for both Morocco and Norway.*
   b) *Using the pyramids answer the following questions.*
      i) *How does population structure change between 1995–2025?*
      ii) *How do the projections for Morocco and Norway compare?*
      iii) *Are the pyramids progressive or regressive?*
      iv) *What factors may help to explain the differences?*
      v) *Calculate the dependency ratios for each pyramid and describe their changes.*
   c) *Use all the information here and the results of any other research to produce a wall display and prepare a presentation for the rest of the group.*

POPULATION DENSITY AND DISTRIBUTION

General characteristics

Population density is the relationship between area and population numbers. This can be given at a global level, showing the average densities for the major regions (see Figure 9.14) or for the districts making up a county or enumeration districts making up the ward in a town or city (see Figure 9.15). Global population density has shown a significant increase in the last 100 years, in line with population growth. The global trends do, however, hide some interesting local variations: some urban areas have shown increases while other inner city areas have recorded decreases; in EMDCs rural areas have recorded a density decrease while in ELDCs these areas have usually continued to become more populated.

	1950	1960	1970	1980	1990	2000
WORLD	18	22	27	33	39	46
AFRICA	7	9	12	16	21	29
ASIA	50	60	76	94	113	134
EUROPE	76	82	89	93	96	98
NORTH AMERICA	8	9	11	12	13	14
OCEANIA	1	2	2	3	3	4
SOUTH AMERICA	6	8	11	14	17	20

Figure 9.13 *Population density: major regions of the world, (persons per km^2)*

Q7. *Using Figure 9.13 describe and explain the regional variations in population density in 1990. Identify the major trends in population density and the possible impact on society.*

Norway	1995			2010			2025		
Age group	Both sexes	Males	Females	Both sexes	Males	Females	Both sexes	Males	Females
All Ages	4271	2111	2160	4415	2191	2225	4501	2232	2269
0–4	265	136	129	244	125	119	238	122	116
5–9	262	134	128	258	133	126	241	123	117
10–14	259	133	126	272	140	132	243	125	118
15–19	274	140	134	271	139	132	250	128	121
20–24	324	166	158	267	137	130	263	135	128
25–29	337	173	164	284	136	128	277	142	134
30–34	316	161	154	279	143	136	276	142	134
35–39	315	161	154	329	168	161	272	140	132
40–44	305	157	149	339	174	165	267	137	130
45–49	313	161	152	315	161	154	279	143	136
50–54	231	117	114	309	157	152	324	165	158
55–59	188	93	95	292	148	144	329	168	161
60–64	184	89	94	289	145	144	295	148	147
65–69	189	89	100	203	98	105	276	134	142
70–74	191	84	107	151	68	83	241	113	128
75–79	149	60	89	128	53	74	209	92	118
80+	169	57	112	207	68	139	221	74	147

Morocco	1995			2010			2025		
Age group	Both sexes	Males	Females	Both sexes	Males	Females	Both sexes	Males	Females
All Ages	28301	14173	14127	37586	18848	18739	45647	22890	22758
0–4	3995	2037	1958	3811	1941	1871	3777	1925	1852
5–9	3693	1880	1813	3866	1968	1899	3632	1849	1783
10–14	3295	1675	1619	3975	2023	1952	3729	1897	1833
15–19	3039	1548	490	3898	1982	1915	3764	1912	1852
20–24	2689	1369	1319	3634	1845	1788	3827	1943	1884
25–29	2334	1187	1147	3231	1638	1593	3927	1992	1935
30–34	1997	979	1018	2967	1507	1461	3841	1946	1894
35–39	1773	881	892	2616	1328	1288	3573	1808	1785
40–44	1369	684	685	2260	1145	1115	3166	1599	1567
45–49	999	490	509	1914	932	982	2885	1457	1428
50–54	720	334	387	1670	821	849	2507	1261	1246
55–59	731	320	410	1253	614	638	2113	1053	1060
60–64	561	266	295	872	414	457	1720	814	906
65–69	507	229	278	581	256	325	1399	657	742
70–74	272	136	137	512	209	303	925	425	500
75–79	198	97	101	302	132	171	509	221	288
80+	129	60	69	224	93	131	354	133	221

Figure 9.14 Population structure: Norway and Morocco (medium projection)

A fundamental fact of population geography is that people are not evenly spread across the face of the earth (see Figure 9.16). In 1993 there were an estimated 5.51 billion people occupying the globe with a mean density of 42 per km^2 (the figure for 1980 was 33 per km^2). Figure 9.16 on page 238 reveals how densities vary: Oceania for example has densities one-tenth of the global figure while Antarctica is virtually uninhabited. On the other hand Asia and Europe show densities that are three times the average. At this scale it is also important to consider the distribution of people within each continent: major concentrations avoid the continental interiors with 65 per cent of the world population living within 500 km of the coast and 75 per cent within 1000 km. Also significant is the fact that less than ten per cent of the world population can be found in the Southern Hemisphere. This same figure applies to the area between the equator and 10° North. Nearly 50 per cent of the world's population can be found between 20° and 40° North, and 30 per cent between 40° and 60° North. Latitude therefore

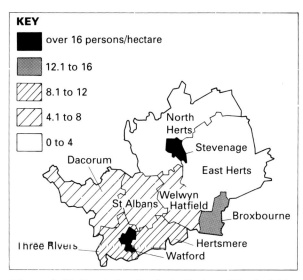

Figure 9.15 *Population density in districts of Hertfordshire*

has a significant influence. Europe, CIS and Asia account for 75 per cent of the world's population. When the share of land in the economically less developed world is considered, then density is far less favourable: 75 per cent of the population here is squeezed onto 63 per cent of the land surface. This helps to explain some of the population problems outlined at the start of this section.

Africa and Europe show striking contrasts in population distribution. In Africa highest densities are found in coastal locations, particularly in West Africa and along stretches of the east shoreline. The country with the highest density figure is Burundi with 264 persons per km^2 in 1990. Nigeria is next with 162, South Africa has 36 and Mauritania only three persons per km^2. In Europe highest densities are found in the core areas of Netherlands, northern Germany and France, with much lower concentration in peripheral areas like Spain and Turkey. All members of the EU have densities higher than Burundi. Even in these countries concentration is greatest around the Mediterranean fringe creating pressures on the sea and land.

Explaining distributions

The number of people in a given area and their distribution within it are the result of a number of interacting factors (see Figure 9.17). Each of the factors may affect population distribution in different ways. For example, they may have an indirect effect via population change (natural increase or migration); they may influence distribution through the carrying capacity of land and, since carrying capacity varies so much from place to place it is bound to be an important factor; and, they may exercise a direct influence on population distribution

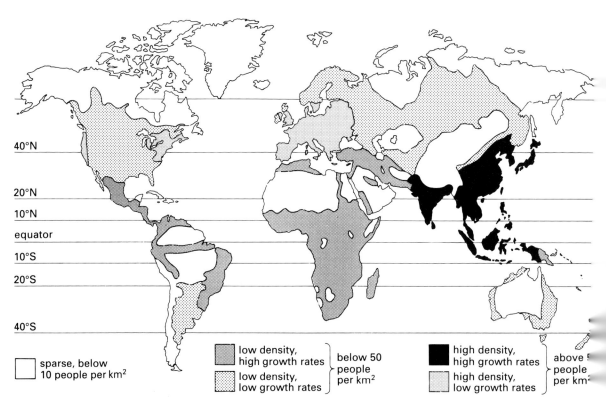

Figure 9.16 *Population distribution and density*

Factors	Population change	Carrying capacity	Population distribution
Physical	Natural disasters and hazards. Unreliability of physical environment.	Physical potential for food production. Availability of natural resources.	Limiting effects of physical extremes. Attraction of physical optima.
Biological	Age and sex structure. Changes in morbidity.	Incidence of pests reducing agricultural productivity. Richness of native flora and fauna.	Contagious diseases and epidemics. Human physiology.
Economic	Improving living standards. Investment in medical and welfare services.	Degree of economic development. Investment in resource exploitation.	Location of economic activities. Productivity of agriculture.
Social	Attitudes concerning children, family structure, etc. Disposition to migrate in search of better opportunities.	Willingness to adopt new technology. Life styles.	Spatial diffusion of urbanism. Ethnic identity and segregation.
Political	Government promotion of family planning. Oppression, inducing migration.	Government support of the economy. Regional aid programmes.	Political stability and security. Colonialism.
Technological	Increasingly effective disease and death control. Greater personal mobility.	Introduction of improved methods of food production. Ability to exploit previously marginal resources.	Improved accessibility of remote areas. Ability to cope with difficult environments.
Historical	Evolutionary cycles of occupation. Traditional seasonal migrations.	Persistence of the traditional economy. Immigrant skills.	Age of peopling. Persistence of traditional patterns.

Figure 9.17 *Explaining population distributions*

National contrasts

Japan

Japan, located in the economic heart of South East Asia, is generally regarded as a superpower. Although not much larger than the UK (377 800 km^2 compared to 274 800 km^2) Japan has a population twice that of the UK. The last available Japanese census was conducted on 1 October 1990 and put Japan's population at 123.6 million with a mean density of 332 per km^2. The 1980 population total was 117.1 million and the new figure represented a 5.6 per cent increase, making Japan one of the most densely populated countries in the world.

Japan is composed of four main islands and hundreds of smaller ones. Ironically much of the land area is too steep and mountainous to build on and so most of Japan's population is squeezed onto the thin coastal plain. So great are the pressures, that communities around the shoreline are continually reclaiming land from the sea. Properties are frequently redeveloped when they are 30–40 years old, often to accommodate the desire for higher housing densities.

Norway

Norway and its Scandinavian partners occupy a unique position abreast of the Arctic circle. With a total of 323 900 km^2, Norway is close in land area to Japan but has a population of only 4.2 million (1990), and an average density of only 13.1 per km^2. The contrast with Japan is striking and the pattern of population distribution is equally revealing.

JAPAN
Date of census: 1 October 1990
Total population: 123 612 000

Prefecture	Population 1990	Density persons per km²
Aichi	6690	1300
Akita	1227	106
Aomori	1483	154
Chiba	5555	1078
Ehime	1515	267
Fukui	824	197
Fukuoka	4811	969
Fukushima	2104	153
Gifu	2067	195
Gumma	1966	309
Hiroshima	2850	336
Hokkaido	5644	72
Hyogo	5405	645
Ibaraki	2845	467
Ishikawa	1165	278
Iwate	1417	93
Kagawa	1023	546
Kagoshima	1798	196
Kanagawa	7980	3310
Kochi	825	116
Kumamoto	1840	249
Kyoto	2603	564
Mie	1793	310
Miyagi	2249	309
Miyazaki	1169	151
Nagano	2157	159
Nagasaki	1563	382
Nara	1375	373
Niigata	2475	197
Oita	1237	195
Okayama	1926	271
Okinawa	1222	540
Osaka	8735	4640
Saga	878	360
Saitama	6405	1687
Shiga	1222	304
Shimane	781	118
Shizuoka	3671	472
Tochigi	1935	302
Tokushima	832	201
Tokyo	11855	5430
Tottori	616	176
Toyama	1120	264
Wakayama	1074	228
Yamagata	1258	135
Yamaguchi	1573	257
Yamanashi	853	191
Japan	123612	332

Figure 9.18 *Population in Japan, 1990*

Q8. Using the 1990 census for Japan (Figure 9.18) and the outline map (Figure 9.19) produce a choropleth map to show the population distribution and density in 1990. Use these maps to produce a report on population in Japan. In your report try and explain population distribution and use a Lorenz Curve to search for any spatial concentration in the rate of population change. The Lorenz curve can be calculated by working out the land area of each prefecture for 1990 (divide the 1990 population figure by the density figure to give the land area in km²). Sort the prefectures into rank order of population and work out the percentage of each prefecture's share of the total. Add these cumulatively. Repeat for the land area data keeping the ranking the same. Graph your results.

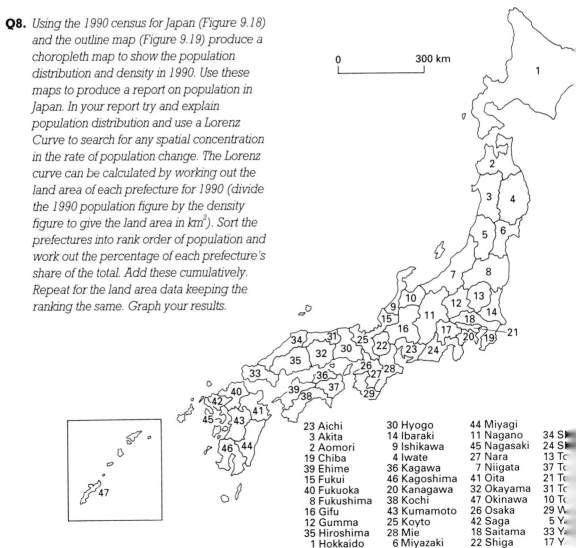

23 Aichi 30 Hyogo 44 Miyagi
3 Akita 14 Ibaraki 11 Nagano 34 S...
2 Aomori 9 Ishikawa 45 Nagasaki 24 S...
19 Chiba 4 Iwate 27 Nara 13 To...
39 Ehime 36 Kagawa 7 Niigata 37 To...
15 Fukui 46 Kagoshima 41 Oita 21 To...
40 Fukuoka 20 Kanagawa 32 Okayama 31 To...
8 Fukushima 38 Kochi 47 Okinawa 10 To...
16 Gifu 43 Kumamoto 26 Osaka 29 W...
12 Gumma 25 Koyto 42 Saga 5 Y...
35 Hiroshima 28 Mie 18 Saitama 33 Y...
1 Hokkaido 6 Miyazaki 22 Shiga 17 Y...

Figure 9.19 *Prefectures in Japan*

240 ALL THE PEOPLE

Q9. Using Figures 9.20 and 9.21 produce an isoline map to show population distribution in 1985 and 1990 in Norway (use the points in the middle of each region for locating the data). Using an atlas try to explain the pattern of population. Calculate the population densities for the main regions or Fylkes. You will need to find the land area of each region. Why are the densities so low in Norway compared to Japan?

Norway
Date of Census: 3 November 1990
Total population: 4 247 553
The population totals are given for the Fylke and any Kommune with over 10 000 inhabitants. The population at 1 January 1985 is also shown.
The 1985 and 1990 figures are not always comparable because of boundary changes/amalgamations. The Fylke are shown, as normally in Norwegian listings, in geographic order south to north.

AREA	1985	1990
NORWAY	4 159 335	4 247 553
ØSTFOLD	234 941	238 296
Halden	25 876	25 873
Sarpsborg	12 069	11 790
Fredrikstad	27 125	26 546
Moss	24 830	24 683
Borge	11 291	11 966
Skjeberg	13 495	14 279
Askim	12 500	12 864
Rune	18 511	18 324
Onsoy	12 488	12 914
Rygge	11 540	12 037
AKERSHUS	393 217	417 653
Vestby	10 657	11 266
Ski	20 849	22 337
Ås	11 410	11 946
Sogn	9 223	10 298
Nesodden	11 243	13 189
Oppegård	18 603	20 669
Bærum	82 918	90 333
Asker	37 767	41 848
Aurskog-Holand	12 438	12 582
Sorum	9 905	11 265
Fælingen	13 387	13 781
Sorenskog	23 779	26 454
Skedsomo	32 339	34 110
Nittedal	14 945	16 177
Ullensaker	17 506	18 125
Nes	14 729	15 703
Eidsvoll	15 643	16 689
OSLO	447 351	459 292
HEDMARK	186 355	187 275
Hamar	15 693	16 315
Kongsvinger	17 467	17 469
Ringsaker	30 419	31 377
Stange	17 790	17 616
Elverum	16 903	17 428

AREA	1985	1990
OPPLAND	181 791	182 578
Lillehammer	22 012	22 850
Gjøvik	25 957	26 207
Østre Toten	14 131	14 336
Vestre Toten	13 626	13 366
Gran	12 514	12 641
BUSKERUD	219 967	225 172
Drammen	50 749	51 880
Kongsberg	20 913	21 185
Ringerike	26 870	27 384
Modum	12 160	12 243
Øvre Eiker	14 113	14 801
Nedre Eiker	17 906	18 901
Lier	18 113	18 961
Røyken	12 687	14 393
VESTFOLD	191 600	198 399
Borre	8 940	22 568
Tønsberg	8 891	31 551
Sandefjord	35 011	36 095
Larvik	8 070	38 223
Nøtterøy	17 183	18 031
TELEMARK	162 547	162 907
Porsgrunn	31 402	31 268
Skien	46 656	47 870
Notodden	12 622	12 426
Bamble	13 162	13 784
Kragerø	10 904	10 817
AUST-AGDER	94 688	97 333
Arendal	12 051	12 439
Grimstad	14 631	15 656
VEST-AGDER	140 232	144 917
Kristiansand	62 197	65 543
Mandal	12 218	12 496
Vennesla	11 092	11 544
ROGALAND	323 365	337 504
Eigersund	12 126	12 409
Sandnes	39 678	44 798
Stavanger	94 193	98 109
Haugesund	27 014	27 736
Hå	12 660	13 022
Klepp	11 289	11 854
Time	11 013	12 051
Sola	14 420	15 944
Karmøy	33 888	35 087

AREA	1985	1990
HORDALAND	399 702	410 568
Bergen	207 416	212 944
Stord	13 591	14 632
Kvinnherad	13 159	13 093
Voss	124 059	14 082
Os	11 499	12 768
Fjell	12 200	14 912
Askoy	18 036	18 598
Lindås	11 709	11 863
SOGN OG FJORDANE	106 116	106 659
Flora	9 344	10 049
MØRE OG ROMSDAL	237 290	238 408
Molde	21 310	22 251
Krisitansund	17 818	17 190
Ålesund	35 008	35 862
Orsta	10 289	10 248
SØR-TRØNDELAG	246 824	250 978
Trondheim	134 075	137 846
Orkdal	9 960	10 136
Melhus	11 834	12 505
NORD-TRØNDELAG	126 692	127 157
Steinkjer	20 590	20 665
Namsos	11 847	11 909
Stjordal	16 717	17 321
Levanger	16 278	16 829
Verdal	13 131	13 503
NORDLAND	242 268	239 311
Bodø	34 013	36 890
Narvik	18 865	18 609
Vefsn	13 282	13 410
Rana	25 251	24 650
Fauske	10 093	10 001
Vestvågoy	10 846	10 566
TROMS	146 736	146 716
Harstad	21 760	22 375
Tromso	47 753	51 218
Lenvik	11 204	10 843
FINNMARK	75 667	74 524
Alta	13 928	15 170
Sor-Varanger	10 073	9 671

Figure 9.20 Population distribution in Norway, 1990

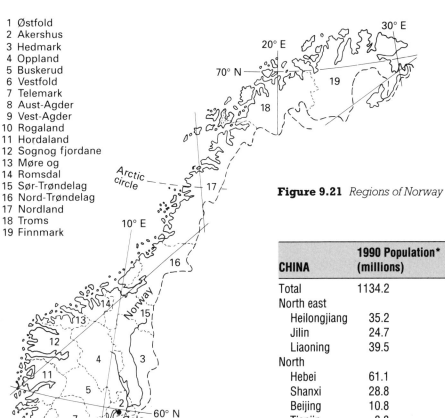

Figure 9.21 Regions of Norway

1 Østfold
2 Akershus
3 Hedmark
4 Oppland
5 Buskerud
6 Vestfold
7 Telemark
8 Aust-Agder
9 Vest-Agder
10 Rogaland
11 Hordaland
12 Sognog fjordane
13 Møre og
14 Romsdal
15 Sør-Trøndelag
16 Nord-Trøndelag
17 Nordland
18 Troms
19 Finnmark

China

China's population was put at 1.13 billion in the 1990 census, the fourth national population census (see Figure 9.22). This was a marked increase from the 1982 figure of 1.01 billion, reflecting a continuing upward surge in China's demographic transition. The people of China are not evenly distributed across this vast country, nor do they all belong to one ethnic group. Although the bulk of the Chinese are *han*, there are over 91.2 million people that belong to minority groups. China's ethnic minority includes 18 officially recognised groups, including the *zhuang*, the largest of the minorities, and the *hui*, an all Moslem group.

China's changing pattern of population distribution reveals the impact of surging urbanisation. In the context of China's economic development and modernisation the effects of population redistribution can operate at two levels. Redistribution can include rural-urban migration and also, the outpouring of people from the agricultural sector into the industrial and service sectors. The increase in urbanisation has been dramatic and can be attributed to two factors: natural increase and migration. The latter has been encouraged by the de-collectivisation of agriculture and economic deregulation. Two of China's leading

CHINA	1990 Population* (millions)	% increase since 1982	Density[1] 1990
Total	1134.2	12.5	118
North east			
Heilongjiang	35.2	7.8	78
Jilin	24.7	9.3	132
Liaoning	39.5	10.5	270
North			
Hebei	61.1	15.2	325
Shanxi	28.8	13.7	184
Beijing	10.8	17.2	644
Tianjin	8.8	13.2	777
Shandong	84.4	13.4	539
Henan	85.5	14.9	572
East			
Anhui	56.2	13.1	404
Jiangsu	67.1	10.8	654
Shanghai	13.3	12.5	2118
Zhejiang	41.4	6.6	407
Central			
Hubei	54.0	12.9	290
Hunan	60.6	12.3	286
Jiangxi	37.7	13.6	226
South east			
Fujian	30.0	16.1	248
Guangdong	62.8	17.2	353
Guangxi	42.2	16.0	178
Hainan	6.6	15.7	193
South west			
Guizhou	32.4	13.4	184
Sichuan	107.2	7.5	188
Yunnan	37.0	13.6	94
Xizang (Tibet)	2.2	16.0	2
North west			
Nei Monggol	21.5	11.3	18
Shaanxi	32.9	13.8	160
Ningxia	4.7	19.5	90
Gansu	22.4	14.3	49
Qinghai	4.5	14.4	6
Xinjiang	15.2	15.9	9

* Population totals do not include immigrants without official clearance and 3.2 million persons serving in the armed forces
[1] Persons per km²

Figure 9.22 China provincial population data, 1990

Figure 9.23 *China province map*

metropoli illustrate the effects of these reforms: Beijing, the capital, reported a 17 per cent increase in population, and Shanghai a 13 per cent rise between 1982 and 1990; during the same period the rate of natural increase was eight per cent in Beijing and five per cent in Shanghai; in-migration therefore accounted for the lions share of population growth.

Q10. *Using appropriate statistical and cartographic techniques produce a word processed report on China's population distribution. Include in your report an explanation of the pattern of distribution and an investigation into the relationship between density and population increase. (Use Spearmans Rank Correlation Coefficient to test the hypothesis that 'population increase is significantly related to density'.)*

PEOPLE ON THE MOVE

Population change in a region or country can result from a natural increase or decrease or from the effects of migration. Migration can be thought of as population movements. Under this general heading, all sorts of movement can be included from a shopping visit, to moving from one flat to another in the same town, to the mass movement of peoples across international frontiers. Movements can also differ with respect to time: some may be daily (commuting) or seasonal (nomadic pastoralism) or semi-permanent (shifting cultivators) or permanent (Jews into Israel). To avoid these confusions the UN has recommended that the term migration should only be used to describe those movements which involve people moving from one administrative area to another and which result in a permanent change in residence. The other types of movement outlined above are now referred to as **circulation** (see Figure 9.24). Circulation therefore involves circular flows which are often short-term and repetitive; migration results in displacement. In population geography mobility is the umbrella term used to describe all types of **population movement**.

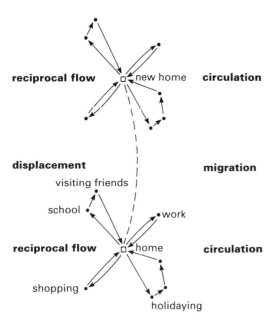

Figure 9.24 *Circulation and migration*

Out-migration: the High Atlas and Massif Central

Both the Massif Central and the High Atlas have a long history of population loss (see Figures 9.26 and 9.27). The current population of the Massif is about half of what it was 150 years ago: in the Lozère Department the population peaked in 1851 (144 765) and again in 1881 (143 565) but since then has fallen consistently, reaching 72 814 in 1990. The problems associated with such a massive out-migration dominate the social and economic issues in the Massif. In the High Atlas mountains of Morocco out-migration is a more recent phenomena. In one region, the Ait Mizane, young *berbers* have moved with increasing frequency to the cities, threatening the long-term viability of agricultural systems in the mountains. In 1989 a total of 119 people left the villages of the Ait Mizane.

In the eighteenth and early nineteenth centuries the peoples of the Massif depended on three economic activities for their survival: agriculture, craft industries and mining. Much of the agricultural activity was subsistence and farm workers often had to supplement their meagre incomes with additional jobs. Some worked on the grape harvests in Languedoc while others travelled as far as Scotland and Ireland to work in the fisheries. A series of harsh summers and poor harvests in the 1840s severely disrupted the subsistence equilibrium and the

Figure 9.25 *High Atlas village, Morocco*

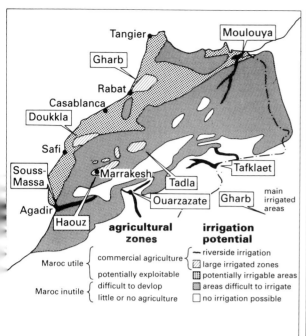
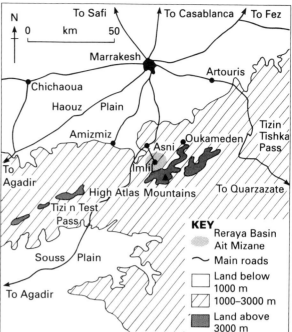

Tourism and migration: Reraiya Basin 'Ait Mizane'

Village/clan	Number guides	Number shopkeepers	Migrants to: Casablanca	Marrakesh	Rabat	Others	Abroad	Total
Ait Souka	5	0	12	8	4	9	3	86
Arrhen	5	0	13	1	0	1	0	15
Arremdt	31	5	24	14	4	6	4	52
Ashain	7	0	1	2	2	2	0	7
Imlil	4	1	5	4	1	3	0	13
Mzig	10	3	22	9	1	0	0	32
Taigadirt	8	3	7	3	0	1	0	11
Tairgaimoula	6	3	8	7	1	2	0	18
Talawal	0	0	0	0	0	0	0	0
Tamatoirt	1	1	4	1	7	1	1	14
Taourirt	1	0	0	1	0	0	0	1
Totals	**78**	**16**	**96**	**50**	**20**	**25**	**8**	**119**

Note: The revenue tourists bring into the area and the income sent back from migrant workers is an essential part of the economy of the area. The young are increasingly less interested in traditional agriculture. Guides are becoming the most important people in the villages.

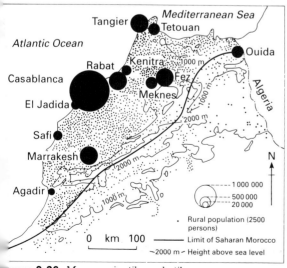

Population change: Reraiya Basin 'Ait Mizane'

Village/clan	Population 1971	%	Population 1989	%	Rate of change
Ait Souka	134	6.9	234	10.8	+3.9
Arrhen	171	8.8	180	8.3	−0.5
Arremdt	618	31.9	732	33.7	+1.8
Ashain	107	5.5	105	4.8	−0.7
Imlil	164	8.5	149	6.7	−1.8
Mzig	362	18.7	290	13.4	−5.3
Taigadirt	50	2.6	121	5.6	+3.0
Tairgaimoula	80	4.1	130	5.9	+1.8
Talawal	45	2.3	32	1.5	−0.8
Tamatoirt	146	7.5	160	7.4	−0.1
Taourirt	61	3.1	37	1.7	−1.4
Totals	**1938**		**2170**		

Figure 9.26 *Morocco inutile and utile*

(a) Population change in Paris Basin 1975–1990

Départements	Total population '000		
	1975	1982	1990
Paris City	2999.8	2176.2	2152.4
Inner suburbs (Paris region)			
Hauts-De-Seine	1438.9	1387.0	1391.7
Seine-Saint-Denis	1322.1	1324.3	1381.2
Val-De-Marne	1215.7	1193.7	1215.5
Outer suburbs (Paris region)			
Essones	923.1	988.0	1084.8
Val-D'Oise	840.9	920.6	1049.6
Seine-Et-Marne	755.8	887.1	1078.2
Yvelines	1082.3	1196.1	1307.1
Total	9878.6	10073.0	10660.6

(b) Population change in Massif Central 1801–1990

Year	Population	Year	Population
1801	126 565	1891	135 517
1806	143 201	1896	132 151
1821	133 915	1901	128 866
1826	100 711	1911	122 738
1831	140 310	1921	108 822
1836	141 710	1926	104 738
1841	140 710	1931	101 849
1846	143 310	1936	98 480
1851	144 765	1946	90 523
1856	140 810	1954	82 391
1861	137 362	1962(a)	80 891
1866	137 265	1962(b)	81 863
1872	135 101	1968	77 258
1876	138 919	1975	74 825
1881	143 565	1982	74 294
1886	141 261	1990	72 814

Figure 9.27 *Population change: Paris Basin and Massif Central*

resulting famines pushed the traditional agricultural system to breaking point. Resistance to disease was lowered and infant and adult mortality rates soared. Out-migration, stimulated by the famine and encouraged afterwards by the rural poverty, combined with a temporary excess of births over deaths to produce a very serious rural depopulation. Simultaneously, craft industries underwent a sudden decline as a result of competition from cheap factory produced goods from other, rapidly industrialising areas. In the latter half of the nineteenth century, road and rail construction reduced the isolation of the region and partially integrated it into the national economy. The effect of this was to highlight the inefficiencies of the three economic activities, which declined still further. People were also provided with a cheaper and easier means of escape. The wave of out-migration continued into the early twentieth century as the bright lights of Paris, Lyon and the other cities attracted more people. The spread of primary education and compulsory military service made the situation worse as children learnt about the attractions of city life and young men experienced new urban environments for the first time. Today a visit to many parts of the Massif reveals evidence of out-migration, e.g. in Lozère. There is evidence of deterioration and neglect everywhere, with tumbling terraces and deserted farmsteads. The farming communities are rapidly ageing and the economic and social life of many villages is threatened.

The story in the High Atlas is very similar, although the problems of this mountainous area have been exacerbated by the colonial policies of the French, who ruled Morocco from 1912–56, and subsequently from neo-colonialism. The French neatly divide Morocco into two parts: *Maroc Utile*, the well-watered and favoured lowlands and *Maroc Inutile*, the drier and more mountainous areas. French investment in agriculture and industry went to Maroc Utile while Maroc Inutile, had until fairly recently changed little from medieval times. The pressures in High Atlas to migrate come from several factors: the subsistence nature of the economy; natural hazards such as rock avalanches and flooding; risks from disease and poor health care; isolation of many communities (some villages can only be reached on foot or by mules); poor educational opportunities and deforestation and desertification.

In-migration: the Paris Basin and Maroc Utile

Maroc Utile has long been the goal of many migrant moving from the deprived areas of Morocco. In som cases this has been only one step on their move into Europe, and some Moroccan migrants have made their way to the Paris Basin to swell the large numbers of migrants attracted to this area. Most of the migrants from the High Atlas will find permanen locations in the coastal cities, Casablanca attracting the largest percentage (see Figure 9.26b). Opportunities exist for migrants in the informal sect (unregulated employment without salaries or holidays) and also as labourers on the large agricultural estates. Migrants often live in squalid conditions in the shanty towns of these cities where the quality of life may often be better than the basic accommodation and services available in the mountains.

The Paris Basin has historically acted as a magnet on the rest of France, drawing people from a wide area (see Figure 9.27). The population currently stands at 8.6 million which is an increase of eight per cent from 1960. Most migrants are drawn from the immediate area, particularly the north and west. In addition large numbers of French migrants, from as far away as the Mediterranean and the Massif Central, have made their way to the capital. Attractions in the Paris region are numerous: problems from the overpopulation of this city are less well-known and include pollution, congestion, high costs of living, ethnic crime and racism.

Migration: causes

There are some basic terms used in the study of migration and it is useful to define these before the causes and theories of migration are discussed. People involved in the same migration form a migration stream. When there is a two-way flow in migration the stronger migration is called the dominant migration and the weaker, the reverse or counter migration. Migrants who cross international boundaries are referred to as immigrants in the country of destination and emigrants in the country of origin.

Traditional views of migration group the causal factors into 'push' and 'pull' categories. Another way of looking at migration is to consider factors under social, political, economic and physical categories (see Figure 9.28). Generally migration takes place because of the operation of two sets of forces, one operating in the migrants home area, and one set operating elsewhere. Push forces are those which create pressures for individuals, and eventually persuade them to move away. These might include things like natural disasters, low wages, ethnic or religious persecution, or civil war. Pull factors or forces attract the individual migrant to a new location. Typical factors here may be good social, educational and welfare services and political freedom.

> **Q11.** *Use the information on out-migration from the Massif Central and High Atlas and immigration into the Paris Basin and Maroc Utile to construct a list of push and pull factors.*

Sometimes groups make the decision to migrate, but often it is an individual choice. Much individual migration is voluntary: the decision to move taking place as a decision making process. This is often influenced by personal factors. For example, some individuals will need a strong stimulus to move (a natural disaster) while others will migrate for economic or social reasons. This individuality is known as migration elasticity.

The information that an individual holds about certain destinations is very important in understanding and explaining migration. The quality and quantity of information make up what is known as the information field; this varies greatly from one person to another and the size can be related to age, background, experience and education. A common characteristic of information is that it tends to show a distance decay effect. In some situations, knowledge about far away destinations is passed on by people who have already migrated or who are return migrants along the information chain. After the 1939–45 war these chains were evident in Indian migration to the UK.

> **Q12.** *Construct a flow diagram to show the information field of a migrant from the Massif Central and Maroc Utile. How do these fields differ? Explain your answer.*

Migration models

Migration models have been developed by geographers to try to predict and explain population movements. They vary from the mathematical gravity model to the descriptive general migration model.

The gravity model

This model is based on the work of EG Ravenstein in the 1880s who put forward the laws of migration. One of the best known laws is that migrants travel short distances, and with increasing distances the number of migrants decrease. This distance decay effect is

	Push	Pull
physical	inaccessibility harsh climate natural disaster	scenic quality fertile soils lack of natural hazards
economic	unemployment poverty high rents heavy taxes	high living standards good wages promotion resource exploitation
social	discrimination lack of housing bereavement growth of family	good welfare services relatives and friends marriage higher education
political	civil unrest persecution planning decision	freedom of speech propaganda political asylum

Figure 9.28 *Factors affecting migration*

one of the features of the gravity model which states that: the volume of migration is inversely proportional to the distance travelled by the migrants and directly proportional to the relative sizes of the two places. The formula is shown below:

$$Mij = \frac{PiPj}{Mij^2}$$

where M = migration, P = population and D = distance between i and j.

The stepwise model

This model was developed around another of Ravensteins laws that *'migration occurs in stages and with a wave-like motion'*. In this model the capital, or largest city, attracts migrants from the smaller cities, who in turn draw people from the towns, which in turn draw migrants from villages in rural areas. Migration to the leading city can therefore have a ripple effect that can stretch right across a country (see Figure 9.29).

Lee's general model of migration

This model emphasises the notion of positive and negative factors in the origin and destination of all migrants (see Figure 9.30). For migration to take place, the positive destination factors must outweigh the positive origin factors. During the move to the final destination, possibly the capital city, intervening opportunities present themselves to the migrant, slowing their passage or even halting the flow.

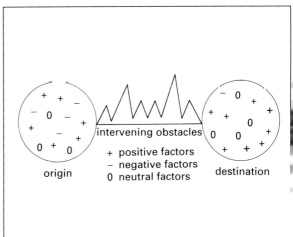

Figure 9.30 *Lee's general migration model*

The Reilly model

This model uses the ideas behind the gravity model to try to identify the migration area, the zone from which potential flows will take place. Using the formula below it is possible to calculate the break point between two cities which marks the migration boundary.

$$\text{Break point} = \frac{\text{km between city x and city y}}{1 + \sqrt{\frac{\text{population of x}}{\text{population of y}}}}$$

The Huff model

Huff's model uses probability theory to try to assess the relative pull of urban centres on potential migrants. The model is based in part on the gravity model but then uses information on the other possible destination centres to estimate the likelihood of migrants moving to a city. The model can be divided into two parts: a numerator (the usual gravity formula) and a denominator which takes into account the size and distance away of the other centres which a migrant may select from.

$$\text{Probability of migrant moving to city } 1 = \frac{\text{attraction of centre } 1}{\text{total attraction of all centres}}$$

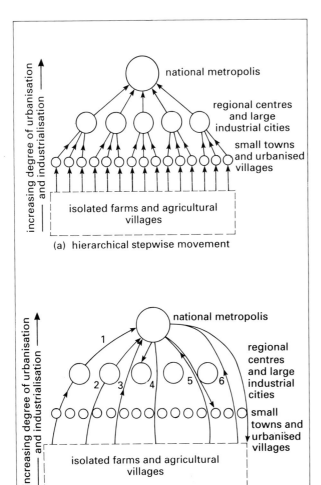

Figure 9.29 *Stepwise migration model*

$$= \left\{ \frac{\dfrac{S_1}{D_{A1}}}{\displaystyle\sum_{j=1}^{4} \dfrac{S_j}{D_{ij}}} \right\} \begin{array}{l} \text{numerator} \\ \\ \text{denominator} \end{array}$$

where S_1 = population size of centre 1.
D_{A1} = distance from migrants home to centre 1.
S_j = size of other centres for possible migration (4 in this example).
D_{ij} = distance from migrants home to each centre.

N.B. It is possible to substitute time or cost for distance and employment opportunities (rate) for population size.

The diffusion model

This is a simulation model which uses the Monte Carlo technique (computer simulation model incorporating a chance element. The working of the model is like the spinning of a roulette wheel!). The ideas behind this model were developed by Torsten Hagerstrand in 1953 and can be used to identify the spatial stages in the spread of migrants.

The scale of migration

Migration varies considerably across space and time. In 1990 the UN estimated that some 50 million people, one per cent of the world population, lived in a country other than their country of origin. In 1992 the World Bank estimated international migrants of all kinds at 100 million. Refugees in 1991 totalled approximately 17 million, 87 per cent of them in ELDCs. Of the remaining international migrants some 25 million were in sub-Sahara Africa, and 13–15 million each in western Europe and North America. An additional 15 million migrants are in Asia and the middle East, where a few countries have particularly heavy migrant concentrations.

At the national scale migration is responsible for the rapid urbanisation in ELDCs. Most of the growth of the world's top 20 agglomerations can be attributed to this process: in 1950, 17 out of the 20 top cities were in more developed countries: in 1990 it was only five! The movement from rural to urban areas accounts for about 60 per cent of the urban growth in ELDCs. At the global scale, transnational migration is a more complex picture with numbers often far greater than those reported. Migrants are usually in their peak years of fertility and, concentrating in a few areas, can increase their visibility and perceptions of cultural difference by living in segregated ethnic communities.

Population growth helps to drive international migration. The countries producing the most international migrants are Mexico, Turkey and Morocco. Today these nations have relatively modest rates of population growth and the migration surge results from the increase in the number of young people in their peak years of fertility, and this has the potential for huge increases in the future. The demographic impetus towards migration will therefore rise in the next three decades. If this is projected into the future and combined with the pressure of population, economic imbalances and improved communications, then the picture by the end of the decade could be mass migration from richer to poorer nations.

Figure 9.31 *Refugees in Somalia, Africa*

Figure 9.32 *Urban slum area in Delhi, India*

The politics of migration

Some 15 million people entered western Europe as migrants between 1980 and 1992. Foreign residents in EU countries totalled 13 million, four per cent of the population in 1990. This figure excludes large numbers of the 'foreign born' who have become naturalised. Of these 13 million, around eight million come from countries outside the EU. Of these, 50 per cent originate in North Africa, Turkey and Yugoslavia. Migrants from Asia and sub-Saharan Africa are increasing as a proportion of the overall international migrants. In the former West Germany the foreign resident population had risen from 4.5 million in 1980 to 5.2 million in 1990 (8.4 per cent of the total population). The largest migrant groups were Turks 1.7 million, Yugoslavs 652 000, and Poles 241 200. In the UK foreign citizens totalled 1.9 million (3.3 per cent of the population). If this figure is combined with the Caribbean and Asian ethnic minorities then the real figure is closer to 4.5 million or eight per cent of the population. France, on the other hand has a smaller proportion of foreign residents, mostly Algerians, Moroccans and Portuguese. In Sweden, Iranians and Lebanese are the largest groups of current migrants. In neighbouring Norway it is Pakistanis, while in Spain and Italy the increasingly dominant group are Filipinos. Eastern and central Europe – Hungary, Czech Republic, Slovakia and Poland – have recently become receiving countries for migrants and 'asylum seekers'.

The arrival of migrants in many Indian cities has created an urban crisis which needs all the ingenuity of the urban planners to solve. In 1990 it was estimated that 'squatter families' made up 40 per cent of the population in Calcutta, 42 per cent in Bombay and 30 per cent in Delhi. This group is the most at risk from disease, industrial contamination, poor sanitation and natural hazards. These illustrate clearly the case for the unsustainable city, the emerging phenomena of the world's top 20 agglomerations.

Q13. *Collect evidence covering two to three months from national newspapers on the issues of European migration. How does the media portray the refugee and asylum seeker? How might governments react to the migration crisis in the future?*

POPULATION PROBLEMS AND SOLUTIONS

There are no clear cut solutions to the population problem: the Aids pandemic is currently the closest the planet will come to a major change in the global death rate; a nuclear holocaust is less likely and fertility may never fall fast enough to prevent the population crisis. Issues of global migration and rapid urbanisation just add to the difficulties facing the planet (see Figure 9.34).

Q14. *Using Figure 9.33, produce three sets of arguments which you think justify the views expressed in each cartoon. Which views do you agree with and why?*

This section examines some of the possible solutions to the population problem: some have been tried in the past, others are more recent examples.

Figure 9.33 *Causes of famine*

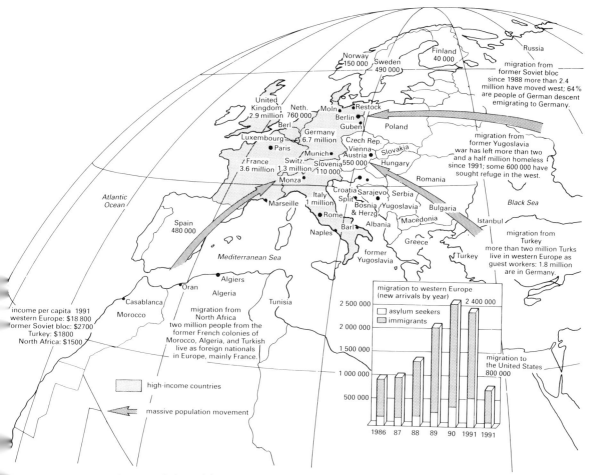

Figure 9.34 *A growing population crisis*

Population and development – the scarcity puzzle

The link between resources and population growth has been explored in an earlier section. The resource issue is part of an interesting debate. Some researchers, for example, Klieber, argue that the world could easily support a population of 50 billion if patterns of food consumption, diets, industrialisation and urbanisation radically changed. In this scenario, low energy agricultural systems would predominate, producing high energy organic foods; industrial and urban areas would follow sustainable policies with significantly lower levels of consumption and production. The issue of food production, consumption and starvation clearly illustrates the scarcity puzzle. In a world of plenty, where Europe and North America can produce vast surpluses, large parts of Africa, central and southern Asia suffer from food shortages. Scarce food, a fundamental resource, is linked to a complex set of factors which includes population growth and migration (see Figure 9.34).

Q15. *Study Figure 9.35 on page 252.*
   a) *Comment on the pattern of debt shown.*
   b) *What influence will debt patterns have on*
      i) *attempts to increase agricultural production;*
      ii) *the relationship between population growth and food supply?*
   c) *Use Spearmans Rank Correlation Coefficient to investigate the relationship between population, GNP and debt.*

The politics of population change

Population change is influenced by a range of political factors. In a number of traditional societies the ruling elite make decisions about population growth and the role of women, sometimes imposing rules such as the spacing of children (reducing growth rates), or maintaining high fertility levels as status symbols. Sometimes women are subjected to practices of male domination such as female circumcision. In more advanced economies the political hand is evident in the demographic transition: pro-natalist policies encourage growth; anti-natalist policies slow the birth rate, while some governments actively participate in forced migration.

Country	Population estimate, mid-1990 (millions)	Natural increase rate, circa 1990 (%)	Total debt outstanding, 1988 (US$bn)	Ratio of debt to GNP, 1987 (as %)	Ratio of debt interest to value of exports, goods & services, 1987 (as %)
	(1)	(2)	(3)	(4)	(5)
Bolivia	7.3	2.6	5.7	133.7	44.4
Brazil	150.4	1.9	120.1	39.4	28.3
Chile	13.2	1.7	29.8	124.1	27.0
Costa Rica	3.0	2.5	4.8	115.7	17.5
Ecuador	10.7	2.5	11.0	107.4	32.7
Ivory Coast	12.6	3.7	14.2	143.6	19.7
Jamaica	2.4	1.7	4.5	175.9	14.2
Mexico	88.6	2.4	107.4	77.5	28.1
Morocco	25.6	2.6	22.0	132.4	17.3
Philippines	66.1	2.6	30.2	86.5	18.7
Venezuela	19.6	2.3	35.0	94.5	21.9

Figure 9.35 *Population and development*

Politics and population growth

In the section on population structure (page 235) several countries were identified as pursuing pro-natalist policies. The basic principle behind this strategy is to view population growth as a way of achieving development: *'necessity is the mother of invention'* (see Figure 9.37). Australia is a nation which has encouraged population growth by trying to raise fertility levels and also by encouraging selective in-migration. More recently, two other nations, Singapore and Romania, have become well-known for their pro-natalist approaches. The Singapore Government have encouraged growth because of serious labour shortages feared in the future (see Figure 9.36). In Romania population growth was seen as a way of achieving communist style development: the regime banned artificial forms of contraception and abortion. The resulting child boom led to children being abandoned and condemned to a horrific life in the state orphanages.

Singapore: go forth and multiply
The Singapore Government's recent enthusiasm for population growth is spurred on by the prediction of serious labour shortages in the next century. In the Chinese 'Year of the Monkey' Singapore hopes to spur couples to greater efforts. The slogan of the 1980s *'Stop at two'* has been replaced by the new *'Go for Three'*.
Unfortunately the results have not been that encouraging – births are down, divorce is up and marriages are taking place even later. Officials have been frustrated by a society who cannot easily be convinced about the joys of parenthood. The 'Year of the Goat' in 1991 saw the birth rate falling by 2.9 per cent; the 1988 'Year of the Dragon' was much more favourable with birth rates soaring by one fifth. Can the monkey work some magic?

Figure 9.36 *Singapore population control*

Politics and population control

There are numerous examples of the excesses of government population policies. Perhaps the best known are the Chinese two and one child policies developed in the 1970s and 1980s and partially abandoned in the 1990s, particularly in rural areas. The success of the one child policy was limited to urban areas, particularly where communities depended on one or two main industries. In these locations the factory became the dominating force in birth control and considerable peer pressure was exerted so that factories could stay within their allocated 'quota' of babies. Workers had to seek 'permission' to try for a child and benefited from free factory nurseries and the use of factory showers if they followed the rules. Outside the factory walls extra assistance came from the 'granny police' who snooped on women of all ages and reported signs of broodiness. The Chinese propaganda machine and the use of financial incentives helped to maintain the momentum of falling fertility. Women who agreed to be sterilised after their first child were able to obtain a 'glory certificate' which entitled their child to a payment of £20 per year, free education and priority for a university place.

In India the 1960s' population programme was less successful and caused lasting damage to future birth control policies. The use of male and female sterilisation was the centre piece of the programme. In the early years the policy was pursued slowly but in subsequent years sterilisation was so vigorously pursued that it, and the birth control programme, became known as the 'terror'. Large numbers of people were sterilised against their will, some bribed with transistor radios.

Politics and migration

Forced migrations are often the result of political intervention. The refugee and asylum seeker problem in Europe described in the migration section (page 250) is a good example. Perhaps less well-known are government attempts to reduce the effects of high population density by forcing people to move into low density areas. The transmigration policy pursued by the Indonesian Government does just this. People have been moved from the islands of Java, Bali, Madura, and Lombok to the 'outer islands' such as Sumatra, Sulawesi, Kalimanten and Iran Jaya. The policy, which started in the 1930s continues today with around 25–30 000 transmigrants each year. In 1986 it was estimated that some three million people (700 000 families) had participated in the programme. Not all the planned movements have been successful. In Iran Jaya the policy of transmigration increased Indian hostility towards the government and helped to start a guerrilla war: Indian blowpipes regularly fire poisonous darts at government helicopters!

Q16. *Carry out some library research on transmigration in Indonesia. Use this information and an annotated sketch map to evaluate the overall success of the policy.*

Managing populations

Much of the recent work on population management has moved away from enforced policies of birth control or investment in large-scale development programmes in the naive hope that development is the best form of contraceptive. Managing growth, in a sustainable approach, can be positive in enriching human resources (see Figure 9.37). There are of course high costs involved if management is to be successful. Today's population programmes are on a much smaller scale and are geared towards giving women the ability to control their own fertility (see Figures 9.37 and 9.38, page 254).

Policies now recognise the key importance of giving women improved status so that the issue of population and development can be dealt with effectively. In some countries this has meant the identification of traditional male attitudes as the obstacles (see Figure 9.39, page 254).

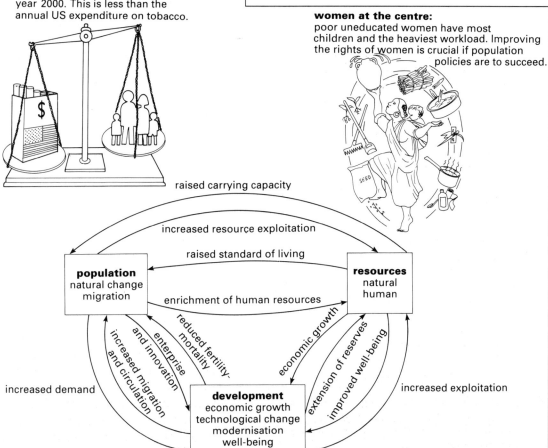

Figure 9.37 *Population, resources and development*

Meeting local needs in Nepal and Bangladesh
Population Concern, a registered charity based in London, is involved in a number of population projects worldwide. In Bangladesh, projects have tackled two areas: family planning services for industrial workers in Dhaka and community development linked to family planning among tribal groups in south-east Bangladesh. Future projects are planned in Jessor, west Bangladesh and Sylhet in the north-east. In Jessor, Population Concern will work in partnership with the Jagorni Chakra to develop mother and child healthcare and family planning. In Sylhet Population Concern has joined forces with the Voluntary Association for Rural Development (VARD) who work with poor local villagers, especially women, on the provision of education and skills training. In Nepal, Population Concern is supporting the Nawalparasi Women's Development Project through the local Family Planning Association. A mid-term evaluation of this project has shown an increase in the local contraceptive prevalence rate and in the numbers of women taking part in literacy and skills training courses.

Figure 9.38 *Nepal and Bangladesh*

Machismo: a major obstacle to population control in Mexico
Studies in Mexico have shown a link between family planning, religion and machismo. The latter has no one simple definition. In Mexico it has been described as a show of aggressive masculinity and power over women. Recent evidence shows that the 'machismo factor' has a strong influence on women's reproductive behaviour and contraceptive use. At family planning clinics five to ten per cent of women attend without their husband's knowledge, and internal contraceptive methods, notably the IUD, are chosen so that their husbands will not find out. Women often did not return for the necessary check ups for fear of being found out. Some men will not allow women to use family planning because children guarantee manhood. If they have few children they are *'poco hombre'*.
Fertility remains high amongst the poorest and least educated women, and rates of infant mortality and the number of complications relating to child bearing remain unacceptably high. Machismo has to be taken into account in family planning programmes.

Figure 9.39 *Mexican machismo*

PROJECT SUGGESTIONS

1 Global population change

AIMS: to collect population data from a variety of countries and examine population trends.
EQUIPMENT: census data World Bank.
RECORDING AND ANALYSIS: use IT to select projectable information and investigate population change over a ten-year period.

2 National migration trends

AIMS: to investigate current and past migration trends.
EQUIPMENT: IT.
TECHNIQUES: use secondary data sources from the UK 1991 and 1981 census.
RECORDING AND ANALYSIS: produce flow line charts to show population movements.

3 Local study

AIMS: to investigate population movements in the local community.
EQUIPMENT: questionnaire.
TECHNIQUES: use of local secondary data and comparison with primary data collected using questionnaire.
RECORDING AND ANALYSIS: produce local flow line maps to show circulation and in-migration and choropleth maps to show population change.

4 Managing populations

AIMS: to investigate policies which have tried to manage populations.
EQUIPMENT: IT, recording sheets, questionnaire.
TECHNIQUES: use of CD-ROM, library research, interviews with strategic planning department of local council and Population Concern.
RECORDING AND ANALYSIS: annotated maps at global, regional and local scales.

5 Population and industry

AIMS: to investigate the importance of demographic information in industrial location and planning.
EQUIPMENT: IT, questionnaire.
TECHNIQUES: use of local or regional industrial directories and postal questionnaire to stratified sample of industries.
RECORDING AND ANALYSIS: graphics to show results of questionnaire survey and Chi-Squared analysis of differences between types of industry.

GLOSSARY

Bio-diversity ecological richness in natural environments such as tropical rain forests.

Carrying capacity ability of environment to sustain use by people without environmental deterioration.

Census the collection of information on population.

Circulation the temporary short-term movement of people that does not involve a change in home location.

Crude birth rate the number of live births per year as a ratio of the total population.

Crude death rate the number of deaths per year as a ratio of the total population.

Demography the study of population.

Dependency ratio the number of dependants in a population usually defined as those below 16 years and above 65 years.

Extended family the very large family unit found in ELDCs where parents and several children live together with grandparents, uncles, aunts and cousins.

Fertility the occurrence of live births.

Infant mortality the deaths of infants before their first birthday per thousand live births per year.

Life expectancy the average number of years that a person can expect to live from birth.

Migration the movement of people that involves a permanent change in home location from one region or country to another.

Mortality the number of deaths in a population.

Natural increase the difference between births and deaths, usually expressed as a percentage rate of change.

Nuclear family the name given to the small family unit (parents and two children) associated with families in EMDCs.

Overpopulation when the population of a country exceeds the ideal or optimum.

Population density the number of people per unit area of land, usually given as km^2.

Population movement the movement of people over space and time to include migration and circulation.

Population structure the composition of a population based on age, sex, ethnicity, religion or class.

Renewable resources resources that are not finite and can be recycled or reproduced, e.g. water, wind etc.

Replacement level the rate of natural increase needed for a country's population in order for it to maintain its present size in the future.

Vital rates rates of those components such as birth, fertility, marriage and death which indicate the nature and possible changes in a population.

REFERENCES

W F Hornby and M Jones (1993), *An introduction to population geography*, Cambridge University Press

R Hall (1991), *World population trends*, Cambridge University Press

M E Witherick (1990), *Population geography*, Longman

USEFUL ADDRESSES

UNFPA – United Nations Population Fund
220 East 42nd Street
New York
NY 10017 USA

POPULATION REFERENCE BUREAU, INC
1875 Connecticut Avenue NW
Suite 520
Washington DC 20009
USA

SIMON POPULATION TRUST
99 Gower Street
London
WC1E 9A2

POPULATION CONCERN 231 Tottenham Court Road, London, WIP 9AE

index

ablation 19, 24, 32
abrasion 25, 32
accumulation 19, 20, 24, 32
agglomeration 133, 136, 143, 146, 154, 159, 177, 222
agribusiness 97, 167, 171, 177, 179
air mass 74, 76, 94
air pollution 65–70, 84–7, 94
albedo 21, 32, 70, 94
anticyclone 82–4, 94
arête 24, 32
atmosphere 63–96
bedding plane 15, 18
biodiversity 225, 227, 255
biome 97–9, 114–21, 126
birth rate 197, 228, 231, 235, 255
Boserup 164, 231
brown earth 111, 112, 126
CAP 116, 118, 178
census 191, 229, 235, 240, 255
chernozem 115, 126
cirque 19, 20, 21, 23, 24, 25, 26, 32
climatic climax 103, 126
cold front 76, 78, 94
comprehensive redevelopment 208, 212, 222
compression flow 26
conurbations 146, 198, 222
counterurbanisation 146, 154, 159, 197–9, 222
crevasse 23, 27
cumulative causation 144, 159
death rate 197, 228, 231, 235, 250, 255
depressions 76–81, 95
denudation 5, 18, 43, 61
de-industrialisation 140, 141, 159
discharge 34, 42, 44, 46, 50, 60, 61
drainage basin 34–7, 41, 49–50 61
drainage pattern 39, 61
drought 35, 52, 53, 83, 164, 168
dynamic equilibrium 35, 46
ecosystem 97, 105, 110, 113, 125
enterprise zones 143, 147, 159
export processing zones 150, 159
erosion 6, 18, 25, 32, 46–7, 116, 164, 179, 181
famine 35, 164, 166, 235, 246, 250
fault 5, 18
ferralsol 110, 111, 126
fertility 108, 118, 224, 229, 235, 255

flood 34, 47, 50–1, 57–60
footloose 130, 155, 159
geopolitics 38
gleying 120, 121, 126
grade 46, 48, 61
green belts 208, 216, 222
greenfield sites 139, 154, 159
Greenhouse Effect 63, 65, 66, 95
Green Revolution 164, 167, 168, 200, 222
Hadley Cell 73, 95
humus 12, 18, 102, 115, 126, 169, 179
hydrograph 36–8, 50, 51, 57, 61
hydrological cycle 34, 35, 37, 49, 61, 112
hydrosere 103, 126
ice ages 22, 32, 66, 109, 111
igneous 8, 18
industrial revolution 136, 145, 147, 194
insolation 63, 64, 67, 70, 71, 95
intrusion 8, 18
ITCZ 35, 73, 75, 95
jet stream 74, 95
kondratiev waves 136
lapse rates 89–90, 95
latent heat 89
laterite 115, 126
leaching 101, 126
lithosere 103, 126
load 43, 61
loam 107, 124, 126
magma 8, 18
Malthus 163, 164, 231
manufacturing industry 130–50, 158, 206
mass movement 5, 14–17, 18, 30–1, 43, 47
metamorphic 5, 18
migration 196, 197, 200, 222, 224, 235, 236, 242, 250, 253, 255
moraines 26, 32
mortality 223, 229, 230
multiplier effect 137, 159
névé 19, 32
new towns 143, 208, 222
NICs 139, 147, 148–51, 159
outwash 27, 32
overpopulation 225, 226, 227, 231, 251, 255
ozone 68–70, 95
peat 179, 190

periglaciation 29–31, 32
permafrost 29, 32
photosynthesis 99, 126
pioneer species 104, 126
plagioclimax 104, 126
pleistocene 22, 32
podzol 109, 111, 126
polar front 73, 95
population density 163, 236–8, 255
population distribution 152, 237–43
population growth 222, 223–36, 242, 251–4
porous 15, 18, 31
psammosere 103, 126
rejuvenation 46, 48, 61
resources 33, 131, 164, 225, 227, 228, 251, 253, 255
river flow 35–43, 52
river regime 36, 37, 61
Rossby waves 74, 95
salinisation 167, 190
science parks 141
sedimentary 5, 15, 18
sphere of influence 152, 193
service industry 129–30, 151–8, 206 216
smog 70, 84–6, 95
soil fertility 108, 126, 161, 167–9, 176, 226
soil structure 108, 118, 126
soil texture 106, 107, 124, 126
solonchak 119, 126
succession 103–105, 127
striations 25, 32
tectonics 5, 14, 18
till 26, 30, 32, 49
transnationals 130, 135, 137, 142, 148, 150, 171–3, 190
trophic level 163, 190
troposphere 68, 82, 95
urbanisation 87, 145, 192–200, 208–16, 222, 251
urban climates 87–8, 93, 95, 96
urban planning 192, 194, 201, 206, 208–19
urban structure 201–207
varve 27, 32
water balance 9, 50
water borne disease 54, 55
water supplies 33, 38, 52–7
water quality 35, 38, 52–7, 61, 123
weathering 5–13, 18, 31, 43